Light is a Messenger

Light Is a Messenger

The life and science of William Lawrence Bragg

Graeme K. Hunter

OXFORD
UNIVERSITY PRESS

OXFORD
UNIVERSITY PRESS
Great Clarendon Street, Oxford OX2 6DP
Oxford University Press is a department of the University of Oxford.
It furthers the University's objective of excellence in research, scholarship,
and education by publishing worldwide in
Oxford New York
Auckland Bangkok Buenos Aires Cape Town Chennai
Dar es Salaam Delhi Hong Kong Istanbul
Karachi Kolkata Kuala Lumpur Madrid Melbourne Mexico City Mumbai
Nairobi São Paulo Shanghai Taipei Tokyo Toronto

Published in the United States
by Oxford University Press Inc., New York

First published 2004

A catalogue record for this title is available from the British Library
Library of Congress Cataloging in Publication Data
(Data available)
ISBN 0 19 852921 X (Hbk)
10 9 8 7 6 5 4 3 2 1
Typeset by Newgen Imaging Systems (P) Ltd., Chennai, India
Printed in India
on acid-free paper by
Thomson Press (I) Ltd.

"The light is a messenger, carrying a story about the form of the object. . ."

William Lawrence Bragg,
Mackenzie Davidson Memorial Lecture, November 14, 1928.

To Francine

Contents

Acknowledgements

Funding for the initial stage of my research on William Lawrence Bragg was provided by a grant from the Social Sciences and Humanities Research Council of Canada. The Burroughs Wellcome Fund provided a travel grant that funded my research trips to the United Kingdom. Finally, Associated Medical Services, Inc, provided me with a Hannah Institute for the History of Medicine grant to cover costs of transcribing interviews and photocopying.

I am very grateful to members of the Bragg family for their cooperation with this project. Stephen Bragg and Patience Thomson were interviewed and Margaret Heath provided information by letter. The Bragg family permitted me to quote from William Lawrence Bragg's and William Henry Bragg's personal papers in the Royal Institution archives but they do not necessarily agree in any way with my assessment of Bragg's life and career. I am also indebted to the following former colleagues of Bragg who agreed to be interviewed or provided me with information by letter or e-mail: David Blow, William Cochran, Francis Crick, Peter Duncumb, Antony Hewish, Peter Hirsch, Louise Johnson, Aaron Klug, Mick Lomer, Anthony North, John Nye, Peter Pauling, the late Max Perutz, Brian Pippard, Michael Whelan, and Michael Woolfson.

Frank James, Keeper of collections at the Royal Institution, was a gracious host during my research trip to the RI in 2000 and has provided me with much assistance since. Margaret Woodall's magnificent catalogs of the William Lawrence and William Henry Bragg papers made this project possible. I am also grateful to the staffs of the Modern Papers Room of the Bodleian Library, Oxford University; the Churchill Archive Centre, Cambridge; and the Royal Society Library; and to Jonathan Smith of Trinity College Library, Chris Petersen of Oregon State University Library, Robert Cox of the American Philosophical Society, and Carol Beadle of the Medical Sciences Video Archive at Oxford Brookes University. For assistance in obtaining the photographs used in this book, I thank the aforementioned Frank James, Lesley Black of the Bridgeman Art Library, Annette Faux of the MRC Laboratory of Molecular Biology, and Heather Lindsay of the American Institute of Physics.

Horace Judson, Robert Olby, and David Edge kindly agreed to give me access to transcripts or recordings of their interviews with Bragg and other scientists. John Jenkin provided me with valuable advice on sources of information about W. L. Bragg's early years.

Vincent Yeung assisted with the early stages of the research, and Bev Goldsmith and Brenda Sefeldas transcribed the interviews I conducted. Several people helped me in small but important and much-appreciated ways, including Anders Bárány, Frank Beier, David Hunter, Robin Marshall, Les Murison, and Gordon Squires.

Finally, I am very grateful to Dr Durward Cruickshank for his detailed and insightful comments on parts of this manuscript. Dr Michael Woolfson also read a draft of the manuscript and made many useful suggestions.

Introduction

When William Lawrence Bragg won the 1915 Nobel Prize in physics, he was 25 years old; to the present day, he remains the youngest person ever to win a Nobel Prize. Because he shared the Prize with his father, William Henry Bragg,[a] many people assumed that Bragg's role in the research recognized by the Nobel Foundation was a subordinate one. Not so: it was actually the son who, in the autumn of 1912, realized how to interpret the diffraction pattern that Max von Laue, Walter Friedrich and Paul Knipping had obtained six months earlier by irradiating a zincblende crystal with a beam of X-rays. It was also the younger Bragg who carried out most of the seminal studies of 1913–14 that showed how X-rays could be used to determine the ways that atoms and molecules are arranged in crystalline solids. On the basis of these events alone, William Lawrence Bragg would be worth studying as a remarkable example of precocious scientific genius.

In retrospect, it may seem odd that the Braggs were awarded the 1915 physics Nobel in preference to Max Planck and Albert Einstein, among other better-known physicists. It could well be argued, however, that the scientific impact of X-ray crystallography has been as great as those of quantum theory and relativity, and the impact on everyday life even greater. But the impact of Bragg's work has very largely been on chemistry and biology rather than on physics. Significantly, the Nobel Prizes that have been awarded for X-ray crystallography in the 1960s and since were not in the physics category, but rather in the chemistry and physiology categories. Within his lifetime and with his active participation, Bragg's 1912 discovery had revolutionized chemistry and helped create a new discipline—molecular biology.

After Bragg died in 1971, one of his most distinguished protégés, David Phillips, composed a lengthy and detailed biographical memoir for the Royal Society. In 1990, the centenary of Bragg's birth, Phillips and John Thomas edited a volume of reminiscences entitled *Selections and Reflections: the Legacy of Sir Lawrence Bragg*, which included contributions by two of Bragg's children and many of his former colleagues. Until now, no biography of Bragg has been written, and the 1990 *Festschrift* remains the only book devoted to Bragg's life and career.

[a] William Henry Bragg will be referred to as "WHB" to avoid confusion with his son, William Lawrence Bragg, who will be referred to as "Willie" or "Bragg."

Considering the number of books that have been published on other eminent twentieth-century physicists, such as Einstein and Ernest Rutherford, it is curious that there has been until now no biography of William Lawrence Bragg. Two possible explanations come to mind. One is that Bragg did not have a flamboyant personality—it almost goes without saying that he lacked the charisma of Rutherford or the cult status of Einstein. The second is that Bragg's career, which began with mathematics and ended with proteins, does not fit into any standard academic pigeonhole and presents a considerable challenge to those (such as the present writer) whose professional expertise is in only one of the relevant areas. As Jack Goldsmith wrote to Bragg's wife in 1965: "Who is going to write his biography? Perhaps a symposium of writers is the only possibility."[1]

The Bragg scholar does not, at first sight, face any lack of raw material. The William Lawrence Bragg archive at the Royal Institution has an extensive collection of documents bearing on many aspects of Bragg's life and career. This includes his unpublished autobiography, apparently written over many years and covering the period 1890 to 1951, and a memoir written by his wife Alice. However, the vein is not as rich as it may appear. Bragg was an intensely private man who did not often commit his innermost thoughts to paper, or even confide in close friends and colleagues. The correspondence in the RI archive reveals little about Bragg's opinions on the scientific events in which he participated, far less about his personal feelings; even his closest scientific collaborator, Max Perutz, claimed to know little of Bragg's views on their fellow scientists. With the possible exception of his sister Gwendy, his wife Alice was probably the only person who truly knew Bragg. The autobiography, although an invaluable source of information, is not always a reliable one. It appears to have been dictated by Bragg and never edited; as a result many names are spelled phonetically and there are many factual errors and duplications. To take a particularly egregious but by no means isolated example, Bragg claimed that he entered Trinity College in 1908 and graduated in 1911, when in fact the correct dates are 1909 and 1912, respectively. The autobiography, apparently written as a family history, also contains little information about Bragg's research activities. Because of these factors, the Bragg scholar has to reconstruct Bragg's key relationships, such as those with his father and with Linus Pauling, from pieces of circumstantial evidence. To use a crystallographic analogy, the present work is a product of the trial-and-error method rather than the result of a Fourier synthesis.

There is something of a tradition that biographies of British crystallographers have been written by individuals who are neither professional scientists nor professional historians—this was the case for books about John Desmond Bernal and Dorothy Hodgkin, as well as both lives of Rosalind Franklin and Gwendy Caroe's biography of WHB. Aside from their scientific achievements, Franklin is of interest because of sexual politics, and Bernal because of sex and politics. Franklin and Hodgkin were both pioneers in a male-dominated world. Bragg, in contrast, was a non-political man who blazed no trails—indeed, was often thought of, both personally and professionally, as a bit of a throwback—and had a very conventional personal life. This is not to say that the non-scientific aspects of Bragg's life are entirely devoid of interest or historiographical significance. His relationship with his father affected Bragg's career at many points, as did his depressive disorder; his rivalry with Linus Pauling had both personal and scientific elements. But even if one considered Bragg's personal life sufficiently interesting on which to base a biography, one would face the solid wall of privacy that he built around himself. This book is therefore a scientific biography rather than a biography of a scientist, if such a distinction can be made.

Bragg's scientific work touched upon several disciplines, including chemistry, biology, and mineralogy. Its overarching theme, however, was the use of X-rays to determine the structures of crystals (or molecules in the crystalline state). Crystallography is an esoteric branch of science but not an inherently difficult one to comprehend. Indeed, one of its great attractions to Bragg was the existence of a very simple geometric relationship—firmly established as early as 1913—between the pattern of spots in an X-ray diffraction pattern and the spacing of atoms in the diffracting crystal. There is little doubt that this knowledge—that the structure of the crystal was hidden in its diffraction pattern—encouraged him to attempt ever more complicated crystals.

When Laue, Friedrich, and Knipping first demonstrated that crystals could diffract X-rays, diffraction of visible light by gratings was already a well-understood phenomenon. In the simplest case, light passes through a grating consisting of parallel lines ruled equidistantly on a glass plate. Light rays impinging upon these lines are scattered in all directions; it is only because of this scattering that we are able to see the lines. The wavefronts emanating from each line interfere with one another constructively or destructively at different angles, depending on whether the wavefronts from each line coincide peak-to-peak, peak-to-trough, or something in between. Constructive interference, in which the waves emanating from different lines reinforce one another, will occur at only certain angles, the exact values of which depend upon the spacing between the lines. The lowest-angle diffracted beam is called zero order (light undeviated by passage through the grating), successively higher angles are called first order, second order, etc. (Figure 0.1). If the diffracted light falls upon a plane—a translucent screen or photographic plate, say—it will result in a pattern of parallel light and dark bands, the former representing the different orders of diffracted light. For a given distance between grating and screen, the distance between the light bands of the diffraction pattern depends upon the wavelength of the light illuminating the grating and the distance between the lines ruled on the grating.

If the diffracting object is a glass plate with parallel lines ruled on it at right angles—a cross grating—the screen receiving the diffracted light will display a pattern of spots rather than bands. Each spot represents light diffracted by the entire grating in a particular direction. As in the case of a line grating, there are first-order, second-order, etc., spots. Again, there is a predictable relationship

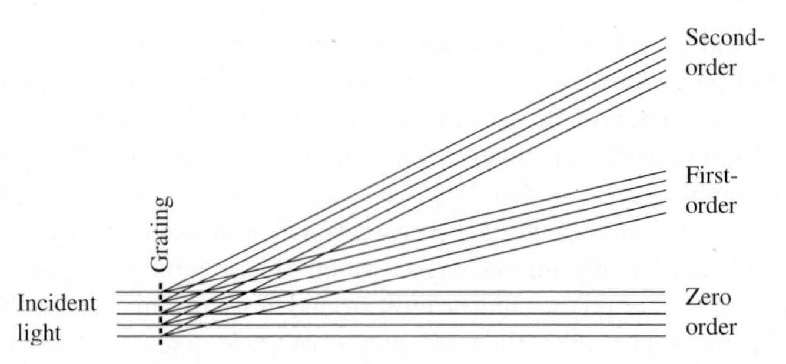

Fig. 0.1 Formation of different orders of diffraction by a line grating. First- and second-order spectra will be formed on both sides of the zero-order spectrum; however, one set is omitted for the sake of simplicity

between the spacing and angles between the lines of the cross-grating and the position of spots in its diffraction pattern. This is also true for more complex gratings.

Not only is the diffraction pattern related to the pattern of the grating, it also contains all the information required to create an image of the grating. If, instead of diverging from one another, the diffracted beams are forced by a lens to converge, a new pattern of bright and dark will result, the bright areas representing points at which light was transmitted through the grating, the dark areas representing points at which light was not transmitted through the grating—in other words, a faithful image of the diffraction grating itself. This is the principle behind optical devices such as the vertebrate eye and the camera. The interchangeability of object and diffraction pattern, optically speaking, is a key aspect of diffraction phenomena. As Charles Taylor put it, "the diffraction pattern of the diffraction pattern is an image of the object again."[2]

Up to a point, the diffraction of X-rays by crystals is very similar to diffraction of visible light by a cross-grating. When a beam of X-rays is shone upon a crystal, planes of atoms diffract X-rays in directions that depend upon the orientation of the planes with respect to the incident beam. Since the number of differently oriented planes is infinite, X-rays are diffracted by the crystal in many directions. Whether the beams diffracted at a particular angle interfere with one another constructively or destructively depends upon the spacing between the planes responsible, just as, in optical diffraction, it depends upon the spacing between the lines of the grating. In this way the crystal, which is actually a three-dimensional lattice of atoms, acts like a visible-light cross-grating, producing, when the diffracted beams are collected on a plane, a pattern of spots.

If X-radiation could be focussed like visible light, the process of X-ray analysis of crystals would be very simple—and Bragg would not have won the Nobel Prize. An X-ray lens placed in the path of the diffracted radiation would reverse the complex process of interference resulting in an image of the crystal lattice. A crystal is, of course, a three-dimensional lattice, whereas the image created by a lens is two-dimensional. Therefore, the image of a crystal that would be obtained by focussing X-rays would be a projection of the lattice. However, this is not a significant limitation to X-ray analysis. The problem is that X-rays cannot be focussed.

Although diffracted X-rays cannot be recombined physically to form an image of the crystal lattice, they can be recombined mathematically. As WHB realized as early as 1915, it is possible to use the mathematical technique of Fourier synthesis to combine the information contained in each diffracted beam to generate an image of the diffraction grating—in other words, a map of all the atoms present in the crystal. WHB realized that the recombination of diffracted beams performed by an eye or a camera represents a superimposition of periodic mathematical functions, which is exactly the technique developed by Jean-Baptiste Fourier. If one knows two characteristics of each of the diffracted beams—its amplitude and its phase—then a Fourier synthesis of these will produce a curve that represents the density of diffracting matter across a plane of the crystal.

Again, there is a catch. The *amplitude* of a wave is related to its intensity, or amount of energy it carries. This can easily be measured; for example, by allowing the diffracted beams to fall upon a photographic plate. The *phase* of a wave represents the positions of peaks and troughs, and can vary from one wave to another by anything between zero (the beams coincide constructively, peak to peak) and 180° (the beams coincide destructively, peak to trough). Herein lies the problem, as no information is recorded about the phases of the beams by allowing them to fall upon a photographic plate or other recording device. This phase problem—the problem of needing to know the phases of the X-rays diffracted in different directions in order to perform a Fourier synthesis—bedevilled attempts to analyze the structures of complex crystals for decades. If the phases were known, the structure of even a protein crystal became merely a matter of number crunching—no trivial thing in the pre-computer age, but obviously doable. The solution to the phase problem for proteins turned out to be to attach one or more heavy atoms to the molecule. The effect on the diffracted radiation is so marked that it becomes possible to locate every atom of the protein relative to the added heavy atoms. To quote Taylor again, ". . . our heavy atom is analogous with the hair or speck of dust that we know is there; if we focus on it we can assume that the rest will be in focus."[3]

The X-ray analysis of crystals can be considered to have occurred in three stages, to each of which Bragg made great contributions. The first stage began in 1912 with the demonstration of the diffraction of X-rays by crystals and was essentially over by 1914, when the outbreak of First World War disrupted scientific research in Britain and Germany. Bragg's first contribution was the realization that each spot on an X-ray diffraction pattern represents radiation scattered by a plane of atoms in the crystal (Figure 0.2). He also realized that there is a simple relationship—"Bragg's law"—between the angle at which the radiation is diffracted, the wavelength of the diffracted radiation and the distance between adjacent planes of that type. During the subsequent two years,

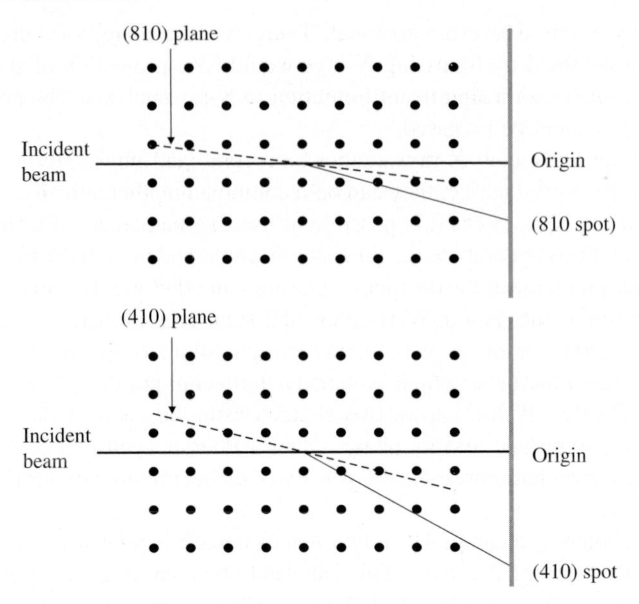

Fig. 0.2 Relationship between atomic plane of crystal and position of corresponding spot on X-ray diffraction pattern. Note that higher-index planes give rise to spots closer to the origin

Bragg and his father developed techniques to solve the atomic structures of the simplest crystals. These were very symmetrical crystals, mostly cubic, in which the position of every atom was defined by symmetry. That is to say, once the size of the cube was determined, solving the structure was simply a matter of ascertaining which atoms occupy a small number of possible positions.

However, even in the first group of crystals that Bragg studied—the alkaline halides, such as sodium chloride and potassium bromide—the situation was more complicated than simply working out spacing between planes of atoms. The complication arises from the fact that planes containing one type of atom are often separated by mutually parallel planes containing a different type of atom. As the two types of plane are parallel, they will diffract radiation in the same direction; as they contain different atoms, the diffracted waves will interfere (usually destructively). This interference between X-radiation diffracted from parallel planes of different compositions can provide valuable information—it allowed Bragg to prove that the alkaline halides were all based on the face-centered cubic lattice—but it vastly complicates the interpretation of diffraction patterns.

In the second stage of X-ray analysis, which lasted until about 1930, Bragg's laboratory remained in the forefront of this new field of X-ray crystallography, solving increasingly complex mineral structures by the mix of careful experimentation and inspired guesswork that he called the "trial-and-error method." This was used to study crystals of lower symmetry than cubic, in which the positions of atoms were not necessarily dictated by symmetry considerations. Determining the structures of these crystals was not merely a matter of deciding which atoms lay at which symmetry-related positions, but rather a matter of finding values for what became known as "parameters"—atomic coordinates not dictated by symmetry. In many of these structures, three parameters, corresponding to their positions along each of the three axes of the crystal, had to be found for some atoms. The number of possible structures therefore became much larger than had been the case for the crystals studied in 1912–14.

In Bragg's trial-and-error method, various pieces of information were put together to limit the possible structures. These included the space group, which is the type of three-dimensional symmetry the crystal possesses; the number of atoms in the unit cell, which is the smallest part of the crystal that repeats throughout three-dimensional space; and the sizes of the atoms involved. Plausible structures generated by these considerations were then tested by determining their theoretical diffraction patterns and comparing these with the patterns actually observed for that crystal. Initially, this comparison was largely qualitative, a matter of whether radiation diffracted in a particular direction was present or absent. Later it became semi-quantitative—whether this radiation was weak or strong. Finally, Bragg's invention of techniques for measuring absolute intensities of diffracted radiation meant that a proposed crystal structure could be tested by comparing the amount of radiation actually diffracted in each direction with the theoretical amount diffracted by that structure. The trial-and-error method was perfectly adequate for crystals of high symmetry, or those with few parameters. Once they had determined the space group of beryl, a moderately complex but highly symmetrical mineral with seven parameters, it took Bragg and his student Joseph West only about 15 minutes to determine its structure. By the late 1920s, however, it was becoming clear that the trial-and-error method was approaching its natural limit. For crystals with more than a hundred parameters, the number of possible structures was simply too large.

The analysis of the most complex crystals would require a method that eliminated the guesswork. To Bragg, it was obvious that the future of X-ray analysis—if it had a future—lay in the Fourier method. As noted above, the limitation of Fourier analysis of X-ray diffraction patterns was that the phases of the diffracted beams could not be measured. However, Bragg realized that the phase problem could be circumvented in some situations. He performed the first two-dimensional Fourier analysis—on the silicate diopside—in 1929. His analysis produced maps of electron density in planes onto which the structure was projected. This may seem a hollow feat when it is realized that the phases used to construct the electron-density map were only known because Bragg had already solved diopside by trial and error. As he pointed out, however, the Fourier method could be of utility in certain circumstances. For certain crystals—those with a symmetry element known as a center of symmetry—the phases have restricted values—they are either 0 (positive) or 180° (negative). Also, the presence of heavy atoms in some crystals means that the phases are all positive.

The third stage of X-ray analysis, which began around 1930, was characterized by the use of Fourier methods to study the structures of complex organic molecules and macromolecules. The lattice structure of the crystal, all-important for inorganic crystals, now became almost irrelevant—what mattered was the structure of the covalently bonded molecule that represented the unit of crystalline pattern. The X-ray analysis of organic crystals had been pioneered by WHB; in the 1930s it was carried forward largely by his students, former students and students of his former students—a true scientific dynasty. As techniques improved, it became possible to analyze by X-ray methods the structures of molecules that could not be determined, even approximately, by chemical means. The two-dimensional Fourier method that Bragg had invented became part of the standard method of X-ray analysis of organic molecules. However, he himself was to make no significant contribution until 1938 when, serendipitously but not coincidentally, he became involved in Max Perutz's 30-year odyssey to solve the structure of the protein hemoglobin. A means of solving the phase problem for proteins was found 15 years later. When the full potential of X-ray analysis was realized in the 1960s by providing solutions to the structures of what Perutz called "the giant molecules of the living cell," it was others—Perutz, John Kendrew, Francis Crick, James Watson, Maurice Wilkins and Dorothy Hodgkin—who shared the Nobel Prizes that were showered on the field. But each of these individuals freely acknowledged the inspirational and intellectual roles that Bragg had played. No 75th birthday present for Bragg could have been more appropriate than the structure of the enzyme lysozyme, solved by his own research team.

The reason for describing the analogy between visible-light and X-ray optics in such detail is that it dominated Bragg's thinking throughout his career. He often discussed it in lectures and review articles, and it clearly underpinned his approach to research. Bragg was motivated by a simple idea: that the pattern of interference of diffracted X-rays could reveal the atomic structure of crystals, no matter how complicated. It is a tribute to how far this idea could be pushed that the first crystals Bragg analyzed have four atoms per unit cell; the last, hemoglobin, has about ten thousand.

It is not to be supposed, however, that the development of X-ray analysis unfolded in a logical series of achievements from sodium chloride to hemoglobin. Bragg himself abandoned the field on two occasions: in the early 1920s, when he became interested in the structure of the atom; and in the 1930s, when his primary research focus was on metal physics. Nor was it preordained that X-ray methods would be able to solve the structures of macromolecules. In fact, most crystallographers thought that Perutz and Bragg's attempt to use X-ray methods to determine the structures of biological macromolecules was completely hopeless. In the absence of a solution to the phase problem, this skepticism was well founded. In fact, one may wonder why Perutz and Bragg kept trying. In the period 1938–53, they were encouraged to continue by the belief that hemoglobin, and other proteins, would have a high degree of internal symmetry that would greatly simplify the analysis. This belief, which seems to have been motivated by an Ockham's razor-like feeling that Nature must be simple, turned out quite erroneous. If they had suspected the real scale of the problem—finding the locations of 2500 non-symmetry-related atoms—surely even the optimistic Bragg and the indefatigable Perutz would have found something more promising to work on. Fortunately for the embryonic science of molecular biology, a solution to the phase problem came along just when the idea of internal symmetry in proteins became untenable.

Because Bragg became such a prominent member of the scientific establishment, it is easy to think of him as the Sir Lawrence of his days at the Royal Institution—a bald, white-mustached, rather portly gentleman with an urbane manner and Edwardian sensibility; playing with his grandchildren or tending his rose bushes when not serving on committees or lunching at the Athenaeum club. The settled and successful senior scientist, though, was only one aspect of Bragg's life. There were also many darker elements. There was the postgraduate student locked in what appeared to him to be a Faustian bargain with his father in which great success would be achieved but the recognition for it denied him; the boy professor struggling to defend his department, career, and reputation against hostile students, colleagues, and administrators; the acknowledged master of X-ray crystallography challenged on his own turf by the younger and more brilliant Linus Pauling; and the sexagenarian derided by his juniors as a bubble-blowing has-been, but whose unparalleled grasp of X-ray optics allowed him to make a series of key breakthroughs in the analysis of hemoglobin.

The life of William Lawrence Bragg is the story of a man who achieved great scientific success and recognition too young for his emotional well-being, but who thereby lived long enough not only to see but also to actively participate in the culmination of the field of research he initiated. Bragg the man remains partly hidden behind his defensive carapace, but Bragg the scientist is what interests us. His career is the story of the development of X-ray crystallography from a technique that could solve the structures of simple salts to one that could provide three-dimensional maps of every atom in the most complex molecules known to man.

1
A shy and reserved person: Adelaide, 1886–1908

On February 27, 1886, William Henry Bragg stepped ashore at Glenelg, South Australia. Only 23 years old, he was the newly appointed Elder Professor of Mathematics and Experimental Physics at the University of Adelaide. WHB spent that night at the Pier Hotel, as the last train for Adelaide had already left. His first sight of Adelaide was therefore the next day, when Dr Alfred Lendon, a local physician of WHBs own age, arrived at the hotel with a carriage to take him visiting.[b]

The colony of South Australia had been founded in 1836. It had obtained self-government in 1857 and was now moderately prosperous from its major exports of copper, wheat, and wool. Adelaide, the capital of South Australia, was located about 6 miles inland from the St Vincent Gulf. The town had been laid out on a novel design in two parts, North and South, the former consisting of 342 one-acre lots, the latter of 700. Dividing North and South Adelaide was a wooded area, the Park Lands, through which ran the River Torrens, a modest stream that had been dammed to create an artificial lake. Another band of Park Lands encircled the whole of Adelaide, preventing the town from undergoing "urban sprawl." North Terrace, which marked the northern boundary of South Adelaide, was the main street of the town. Along it lay the railway station, Government House, the university, and the hospital.

The Adelaide that WHB first beheld that summer day in 1886 was an attractive town of stone and brick buildings, iron-roofed. The streets were gas-lit, there was a municipal water supply, and the first telephone exchange had opened 5 years earlier. Trains linked Adelaide to its ports and to the longer-established colonies of Victoria and New South Wales, and horse-trams provided public transportation within the town.[4]

Lendon first took WHB to "Montefiore," the magnificent North Adelaide residence of Samuel Way, Chancellor of the University and Chief Justice of South Australia. The two young men then rode on to the Observatory, located on the West Terrace of South Adelaide, where they had been invited for dinner

[b] This account of WHBs life in Australia is, unless otherwise indicated, based on Jenkin, J. (1986). *The Bragg Family in Adelaide: a Pictorial Celebration.* University of Adelaide Foundation, Adelaide.

with another prominent local figure—Charles Todd, Government Astronomer, Postmaster-General, and Superintendent of Telegraphs.

Todd was, like WHB, a self-made man. The son of a London grocer, he had in 1848 obtained a job as assistant astronomer at the Cambridge University Observatory. While visiting distant relatives who lived in Cambridge, Edward and Charlotte Bell, Todd made the acquaintance of their daughter Alice. In 1854, Todd moved to the Royal Observatory in Greenwich, as assistant in charge of time-balls. When he was offered the position of Superintendent of Telegraphs and director of the observatory in South Australia, Todd proposed to Alice Bell, and they married in April 1855. The newlyweds arrived in Adelaide in November of that year.

Charles Todd's great achievement was the construction of a telegraph line between Adelaide and Darwin in the Northern Territory. Completed in 1872 after overcoming enormous difficulties, the line connected Australia to a world-wide telegraph network that now allowed news to arrive in hours that previously had taken months by ship. During her husband's frequent absences in the uncharted Australian interior, Alice Todd was raising their large family: Lizzie, Charlie, Hedley, Maud, Gwendoline, and Lorna. One of Todd's surveyors named the site of one of the telegraph repeater stations after her—Alice Springs.[5]

WHB took great pleasure in making the acquaintance of the Todd family, whom he later described as "Such a jolly lot!" Charles was a witty and erudite man, Alice was open and unpretentious, and their uninhibited children immediately dubbed WHB "the Fressor." By the end of his first full day in Adelaide, WHB must have been elated about the prospect ahead of him. As he later wrote: "I was marvellously fortunate in being thrown into a society of the Todds and people like them, so open and kind and good-natured. The whole thing, the going to Australia to a new work and an assured position, the people I met there, the sunshine and fruit and flowers, was a marvellous change for me."[6]

For WHB, Australia offered a new beginning after an early life that had been often difficult.[7] He was born on July 2, 1862, the first child of Robert and Mary Bragg, who had married the year before. Robert Bragg had been born in 1830 and brought up in Birkenhead. Despite the death by shipwreck of his father John, Robert Bragg became a merchant seaman at the age of 16. However, he served only 9 years before an inheritance allowed him to retire from the sea and purchase Stoneraise Place, a farm in Cumberland. There he had married Mary Wood, the daughter of Robert Wood, Vicar of Westward.

The birth of William Henry Bragg was followed by those of another two sons, Jack and James, before Mary Bragg died in 1869. On the death of his mother, William Bragg was taken in by his father's brothers William and James, who lived in Market Harborough in Leicestershire. "The Uncles" could have served Charles Dickens as the models for Betsy Trotwood and Mr Dick: Uncle William owned the local chemist's shop and had a strong personality that had allowed him to also take in WHBs cousin, Fanny Addison; Uncle James was simple-minded and very much dominated by his brother.

WHB attended the grammar school in Market Harborough, which had been re-established by Uncle William, and proved an excellent student. At his father's insistence, he was sent in 1875 to King William's College in the Isle of Man, where another of his uncles was a teacher. There was no tradition of academic prowess in the family—"Braggs are really of yeoman stock," as WHBs daughter later put it. Nonetheless, WHB excelled at King William's: He won prizes for mathematics and geology, was a "praepositor" (prefect) from 1879–81, and Head of the School in 1880–1. Although shy and unsocial by nature, he became actively involved in extra-curricular activities, playing cricket, tennis, and "fives," and joining the Chess Association, Literary and Debating Society, and Histrionics Society. In 1880, he won a Minor Scholarship to Trinity College, Cambridge, but at 17 was considered too young for university. He spent another year at King William's, but his work suffered from lack of competition and a wave of religious hysteria that swept through the school.

WHB entered Trinity College in the summer of 1881, enrolling in the Mathematical Tripos. Inhibited by his shyness and, perhaps, by his humble origins, WHB made few friends among his fellow students. He did play tennis, and according to family legend acquired a scar on his forehead from a hockey stick wielded by the Duke of Clarence (Queen Victoria's grandson). He was more successful academically than socially, winning a mathematics prize and a Foundation Scholarship, which gave him some privileges in the college, in 1882.

The Mathematical Tripos had been established in 1748 as a system of examination for the degree of Bachelor of Arts—the term "tripos" relating to a three-legged stool on which much earlier generations of Cambridge students had sat for oral examinations. In WHBs day, the Mathematical Tripos was in three parts, the first two of which were examined together. Based on their performance in the Part I and Part II examinations, students were classified into three groups: Wranglers (highest), Senior Optimes, and Junior Optimes (lowest); within each group, students were ranked in order of merit. WHB was Third Wrangler (of 35) in the Part I / II Tripos examinations of June 1884 (the Senior and Second Wrangler were both also Trinity men). Only Wranglers were admitted to Part III of the Tripos, based on which they were classified into three "divisions," but not ranked within a division. WHB was listed in Division 1 in the Part III examination of January 1885—the equivalent of a first-class honors degree today.[8]

Like many Cambridge mathematics graduates of the time, WHB went to the University's Cavendish Laboratory to receive training in physics, spending most of 1885 there. On December 1, he was walking to a lecture with the Cavendish Professor of Experimental Physics, Joseph John (J. J.) Thomson, when the latter mentioned a vacant mathematics chair at the University of Adelaide. The deadline was that day, so WHB sent his application by telegraph (to London) as soon as the lecture was over. To his great delight and surprise, as he considered himself too inexperienced for a professorial position, his application was successful. Uncle William's and Uncle James's feelings of joy for their nephew's achievement were mingled with sadness at his impending departure for Australia. Other than "the Uncles," however, WHB was leaving little behind in England. His father had died earlier that year, but they had never been close; as pointed out by the Bragg scholar John Jenkin, WHBs autobiographical notes do not mention the death of Robert Bragg. News of the death of Jack Bragg, who had been an even more gifted mathematical prodigy than his brother, reached WHB on January 13, 1886—the day before he sailed for Australia.

The University of Adelaide had been founded in 1874. It was an ambitious enterprise for a small provincial town of 200,000 people, considering that at that time there were only four universities in England (Oxford, Cambridge, Durham, and London). The creation of the University was made possible by donations of £20,000 each from two Scottish immigrants: the copper magnate Walter Hughes and the landowner Thomas Elder. Hughes' donation was used to endow two portmanteau professorships: Classics and Comparative Philology and Literature, and English Language and Literature and Mental and Moral Philosophy; Elder's to endow professorships in Pure and Applied Mathematics, and Natural Sciences. These professors were supplemented by lecturers in History and Political Science; Botany; Animal Physiology; and Engineering and Surveying.

Teaching at the University of Adelaide began in March 1876, in rented rooms, a dedicated building not being available until 1882. The first Professor of Pure and Applied Mathematics was Horace Lamb, a former fellow of Trinity College Cambridge and Second Wrangler in Mathematics of 1872. He also instituted lecture and laboratory courses in natural philosophy (physics). Lamb left in 1885 to take up the position of Professor of Pure Mathematics at Owens College, Manchester, creating the vacancy that WHB filled.[9]

WHB boarded with Alfred Lendon, who became his closest friend in Australia, and spent many happy hours as an adopted member of the Todd family at the Observatory and on picnics in the hills around Adelaide. A keen sportsman, WHB won prizes for tennis and golf and captained the North Adelaide lacrosse team. He also became an acclaimed performer in amateur plays and musicals.

His happy personal life helped sustain WHB during a very difficult time professionally. Initially a poor teacher, he was responsible for lecturing in mathematics, physics, and acoustics, as well as presenting practical classes in physics. Although he liked to claim that he had learned his physics from textbooks on the boat to Australia, the Mathematical Tripos covered many areas of physics and WHB had, as mentioned above, spent almost a year as a research student in the Cavendish Laboratory. Nor were class sizes large—in 1884, there were four science students at the University of Adelaide, and 10 years later that number had risen to only 13.[10] The problem was, instead, the sheer volume of teaching required of him. In the 1887 academic year, WHB had 672 student contact hours, 168 of which were in the evenings, and set 21 university examinations; the only help he had was provided by a part-time laboratory assistant.[11] According to a story that may be apocryphal, WHB apprenticed himself to a firm of Adelaide instrument-makers in order to learn how to make equipment for practical classes.[c] In 1886, he became Dean and Chairman of the Professorial Board.

The teaching load lightened somewhat in 1888 when Robert Chapman was appointed as an assistant lecturer. WHB interviewed Chapman in Melbourne on his way to Tasmania, where he was vacationing with Gwen Todd and her brother Charlie. While they were in Tasmania, WHB proposed to Gwen, who accepted subject to the approval of her parents. That assent was quickly given. It was not to be an easy engagement, however: Gwen had health problems and WHB was overworked. It was quite possibly the first emotionally intimate relationship that WHB had ever had, and he reacted to it with violent mood swings. In February 1888, he wrote to Gwen: "my engagement has been to me so far a very sweet thing, in spite of its pain";[12] other letters often refer to him being "worried."[13]

The marriage of William Henry Bragg and Gwendoline Todd took place on June 1, 1889, with Lendon serving as best man. The newlyweds took up residence in a semi-detached house on the corner of Lefevre Terrace and Tynte Street in North Adelaide, overlooking the Park Lands. Conveniently for Gwen, who was a talented amateur artist and had attended a college of design before her marriage, the house next door was occupied by Henry Gill, "the leading artist in Adelaide."

Their first child, William Lawrence (Willie), was born on March 31, 1890. A second son, Robert Charles (Bob) was born $2\frac{1}{2}$ years later, on November 25, 1892. WHB thus named his sons after both his biological and surrogate fathers. Willie remembered that his mother was very ill when Bob was born; when he was finally allowed to see her, he asked: "Mummy, do you know that I have got a baby brother?" Another early memory is of being taken out in the pram with Bob, their long hair and fancy clothing prompting gibes from the "larrikins."

The family had a cook and a housemaid, as well as a series of nursemaids. One of the latter was fired for infecting the boys with nits, but Charlotte Schlegel, an accidental immigrant from Schleswig-Holstein, proved more durable, staying with the family (in various capacities) for nearly 30 years. Willie found her to be "neurotic and fierce," as well as fanatical about dress and cleanliness.

The yard at the back of the house was gravelled, with outhouses along one side and a woodpile in the corner for feeding the fires. At the side of the house was a tree in which a previous occupant had built a platform. Eric Gill, the neighbor's son, was Willie's contemporary and "great crony."

At age 5, Willie was sent to a convent school at the other end of North Adelaide.[d] One of his memories of the school was arguing with one of the nuns about how a mirror works—perhaps an early indication of his interest in reflection phenomena!

On Sundays, the family went for lunch at the Observatory, travelling in a horse-drawn wagon. Bragg remembered his grandparents' house as "a

[c] Whether this is true or not, it is certainly the case that WHB was a superb craftsman.

[d] According to Jenkin, the school was probably that of the Dominican Sisters on Molesworth Street.

wonderful place for small boys." It was a rambling two-story house with verandas and balconies at the front, and a circular driveway surrounding a garden with a large Norfolk pine. On one side of the property were the Observatory offices and buildings housing telescopes and other astronomical equipment. At the back of the house were outbuildings, stables, and storerooms. These contained interesting items, such as souvenirs from the building of the telegraph line, including "gorgeous shells from the Tasman Sea." Beyond the back yard was the dome housing the main telescope and buildings containing meteorological equipment. The cellars under the dome contained leftover supplies that Willie and Bob used to make "electrical gadgets." Their grandmother was a "dear placid vague grandmother in her old lady's cap with lace frills," who gave Willie custard as a treat. After lunch, WHB and his father-in-law smoked cigars and the boys were sent out to play in the grounds.[14]

The University of Adelaide had no expectation that its professors would engage in research, and WHB was fully occupied with teaching. However, he took a great interest in scientific developments taking place half a world away. One of these occurred in 1895, when Wilhelm Röntgen discovered a mysterious new form of radiation—so mysterious that he called it the X-ray.

Röntgen, Professor of Physics at the University of Würzburg in Germany, was studying electrical discharge in evacuated tubes. This was done by making a glass tube containing two metal electrodes, pumping the air out of the tube and then sealing it. When the electrodes were connected to an induction coil and charged up so that a high potential difference was created, the tube would discharge by a spark jumping from cathode (negative electrode) to anode (positive electrode). During this process, it had been found that there was a spot of luminescence on the glass wall opposite the cathode; this spot became known as the anticathode. The luminescence of the anticathode was believed to be due to rays that left the cathode at right angles and travelled in a straight line. It had been shown that these "cathode rays" are absorbed in gases and metals in rough proportion to the mass of matter travelled, and inversely proportional to the voltage of the discharge tube. The nature of these cathode rays, however, was in dispute until 1896 when J. J. Thomson showed that they were deflected towards the anode in an electrical field. Cathode rays were therefore composed of negatively charged particles. In a classic experiment, Thomson measured the charge/mass ratio of this new particle—the electron.

Röntgen happened to have a screen made of a fluorescent material, barium platinocyanide, lying on the table in the same room as the discharge tube. He noticed that the screen fluoresced every time a discharge went through the tube. This was due to something that was emitted from the anticathode, which was where the cathode rays struck the tube, but it was clearly not cathode rays that had passed through the glass. The rays that were causing fluorescence of the barium platinocyanide screen had quite different properties than cathode rays or any other form of radiation known. Specifically, these "X"-rays were highly penetrating; Röntgen showed that if he placed his hand on a photographic plate and exposed it to X-rays, the developed plate revealed the bones of the hand but a mere shadow of the flesh. Similarly, he could visualize brass weights within a wooden box. Röntgen quickly established a number of properties of X-rays: They are absorbed in matter in exponential proportion to the mass traversed, but much less so than cathode rays; they ionize air; production of X-rays is increased by placing a plate of heavy metal at the anticathode of the discharge tube. To maximize the emission of X-rays, Röntgen recommended that the cathode be a concave mirror of aluminum and the anticathode a platinum plate within the tube at the focus of the mirror and at 45° to the angle of the cathode rays. The discovery of X-rays caused great public interest; as Paul Ewald put it, Röntgen's visualization of the bones within the flesh elicited "an uncanny memento mori feeling."[15]

When Willie Bragg was 6 years old, he was riding his tricycle in Wellington Square, at the center of North Adelaide, when Bob jumped on him from behind. Willie fell on his left elbow, fracturing it badly. Although X-rays had been discovered only the year before, WHB managed to rig up a primitive radiographic apparatus to visualize the damage to his son's arm—possibly the first diagnostic X-ray to be taken in Australia. Poor Willie did not appreciate his father's ingenuity: "I was scared stiff by the fizzing sparks and smell of ozone, and could only be persuaded to submit to the exposure after my much calmer small brother Bob had his radiograph taken to set me an example." The family doctor, WHBs old friend Alfred Lendon, thought that the arm should be allowed to set stiff in a useful position. However, Gwen's brother Charlie Todd, who had become a physician, believed that the mobility of the joint could be saved. He devised a treatment in which every few days Bragg was anesthetized with ether while the arm was flexed. Not surprisingly, Willie hated these sessions, and started yelling when he heard Uncle Charlie's voice in the house. However, the treatment was successful—although the arm was somewhat deformed and required corrective surgery later in life, it functioned more or less normally.[16]

By 1897, WHBs position at Adelaide University was sufficiently secure that he was able to arrange what was essentially a sabbatical year, to be spent in England studying the school system. A more compelling reason for the trip was WHBs desire to take his family to meet Uncle William, to whom he felt he "owed everything." Gwen's sister Lizzie had married a Cambridge solicitor, Charlie Squires, and moved to England, but she was visiting Adelaide in 1897. It was arranged that WHB and Gwen would leave Australia first, traveling via Egypt and Italy, while Lizzie and the redoubtable Charlotte would follow with Willie and Bob. The ship bearing the latter group stopped in Colombo, Ceylon. Willie, outside Australia for the first time in his life, was fascinated by the exotic sights and sounds of Colombo: Punkah boys in the shops, chanting men paddling outrigger canoes in the harbor, naked children begging in the streets, and lascars operating hydraulic cranes with their feet.

The Bragg family reunited in Marseilles and proceeded on to England, arriving in early 1898. The first stop was Market Harborough, where they stayed with the Uncles in Catherwood House on the market square. Willie found Uncle William to be "rather a vulgar old man" who pinched the boy's bottoms; Uncle James was "a simple character, quite under the domination of Uncle William." Both were retired, but Uncle William owned property, including a brickworks, where Willie "loved seeing the clay oozing out as a long rod, and being cut into bricks by wires." Uncle James had a workshop at the bottom of the garden where he made whistles, kites, and other toys for the boys, and was rewarded by finding his rockery demolished to construct a fort.

Leaving Willie and Bob in the care of Charlotte and Uncle William, WHB, Gwen, and Uncle James went off to tour Wales by bicycle. This form of transportation was so novel that a ferry operator did not have a tariff for bicycles, eventually deciding to charge them the same rate as pigs. On their return from Wales, WHB and Gwen took Charlotte and the boys to Cambridge, where they visited the Squires. After a holiday with the Squires in Hunstanton, the Braggs visited WHBs cousins Will Addison and Fanny Kemp-Smith (neé Addison). While staying with Will Addison, who lived in Croydon, the Braggs went to London, where Willie was impressed by "the Hansom cabs, with the clip-clop of all the horses on the wooden sets, and the adventure of riding behind the apron with the cabbie perched up above." During their visit to the Kemp-Smith's, WHB started to tell the boys a series of bedtime stories; bizarrely, these were about the properties of atoms: "We started with hydrogen and ran through a good part of the periodic table."[17]

The family returned to Adelaide in March 1899, living at the Observatory while WHB designed their new house. Alice Todd had died while the Braggs were in England, and her husband no doubt welcomed the company. Charles Todd laid the foundation stone of the Bragg family house at the corner of Carrington Street and East Terrace, at the south-east corner of South Adelaide and, like their previous house, adjacent to the Park Lands. Perhaps in another acknowledgement of his debt to Uncle William, WHB named his new home Catherwood House. It was, as Willie remembered "an attractive house, of fair size." In its garden, he and Bob were given small plots to cultivate, thereby initiating Willie's lifelong love of gardening. A galvanized shed at the back of Catherwood House was given over to the boys as a workshop. One of WHBs lab assistants taught them the basics of woodworking, which they used to make gadgets. Reading from a book, Willie constructed a motor with a toothed-wheel armature, powered by a bichromate battery, and was amazed that it worked. "Then Bob and I rigged up an electrical bell in the workshop, with a push-button in the nursery, so that Charlotte could summon us when tea was ready." The usefulness of this was limited, however, as Charlotte first had to yell to them to put the battery in the bell. They also made a telephone, clock, and seismograph.[18]

That year, Willie was sent to Queen's preparatory school, on Barton Terrace at the very northern extent of North Adelaide. This proved to be an unpleasant experience for the sensitive 9-year-old. His journey to and from school, involving a long tram ride with walks at either end, provided abundant opportunities for his archenemies the larrikins. School itself provided no refuge: The smaller boys were used as catapult-targets by the larger ones, and the headmaster, Mr Hood, caned the boys for each word over two (out of ten) that they spelled wrongly. Fortunately for Willie, a precocious student, he only had to endure the latter punishment once. However, Willie's immaturity contributed to his own misery. At lunchtimes, he moped alone while the other boys played hockey. Even his budding scholarly abilities could be used against him: "We generally had more than one class in each room, and I remember being in the same room as a very senior class doing Euclid. From what I overheard I realized what it was all about. Somehow Hood must have caught on to what was happening, for he pulled me, a very small boy, out of my class and made me explain the theorems to the large boys while he crowed with delight."

After 2 years at Queen's, Willie's torments eased when he was sent to the Collegiate School of St Peter's, where he entered the fifth form. St Peter's was "the premier Church of England school in South Australia," with 300–400 day-boys and about 75 boarders. The Headmaster was the Reverend Henry Girdlestone, whom Willie remembered as "a vast and impressive man with a china-blue eye and small yellow beard." One of his Girdlestone's impressive characteristics was that he did not believe in corporal punishment; rather, he chastised by saying "Boy, you are a humbug." The year before Willie went to St Peter's, the curriculum had been reorganized into three areas: Classical, scientific, and commercial. Willie took eight of the ten subjects offered: English language, English Literature, French, Latin, Greek, Scripture, Mathematics, and Chemistry. The subjects not taken were Physics and German; when he became a physicist, Bragg would regret his lack of German, as much of the scientific literature was written in that language.

Perhaps foreshadowing his later respect for the humanities, Willie enjoyed Greek and won a school scholarship for Scripture. Girdlestone "took a great interest in our essays and taught us to express ourselves clearly." However, the mathematics teacher was so weak that he had to get Willie and his best friend, Bob Chapman, to explain the answers to the rest of the class. Bob was the son of Robert Chapman, whom WHB had hired as Professor of Applied Mathematics at Adelaide University in 1888.[19] The chemistry teacher, James Thomson, made a more favorable impression: "I remember at our first practical class he said: 'Boys, take up your mortars—now take up your pestles and see how much noise you can make banging your mortars'. We did so. He then said: 'Now that you have found out how much noise you can make, let me never hear that noise again.' "[20] At lunchtime, Willie was allowed to set up the afternoon's experiments while Thomson had a nap in the laboratory. The young Bragg was more interested by Thomson's descriptions of atoms than he had been by WHBs bedtime stories about them, and impressed by the teacher's willingness to answer questions.

Academically, Willie excelled at St Peter's. In 1904, he was top in Mathematics, Chemistry, and French, and top student overall in his class. The following year he passed the higher public examination in Pure and Applied Mathematics and Inorganic Chemistry. Both years he won scholarships.[21] Willie also discovered some athletic abilities. He was a good sprinter, winning the 100-yard handicap race at the 1905 School Sports. Rowing, however, provided his "physical outlet." Girdlestone, who had been a rowing Blue at Oxford, coached the boys in fours on the River Torrens. Every year St Peter's raced Geelong Grammar School of Victoria on the Port River. Unlike Queen's School, there was "very little teasing or persecution" at St Peter's. However, Willie suffered socially from being both immature and bad at team games as well as from being academically precocious: For lunchtime games, he was put in teams with boys from much lower grades; for lessons, he was in classes with much older boys.[22] As Jenkin has pointed out, "In very many ways WLB's school experiences paralleled those of his father."[23]

To escape the heat of the South Australian summer, the Braggs spent their holidays in the Mount Lofty range east of Adelaide or on the coast. On several occasions, the Bragg and Todd families rented part of a boarding house at Port Elliot, a tiny place about 60 miles from Adelaide by train. While Gwen sketched and WHB played golf on a makeshift course in front of the hotel, Willie was entertained by his "great friend and confidante," the teenage Aunt Lorna, who read to him, played games and took him on walks. Their destinations included Victor Harbor, once intended as a major port but now a ghost town, with decaying breakwaters, railway spurs, warehouses, and cranes. Just offshore, the waves crashing on the cliffs of Granite Island provided "a thrilling sight." At Middleton, there was a long beach where the ocean rollers made a deafening noise and the shells were stained blue by a chemical from the seaweed.[24]

Other seaside holidays were spent at Yankalilla, Aldinga, and Port Noarlunga. To get to the latter destination, the Braggs travelled in a horse-drawn mail coach to Willunga, where they were met by Mr Pocock with his dray, who took them the rest of the way. Port Noarlunga consisted of the Pocock's farm and a cottage used by fishermen. Naked except at mealtimes, Willie and Bob explored the reefs and fished in the River Onkaparinga.[25] The summer of 1906–7, when Gwen was again pregnant, WHB hired a horse and buggy so that she could go on excursions. Willie remembered: "Before breakfast Bob and I rode the ponies bare-back into the sea until they swam and we were towed by their manes, and afterwards galloped on the sands till we were dry."[26]

Later in life, Willie retained vivid memories of these seaside holidays. A mile or so out in the St Vincent Gulf, liners were anchored, with small steamboats ferrying passengers and goods ashore. One could also see four-masted grain ships with a chequer pattern along their sides to simulate gun-ports, waiting to be towed up the Port River to Port Adelaide for loading. "There was generally a mirage over the calm sea in the early mornings, and the hulls looked like a picket fence as high as they were long."[27]

In 1903, WHB turned 41. He was a prominent and highly respected figure in Adelaide and very happily settled there. Yet, a chance occurrence would start a chain of events that led within a few years to him leaving Australia and embarking upon a very different career. WHB had been present at the 1888 meeting in Sydney that established the Australasian Association for the Advancement of Science. In 1903, he was asked to serve as President of the Association, which involved giving an address at its 1904 annual meeting, to take place in Dunedin, New Zealand. While he had been in Australia, WHB had followed developments in physics with great interest. As noted above, he had constructed an early X-ray generator; a lecture, with demonstrations, on "Röntgen rays" that he presented in 1896 had been attended by the Governor and the Chief Justice. In collaboration with his father-in-law, WHB in 1897 set up the first "wireless telegraphy" (radio) station in Australia. For his presidential address, he considered discussing the recently discovered electron or the new phenomenon of radioactivity.[e]

In 1896, Henri Becquerel had discovered natural radioactivity by showing that uranium causes darkening of photographic plates. Shortly thereafter, the New Zealand-born physicist Ernest Rutherford found that the radiation from uranium consists of two components, α and β, the latter being much more capable of penetrating matter. An even more-penetrating form of "Becquerel ray," γ, was found in emissions from radium.

The nature of these new forms of radiation was a subject of intense interest to turn-of-the-century physicists. The low penetrability of α-rays and their ability to be deflected by magnetic fields made it likely that they were positively charged particles. β-rays, which had all the properties of cathode rays but were more penetrating, were thought to consist of fast electrons. In her 1903 Ph.D. thesis, the Polish physicist Marie Curie (neé Sklodowska) proposed that α-radiation was analogous to "canal rays," β to cathode rays and γ to X-rays. The latter comparison was based in part upon observations that γ-rays, like X-rays, cause ejection of electrons from metals. Rutherford agreed with Curie on this point, stating in 1904 that γ-rays are hard (highly penetrating) X-rays produced by the ejection of a β-particle (electron) from a disintegrating atom.

In his readings in the area of radioactivity, WHB became intrigued by an apparent anomaly in Curie's studies, in which she reported that all the α-particles emitted by the radioactive decay of radium appeared to travel the same distance in air, rather than the number of particles detected decreasing exponentially with distance from the source. One of the implications of this finding, with implications for the structure of the atom, was that the α-particles must pass through air molecules.

On his return to Adelaide from New Zealand in 1904, WHB was sufficiently interested in the absorption of radiation that he decided to study the phenomenon itself. To that point, "It had never entered my head that I should now do any research work."[28] In the 18 years he had spent in Adelaide, his only published works were three minor papers on electrostatics. However, he had an

[e] The following discussion, and subsequent discussions on the nature of X-rays, is based, unless otherwise indicated, on Wheaton, B. (1983). *The Tiger and the Shark: Empirical Roots of Wave-Particle Dualism.* Cambridge University Press, Cambridge.

excellent mechanic, Arthur Rogers, and a benefactor purchased him the radium he required. It also turned out that WHB was a very talented experimentalist with a gift for instrument design. As his son later put it, "His instrument maker Rogers was a real genius and the α-ray apparatus was a gem. My father aimed at a high standard of perfection in design and construction."[29]

With the help of his first research student, Richard Kleeman, WHB was able to show that α-particles do not obey the exponential law of absorption observed with electrons; instead, the number of particles penetrating matter falls off sharply at a critical thickness. They also found that the ability of an element to absorb α-particles is proportional to its atomic weight. Finally, WHB and Kleeman showed that four different types of α-particle are produced in the radioactive decay of radium, each with a different range of travel in air. This agreed with observations made by Rutherford, professor of physics at McGill University in Montreal. WHB wrote to Rutherford, whom he had met in 1895 when the latter was on his way from New Zealand to Cambridge. After "the necessary three months," to WHBs great joy, a warm reply came from Rutherford, initiating a classic correspondence and lifelong friendship.

Willie entered the University of Adelaide in early 1906, aged 15. The minimum age was formally 16, but exceptions were sometimes made, and WHB may well have remembered his own wasted last year at school. That year, Willie took Physics I, Inorganic Chemistry I, and second-year Pure Mathematics, getting a first-class pass in each. In 1907, he achieved first-class honors in Pure Mathematics III, Applied Mathematics II, Physics II, and Chemistry I (theoretical and experimental). In the third-year, Willie took honors Mathematics, graduating with first-class honors (B.A.).[30] Most of his instruction came from WHB, although he also took many courses from Robert Chapman, father of his school friend Bob. According to David Phillips, Willie also took a course in English and was particularly pleased at winning the University prize for the best English essay "from under the noses of the professionals."[31]

Apart from his academic successes, university was apparently not an agreeable experience for Willie Bragg. As at St. Peter's, he was younger than his classmates, and for Willie that presented insuperable problems: "Although I was 15 when I entered Adelaide University, I think my emotional age was about twelve or less, and my fellow students were mature young men and women. Such a disparity has a cumulative effect. Anyone handicapped in this way is debarred from taking part in the normal activities of his age group, and the very fact that he cannot enter into their plans, schemes, differences of opinion, exercise of authority and so forth, means that he loses the earlier experience which would teach him how to take his part later in life in the world of affairs. He loses touch with what is going on around him and he thinks of the people who guide the course of events as 'they', not as 'we'. He develops a defense mechanism to hide his inexperience from those he meets, and this again makes him shy of asking the questions the answers to which would keep him in touch. He is like a hermit crab with a formidable array of whiskers and claws in front, but with a soft white tail which it has to conceal in a protecting shell."[32] Further alienating him socially from his fellow-students was WHBs insistence that Willie study in his office.[33] This was resented, but apparently silently. Communication between father and son was already difficult.

Although close in age, the Bragg boys were very different in personality—Bob outgoing and athletic, Willie shy and intellectual. Their sister wrote that Willie "took to solitary pursuits, such as shell collecting, being rather dreamy . . . and not good at games like his younger brother"[34] Stephen Bragg, Willie's elder son, never met his Uncle Robert, but gained the impression that "my father was, in some ways, rather a shy and reserved person. Bob was very much the extrovert, as it were, and bouncy member of the party."[35]

Shell collecting became a major interest of the teenage Willie. He accumulated a collection of about 500 different species, which he later donated to the Manchester Museum. The prize specimen was a new species of cuttlefish skeleton that Willie discovered in the summer of 1906–7. He took it to Dr Joseph Verco, South Australia's leading authority, who verified the novelty of the shell and proposed to call it *Sepia gondola*, because of its elongated shape. On seeing Bragg's crestfallen expression, Verco hastily changed his mind, proposing instead the name *Sepia braggi*.[36] Unfortunately for Willie, his father was later to prove less sensitive than Dr Verco to the feelings of a proud young discoverer.

In the seminal period 1904–7, however, WHB was more concerned with his own discoveries. A key development was his meeting with Frederick Soddy, an English physical chemist who visited Adelaide in 1904 after giving a series of lectures in Western Australia. Willie remembered that Soddy "showed us γ rays making a screen glow through a steel plate."[37] On his return to England, Soddy acted as "WHB's agent," helping him with publications and orders and keeping him abreast of scientific developments.[38]

Canal/α-rays were positively charged particles, cathode/β-rays were negatively charged particles (electrons)—but what was the physical nature of X/γ-rays? Röntgen believed that X-rays were longitudinal waves—that is, waves that oscillated in the direction of their propagation. However, there were many similarities between X-rays and visible and ultraviolet light. These latter forms of radiation were believed to be transverse waves, oscillating perpendicular to the direction of propagation. One of the properties of transverse waves is that they can be polarized—resolved into components in which oscillation occurs only in a particular plane. An even more fundamental property of waves is diffraction, the occurrence of interference patterns. Consider, for example, a beam of light passing through a diffraction grating consisting of lines ruled on a glass plate, and then falling upon a screen. Such a screen will exhibit a pattern of alternating light and dark bands. According to diffraction theories, such as that of Ernst Abbé, light scattered by each line spreads out in cylindrical fashion on the far side of the grating. A bright band on the screen occurs where waves from different lines interfere constructively—crest coinciding with crest, or trough with trough. A dark band occurs where waves from different lines interfere destructively—crest coinciding with trough.

Unlike visible light, however, X-rays apparently could not be reflected, refracted, diffracted, or polarized. To explain the anomalous properties of X-rays—in some ways wave-like, in others not—many physicists, including J. J. Thomson, believed that X-rays were impulses, or non-periodic oscillations. The idea was that the collision of an electron with the anticathode of an X-ray tube produces a local disturbance in the electromagnetic field that propagates spherically. A sufficiently high density of such impulses would account for the apparent continuous nature of X-ray beams, but the fact that the impulses are non-periodic would explain their lack of diffraction.

However, several facts that were apparently inconsistent with the impulse theory of X-rays soon became known. In 1899, the Dutch physicists Hermann Haga and Cornelis Wind claimed that X-radiation could be diffracted by passage through a wedge-shaped slit, suggesting that they must be periodic waves. Thomson studied the ionizing properties of X-rays using a discharge electrometer, in which ionization of a gas contained in a tube causes current to flow between electrodes; at maximum current, only $1/10^{12}$ of the atoms in the tube were ionized.

For WHB, the most significant evidence against the idea that X-rays were waves of either the periodic or non-periodic (impulse) variety was the high velocities of electrons ejected from atoms by the impact of X-rays. The amplitude of a wave diminishes with distance from its source as the initial energy becomes distributed over a larger wavefront. When an X-ray ionizes a gas molecule, however, the electron ejected from the molecule has virtually the same energy as the electron that originally created the X-ray—no matter how far the gas molecule is from the X-ray source. To WHB, this could only be explained by assuming that X-rays were particles. According to his view, the anticathode produces X-ray particles in all directions; the density of such particles would of course decrease with distance, but the energy of any individual particle would be conserved and could be transferred to an electron during a collision with an atom.

One problem with the assumption that X-rays were particles rather than waves was that this seemed inconsistent with their high penetrability. However, WHB was struck by the fact that γ-radiation was always associated with α- and β-radiation. He therefore thought that γ-rays may be "neutral pairs," consisting of an α- and a β-particle. The penetrability of X-rays, as well as their lack of deflection in electrical or magnetic fields, could therefore be explained by the electrical neutrality of the particles. WHB further proposed that, during collision with an atom, the relatively massive α-particle is slowed, but the electron continues with the same velocity as before. Willie later remembered first hearing about this hypothesis "some time in 1907" when he and his father were waiting for a horse tram to take them to the Observatory.[39]

The publication of the neutral-pair hypothesis brought WHB into opposition with Charles Barkla, lecturer in physics at University College Liverpool. Barkla had found in 1904 that the secondary X-rays produced when a primary X-ray beam interacts with matter are partially polarized—apparent evidence for their transverse-wave nature. He had also found qualitative and quantitative differences in absorption between primary and secondary X-rays, in disagreement with the neutral-pair theory. Most damaging for WHBs theory was Barkla's observation that the properties of secondary X-rays are characteristic of the substance that is irradiated. Barkla further showed that these characteristic X-rays have two components, which he later called K and L.

In response, WHB reported that when γ-rays strike a metal plate the resulting secondary X-rays are more intense in the direction of incident-beam propagation than in the opposite direction, which he claimed demonstrated conservation of momentum of the γ-ray particles. His trump card, however, was the ionization phenomenon. Describing this in wave terms was analogous to stating that dropping a plank into the ocean from a height of 100 metres could create a wave that would travel outwards for thousands of kilometres and then, encountering a similar plank, send it 100 metres into the air!

WHBs work on α-radiation had gained him a considerable international reputation; the man trained as a mathematician, who worked in far-off Australia and did no research until the age of 41, had in a few short years become one of the leading experimental physicists of his time. In 1907, Horace Lamb, WHBs predecessor at Adelaide and now at Manchester, nominated him for the Royal Society, Britain's most prestigious scientific body. The nomination paper was signed by, inter alia, J. J. Thomson; Rutherford; C. T. R. Wilson, also of the Cavendish Laboratory; and Arthur Schuster, Langworthy Professor of Physics at Manchester. He was elected on the first try. In May of that year, Rutherford left McGill for Manchester, and suggested that WHB succeed him. A serious fire in Montreal meant a delay, at least, in the proposed move. In December, WHB withdrew his name for consideration at McGill—he had already decided that a return to England would be more attractive.

The Cavendish Chair in Physics at Leeds University became vacant when William Stroud resigned in order to take up a business partnership (Barr and Stroud). In January 1907, Soddy suggested WHB as a replacement. Clinton Farr, a former colleague at the University of Adelaide, wrote to Rutherford: "he, more than any other man, has helped to shift the centre of gravity of scientific research a little to the south."[40] When WHB left Australia, the Brisbane Daily Mail wrote: "Australia has sent home her [Nellie] Melba and now is dispatching her Bragg."[41]

The Bragg family was now augmented by Gwendolen Mary, known as Gwendy to distinguish her from her mother, and born on February 26, 1907. They sailed to England on the "Waratah," which was returning from its maiden voyage to Sydney. The captain was very worried about her seaworthiness, and consulted WHB about it. "She had a great castle of decks, and a large extra coal bunker on the top one." On her next voyage, carrying a full load of coal, she was lost with all hands between Durban and Cape Town.[42]

2
Concatenation of fortunate circumstances: Cambridge, 1909–14

The Braggs landed at Plymouth in early 1909. In Leeds, they lived briefly in a house near Shire Oak and then rented a "fine house," Rosehurst, in Grosvenor Road. Gwen was initially "miserable" and "horrified at the grime of Leeds." However, she was a sociable woman and soon made friends. Willie Bragg thought the Leeds years were "the happiest of her life." WHB wrote to Rogers, his former mechanic: "The place itself is grimy, even the suburbs; but you can get out into beautiful country to the North." The Braggs soon acquired a cottage at Deerstones, near Bolton Abbey, "in wild country, surrounded by the moors."[43]

WHB had a harder time than Gwen adjusting to life in Leeds and for several years was "wretched." The physics lab was a poorly heated temporary shed. The dispute with Charles Barkla, which went on for another 3 years, took an emotional toll. Bragg later wrote: "He was disappointed in the progress of his research, and felt that he was not justifying the hopes which Leeds had of him when he was appointed to the Chair."[44]

Bob, who had, like his brother, attended St Peter's School in Adelaide, was sent to Oundle School in Northamptonshire. Bragg followed in his father's footsteps, going to Trinity College Cambridge, where he enrolled in the Mathematical Tripos.

Founded in 1546 by Henry VIII, Trinity was by no means the oldest of the Cambridge Colleges, but one of the largest, richest, and most eminent. Its architectural treasures included the magnificent Great Court, with its fountain and "double-chiming" clock, and the Library, designed by Christopher Wren and containing woodwork by Grinling Gibbons. Trinity had produced many eminent men of letters, including Andrew Marvell, John Dryden, Alfred Tennyson, Lord Byron, and A. E. Housman, as well as the philosophers Francis Bacon and William Whewell. However, it also had a strong mathematical tradition, exemplified by Isaac Newton, appointed Lucasian Professor of Mathematics at Cambridge in 1669 at the age of 26. In recent times, it had become particularly strong in the natural sciences: All three Cavendish Professors to date had been Trinity men.[45]

Like his father, Bragg went to Cambridge for the Long Vacation (summer) term and regretted doing so, considering the time to have been "frittered away." In retrospect, he wished he had used the time to go abroad to learn French or German or take art classes in Leeds—an early example of the conflict he sometimes felt between his scientific and artistic impulses.[46] Even when the autumn term started, Bragg was homesick, writing to his mother: "I miss you and Dad and Sue most horribly, there is a sort of vacant spot in me somewhere that I feel at times, and am feeling a bit just now so I am writing to you hard . . . I am quite a home person I think."[47] According to his sister, he was "lonely and frustrated."[48] Bragg's daughter Patience got the impression that her father—already very shy—may have felt isolated by the fact that the majority of his fellow students at Cambridge were from the English establishment, with public-school backgrounds.[49] (Of the 191 students who entered Trinity College in 1879, 144 came from public schools and only 11 from "Grammar schools etc."[50])

Bragg may also have been taken aback to find that: "My degree at Adelaide puts me on more or less equal terms with 2nd year men here."[51] He took mechanics lectures from Alfred Whitehead, infinite series with Godfrey Hardy and differential equations with Andrew Forsyth. His coach was Robert Herman and his tutor was Ernest Barnes, later Bishop of Birmingham (tutors were assigned by the colleges, coaches were hired privately).

At tennis, at least, the Australians were clearly ahead: Bragg played a close match with "Pym the elder," a tennis Blue.[52] He also ran, "short-distance sprinting being my forte," as it had been at St Peter's. During the winter, he played hockey and lacrosse "extremely badly"; he later attributed his lack of success at team games to "faults of temperament."[53]

In the spring of 1910, Bragg competed in the Trinity scholarships examination. He was suffering from pleurisy and pneumonia and Gwen had to come down from Leeds to nurse him. Because his temperature was still elevated, he was allowed to write the examinations in bed. The Master of Trinity, Montagu Butler, read his essays and commented upon the "brilliant imagination" shown in them. Bragg was awarded a College Senior Mathematical Scholarship, worth £100 a year for 5 years.

At this time, the Mathematical Tripos was in a state of transition. Under the new regulations, the examination occurred in two parts. Students achieving honors in Part I were placed into three classes, but not ranked within the classes. Those achieving honors in Part II were classified as Wranglers, Senior Optimes, or Junior Optimes, and again not ranked within these classes. Bragg wrote the Part I examination in June 1910, and obtained a first-class pass.[54]

Notwithstanding this excellent performance, by the beginning of his second year Bragg had switched from the Mathematical Tripos to the Natural Sciences Tripos—in effect, dropping mathematics in favor of physics. The reason for this momentous decision remains somewhat mysterious. Bragg's statement in his autobiography that WHB "strongly urged" him to switch only deepens the mystery; later in life, Bragg would complain bitterly about his father's unwillingness to advise him on career options. WHB himself had, of course, profited by the change from mathematician to physicist, but if he strongly believed in the superiority of the latter over the former, why did he not intervene earlier in his son's university education? Whatever the reason, WHB had done his son a great favor. Bragg would never be well-regarded by theoretical or mathematical physicists, and it is very hard to imagine that, had he continued in mathematics, he would have achieved anything like the success he did in experimental physics. For the first—and not the last—time, WHB had made a decisive contribution to his son's brilliant career.

For the next 2 years, Bragg worked on Part II of the Natural Sciences Tripos. "[George] Searle gave deadly dull lectures in Heat . . . J.J. gave us stimulating fireworks. I also got very excited over some lectures of [James] Jeans, because they opened up a new world of statistical thermodynamics." After one of the latter lectures, a young man present took Bragg aside and explained where Jeans had gone wrong. The student was the Dane, Niels Bohr. Bohr soon realized that Manchester, not Cambridge, was "the great centre for physics in this country" and moved there "to sit at Rutherford's feet." [55]

Bragg's most influential teacher was the Scot Charles Thomson Rees (C. T. R.) Wilson, a Fellow of Sidney Sussex College and University Lecturer who lectured on optics and demonstrated in the Part II practical class. Wilson's lectures "were the best, and the delivery was the worst, of any lectures to which I have ever been. He mumbled facing the board, he was very hesitant and jerky in his delivery, and yet the way he presented the subject was quite brilliant." [56] Bragg later wrote to Patrick Blackett, who read physics at Cambridge in 1919–21: "I owed a tremendous amount to C.T.R.'s lectures. I remember them vividly but very little of other lectures I attended like you, and I used my notes shamelessly for teaching optics during all my time as a professor . . . His lectures, and talks I had with him when my first ideas about X-ray analysis were brewing, meant everything to me." [57] One of the key elements of Wilson's teaching was the use of amplitude phase diagrams to illustrate diffraction and interference phenomena. Blackett wrote in 1960: "It may be that W.L. Bragg (now Sir Lawrence Bragg), who attended C.T.R.'s lectures just before the war, was aided by them in his later brilliant development with his father, Sir William Bragg, of the application of Fourier analysis to the elucidation by X-rays of complex crystal structures." [58] If so, Wilson's ideas inspired the two greatest achievements of Bragg's research career.

The admiration Bragg had for C. T. R. was reciprocated. Wilson asked Bragg to accompany him to a Royal Society Soireé where his cloud chamber was to be demonstrated for the first time. Inspired by the view from the summit of Ben Nevis, this device by which the path of ionizing radiation could be visualized became an essential tool of nuclear research and earned Wilson a share of the Nobel Prize in physics in 1927.

Bragg's mood improved as he overcame his inherent shyness and started to make friends. With Hugh Townshend (mathematics), Charles Higham (history), Arthur Tisdall (classics), and Brian Gossling (physics), he formed an informal discussion group: "This was the first time in my life that I had [a] simple intimate relationship with a group of kindred spirits, and I revelled in it. I was still a queer fish at whom they often laughed, but our relations were quite easy."

His great friend, however, was Cecil Hopkinson, who was studying engineering. In this, Hopkinson was following a family tradition—his father, John, had been Professor of Electrical Engineering at King's College London at the time of his death, with a son and two daughters, in a mountaineering accident. This tragedy did not deter Cecil from becoming an outstanding skier—he had won the first Kandahar Cup race in 1911—and enjoying other arduous and dangerous pursuits. For Bragg, who had "grown up with no experience of physical adventure," it was—like his relationship with his brother Bob—"the attraction of opposites." "He was the warmest-hearted and most loyal friend it was possible to imagine." When Hopkinson invited Bragg to go skiing with his family to Vermala in Switzerland, Bragg experienced his "usual hesitation in letting myself in for any experience of an unknown nature." Fortunately Bob, who had no such inhibitions, was present, and insisted that he go. On this trip, Bragg for the first time met Cecil's mother, whom he was later to know as "Aunt Evelyn." As there were no lifts, Cecil gave Bragg a couple of days to acclimatize and then took him to the top of the 10,000-foot Wildstrubel.

Bragg also went with the Hopkinsons to their "summer place" at Loch Spelve on the Isle of Mull. It was a farm owned by the Livingstone family, who gave up the main part of the farmhouse to them. The patriarch was "an old tyrant" who would not let his sons or daughters marry (although the eldest boy, then almost 50, had been permitted to get engaged) and would claim to have no English when a problem arose with his guests. They hunted grouse, duck, blackcock, snipe, and hare; however, the shooting was poor as the gamekeeper was a MacPhail, with whom the Livingstones had a traditional feud, so a "ragged bevy of young Livingstones" would drive away the game. Bragg and Cecil also went on sailing trips, where Bragg learned to sail, although he was generally cook. They slept in the bottom of the (open) boat at night.

One summer, Cecil's brother Bertie, Professor of Engineering at Cambridge, hired a large sailing boat at Falmouth for himself, Cecil, Bragg, Russell Clarke, and a cabin boy. Bragg met them at Youghal, Ireland, which the others had reached after a difficult crossing of the Irish Sea, and they sailed to Cork and then Castletownshend. There Bragg developed pneumonia, so was left in the care of the nuns in the infirmary of the Skibbereen workhouse. With one of his visitors, the local schoolmaster, he had a discussion about Charles Darwin's *The Voyage of the Beagle*. This being on the Catholic Church's list of banned books, the nuns organized a special service to pray for the salvation of Bragg's soul. On another occasion, he and some other inmates (all old men) sneaked past the guard on the gate to see the only steamroller in the south-west of Ireland. When he was convalescing, the Townshends of Castletownshend hosted him for three weeks. On his return to England, Gwen met him at Liverpool, the first time she had ever travelled alone. "My mother could not bear ever to be alone . . . It was really counter to her nature to try to think over anything quietly by herself."

Cecil "longed to understand art, he was fascinated by new ideas, he was delightedly amazed by anything quaint or bizarre in points of view, and you could see him muttering and chuckling about such points as he mulled them over. What he gave me was like water in a thirsty land." He got Bragg into adventures which "bolstered up the self-confidence in which I was so sadly deficient."[59]

Cecil Hopkinson also facilitated Bragg's integration into the British establishment. When he went to Cambridge, this was by no means preordained. No doubt there were those among his fellow students who disdained "colonials"—particularly those who were from poor families in the north of England. This kind of attitude may have underlain WHBs statement about Cambridge: "I have always felt a stranger there."[60] Bragg could easily have reacted to the casual snobbery of the public-school set by deciding to be an outsider. Instead, he appears to have enthusiastically embraced the skiing and grouse-shooting lifestyle he was introduced to by the Hopkinsons, and never subsequently questioned its values.

Perhaps because of his new extracurricular interests, Bragg took 2 years to complete Part II of the Natural Sciences Tripos, which was normally a 1-year program—even for students who had taken Part I of the Mathematical Tripos. Practical classes in physics were held at the Cavendish Laboratory, where George Crowe was the "lab-boy." Before the Part II examinations every year, Crowe made out a list of which students he expected to get first-, second-, and third-class honors, and was almost invariably correct. Bragg, whom Crowe considered to be on the borderline between the former two categories, did achieve first-class honors. Students in the Natural Sciences Tripos had never been ranked in order of merit within the classes, but Bragg himself felt that he was "lucky to get a first."[61]

It was not exactly the kind of academic brilliance that often precedes a Nobel Prize-winning career. But Bragg had done enough to be awarded, in the summer of 1912, an Allen Scholarship to conduct post-graduate research towards an M.A. degree at the Cavendish Laboratory. He shared rooms at Trinity with Cecil, who was doing research with his uncle Bertie. This was against the rules, but as a scholar Bragg got quarters with two bedrooms and the cohabitation was arranged by Aunt Evelyn. In the autumn, Bob arrived at Trinity to read engineering.

The Cavendish Laboratory had been established, together with a chair in experimental physics, by Cambridge University in 1871, using funds donated by William Cavendish, seventh Duke of Devonshire and chancellor of the University. Cavendish was related to two great chemists, Henry Cavendish and Robert Boyle, and was Second Wrangler in the Mathematical Tripos of 1829. The first Cavendish Professor was James Clerk Maxwell, also a mathematics graduate of Cambridge (Peterhouse and Trinity). The Laboratory was built in Free School Lane because of the site's proximity to the colleges and lack of traffic, and formally presented to the University in June 1874. It was originally intended to be used for the research endeavors of graduate students.

However, practical instruction in physics was instituted there in 1879. Maxwell died in 1879, and was succeeded by John Strutt, third Baron Rayleigh, a Trinity graduate and Senior Wrangler of 1865. Rayleigh resigned in 1884 and was replaced by the 27-year-old J. J. Thomson, another Trinity graduate and Second Wrangler of 1880.

The research manpower of the Cavendish was greatly increased by a change to University policy allowing the admission of graduates of other universities as research students; if, after 2 years, they submitted acceptable work, they were awarded the degree of Master of Arts. The first of these "outside" students was the New Zealander Ernest Rutherford. Increases in undergraduate and graduate student numbers necessitated expansion of the Laboratory in 1896 and 1908. In the years before the First World War, there were about 30 research students at the Cavendish.[62]

The Cavendish Laboratory was the oldest laboratory for physics research in Britain, and the three incumbents of the Professorship had been among the greatest physicists of the time. The prestige of the Cavendish chair had outstripped that of the much older Jacksonian Professorship of Natural Philosophy, and the Cavendish Professor was head of the University Department of Physics. However, Bragg found it to be "a sad place." Thomson had been in charge for 28 years: His best years as an experimenter were long past, as was his string-and-sealing-wax style of research. C. T. R. Wilson was "the supreme individual artist in experiments" and always worked alone. Searle "had no opinion of researchers and research generally."[63] Thomson gave Bragg a project on the effects of water vapor saturation on ionic mobility, but it was difficult to make any progress as there was very little apparatus available and "practically no workshop facilities at all."[64] He resorted to stealing from a female student the single foot-pump available to blow glassware, and did not return it even when he found her in tears. Crime did not pay, as the results Bragg obtained from his study were "meaningless."

Members of the Cavendish Laboratory might have been "breaking their hearts trying to make bricks without straw," but physics was rapidly advancing elsewhere. WHB, supported by Rutherford, was still promoting the neutral-pair hypothesis of X-rays, although he had abandoned the α-particle, which had been shown to be doubly charged, as the positive part of the neutral pair. Instead, he proposed the "positive electron" or, later, a "quantity of positive electricity" which acted as a "cloak of darkness" for the electron. However, evidence against WHBs hypothesis had accumulated, mainly from studies on the secondary X-radiation generated when a primary X-ray beam impinges upon matter, and shown by Barkla to have properties that were characteristic of the element involved. Barkla's former Liverpool colleague, Charles Sadler, had found that the primary X-ray beam always had greater ability to penetrate matter than the secondary beam. Penetrability was now seen as being directly related to the frequency of the X-rays. If X-rays were impulses rather than periodic waves, they would not have a frequency as such, but the "inverse pulse duration" (the reciprocal of the pulse width) of an impulse was considered the equivalent of the frequency of a periodic wave. The production of characteristic radiation was seen as analogous to the optical phenomenon of fluorescence, in which illumination of certain substances results in emission of light of higher wavelength. For this reason, the secondary radiation started to be referred to as fluorescent X-rays.

In 1910, Richard Kleeman, WHBs former student, showed that electrons ejected from thin metal plates by irradiation with ultraviolet light, like those produced by irradiation with X- or γ-rays, are preferentially scattered in the forward direction; as ultraviolet light was thought to definitely be a transverse wave, this was a further blow to the neutral-pair hypothesis. However, there was still no credible explanation for the efficiency of energy transfer in X-ray ionization. By 1912, WHB had accepted that X- and γ-rays had apparently contradictory properties and that a new theory would subsume both the neutral-pair and impulse hypotheses.[65] As Thomson described the wave-particle controversy, "It is like a struggle between a tiger and a shark, each is supreme in his own element, but helpless in that of the other."[66]

The answer to the conundrum about the nature of X-rays was to come from Munich, where a large number of excellent physicists was then concentrated. At the Institute for Theoretical Physics, Paul Ewald was completing a doctoral thesis supervised by Arnold Sommerfeld, head of the Institute. Ewald's thesis was an examination of how the phenomenon of double refraction of light by a crystal could result from an anisotropic (asymmetrical) lattice of atoms.[f] In January 1912, he sought help from Max von Laue, a theoretician who was a protégé of Max Planck. Their discussion, which took place during a stroll through the English Gardens, set off a chain of ideas in Laue's head: "What are the distances between the lattice points? What happens if you take shorter and shorter wavelengths? Why not try X-rays?" It occurred to Laue that the wavelengths that had been calculated for X-rays were similar to the inter-atomic distances that had been estimated for crystals; if this were the case, then crystals should diffract X-rays in the same way that a line- or cross-grating diffracts visible light.[67]

Laue discussed this idea with Sommerfeld during an Easter-vacation ski trip. However, his proposal that diffraction of X-rays by crystals should be looked for received a negative response. Sommerfeld believed that the collision of cathode rays with the metal of the anticathode produced two components: Pulses with a range of wavelengths that he called *Bremsstrahlung* and thought of as analogous to white light; and homogenous (single-wavelength) waves characteristic of the metal. As crystals were thought to contain an infinite number of different atomic planes, corresponding to an infinite series of diffraction gratings, and if X-ray beams contained a wide range of wavelengths, Sommerfeld quite logically pointed out that a uniform distribution of diffracted radiation, rather than the sharp pattern of spots created by a cross-grating, should be expected.

[f] Double refraction is the ability of certain types of crystal to produce two refracted beams of light, each polarized at right angles with respect to the other, from a single incident beam.

At this time, Munich physicists were in the habit of having lunch at Café Lutz, the marble tables of which were used for writing equations and sketching diagrams during their discussions. Here Laue recruited two experimentalists to help him attempt the diffraction of X-rays by crystals behind the back of Sommerfeld. Walter Friedrich was "the only young physicist at the university with a fair measure of experience with X-rays," having just completed his doctorate in Röntgen's Institute of Experimental Physics, where he had studied the angular distribution of X-rays emitted from a platinum anticathode. Paul Knipping was a current student of Röntgen's. The conspirators believed that the diffracted radiation would be of a specific wavelength that was characteristic of the atom(s) present in the crystal. Barkla had suggested that elements of atomic weight 50–100 would give strong characteristic radiation. Copper has atomic weight 63.5 and crystals of copper sulfate pentahydrate were available in the Institute, so the choice of crystal was made.[g]

The experiment was performed on April 12, 1912. Initially, the photographic plate was placed between the X-ray source and the crystal so that reflected radiation would be detected. When this was unsuccessful, Friedrich and Knipping placed plates all around the crystal and irradiated the crystal again. The plate placed behind the crystal had an intense dark spot corresponding to the position where the undeviated X-ray beam struck it—but it also had a number of elliptical smudges arranged in a ring around the central spot. To ensure that this was the diffraction they were seeking, Friedrich and Knipping ground up the crystal and placed the powder in the path of the X-ray beam—no spots except the central one were obtained.

Laue was not present when X-ray diffraction by copper sulfate was demonstrated—he first learned of it when Friedrich and Knipping reported back to him in Café Lutz. When Sommerfeld was informed, his pleasure that his ideas of the wave nature of X-rays had been confirmed appears to have overcome his displeasure at being disobeyed—he quickly re-assigned Friedrich to the new project. At a May 4 meeting of the mathematical–physical class of the Bavarian Academy of Sciences, Sommerfeld took the precaution of establishing priority for his group by depositing a sealed envelope with details of the experiment. Meanwhile, thin oriented plates of zincblende (ZnS) were ordered. Zinc has atomic weight 65.4, and therefore is in Barkla's preferred range, and was known to be a cubic crystal, which is the class of highest symmetry.

Using these zincblende plates, Friedrich and Knipping were able to direct the X-ray beam down specific symmetry axes of the crystal. In a cube, a line connecting the centers of opposite faces is an axis of four-fold rotational symmetry, because rotation of the cube by 90° (360°/4) around that line brings it to an equivalent position. A line connecting opposite corners of a cube (body diagonal) is an axis of three-fold rotational symmetry, because rotation of the cube by 120° (360°/3) around that diagonal brings it into an equivalent position.

[g] Copper sulfate may have been a good choice physically, but it was a poor one crystallographically; as was known in 1912, copper sulfate pentahydrate belongs to the crystal class of lowest symmetry, triclinic, and therefore the diffraction pattern would be difficult to interpret.

When Friedrich and Knipping directed an X-ray beam down a four-fold axis of zincblende, a four-fold symmetrical pattern of dots was obtained on the photographic plate; when the beam was directed down a three-fold axis, a three-fold symmetrical pattern of dots was produced.[68]

The discovery of X-ray diffraction provided strong support not only for the view that X-rays were waves rather than particles but also for theories of crystal structure that had been developed over about 250 years. As early as the seventeenth century, the remarkable regularity of crystals, notably geological minerals, had inspired scientists such as Robert Hooke and Johannes Kepler to speculate on their internal structure. It was during the nineteenth century, however, mainly in France and the German lands, that a complete analysis of crystal symmetry was carried out.[69]

Because of their plane faces, characteristic symmetries and predictable cleavage behaviors, it was believed that crystals were three-dimensional lattices of atoms or molecules. The defining characteristic of such a lattice was that each lattice point must have an identical environment. One way of categorizing crystals was by consideration of the geometric shapes formed by the lattice points. This involved thinking of crystals as being built up of parallelipipeds whose vertices were the points of the lattice. These parallelipipeds, which became known as unit cells, are related by translation—any unit cell can be superimposed upon any other by movement along one or more of the three major axes of the lattice. The unit cell is therefore the smallest part of the crystal that repeats in three-dimensional space.

In 1848, Auguste Bravais showed that there are only seven unit cells that can be packed together in three dimensions without leaving spaces: Cubic, tetragonal, orthorhombic, rhombohedral, hexagonal, monoclinic, and triclinic. These can be distinguished by the relative lengths of their three types of side and the three angles the sides form with one another. For example, a cubic unit cell has axes of equal length that are all mutually perpendicular, whereas a triclinic unit cell has sides that are all unequal in length and generally form angles that are not equal to 90°.

Bravais also realized that some unit cells could be formed from lattices such that lattice points lay not only at the vertices of the unit cell, but also at its center or the center of one or more of its faces. For example, there are three types of cubic unit cell: Primitive, in which lattice points lie only at the cube corners; face-centered, in which additional lattice points lie at the centers of the cube faces; and body-centered, in which an additional lattice point lies at the center of the cube (Figure 2.1). Considering centered lattices as well as primitive ones, the total number of unit cells was increased to 14.

Another way of characterizing three-dimensional lattices is by the symmetry elements they possess. In the early nineteenth century, six three-dimensional symmetry elements were known: Two-, three-, four-, and six-fold rotation axes, mirror planes, and centers of symmetry. By definition, rotation of an object around a rotation axis by a fraction of a circle brings the object into coincidence. For a three-dimensional lattice, only two-, three-, four-, and six-fold rotation axes, corresponding to rotations of 180°, 120°, 90°, and 60°,

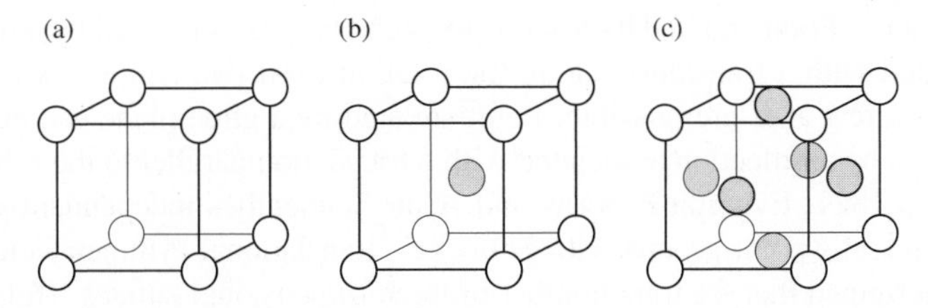

Fig. 2.1 Cubic unit cells. (a) Primitive—one lattice point per unit cell (each point is shared with seven neighboring cells). (b) Body-centered—two lattice points per unit cell. The additional lattice point (shaded) lies at the center of the cube. (c) Face-centered—four lattice points per cell. The lattice points at the centers of the six cube faces (shaded) are each shared by one other cell

respectively, could exist. A mirror, or reflection, plane exists if every point in the crystal has a partner point at the same distance from the plane and directly across from it. A center of symmetry is a one-dimensional version of a mirror plane—for an object with this symmetry element, every point has a partner at the same distance from the center and directly across from it.

Geometric solids possess different combinations of these symmetry elements. The cube has the highest number of symmetry elements: Nine mirror planes, 3 four-fold, 4 three-fold, and 6 two-fold rotation axes, and a center of symmetry. As the octahedron possesses the same combination of symmetry elements as the cube, it was concluded that both these solids were based on an identical three-dimensional lattice—the faces of a cubic crystal represent vertical and horizontal planes of the lattice, the faces of an octahedral crystal represent diagonal planes of the same lattice.

This way of viewing crystals—as three-dimensional arrays of identical parallelipipeds, each of which has characteristic symmetry—was, however, realized to be too simplistic. Another way of representing crystals was as asymmetric units—individual pattern elements—that could be packed, without leaving space, into one of Bravais' types of unit cell. Depending upon how the asymmetric units are arranged within the unit cell, some of the symmetry elements characteristic of that lattice may be lost. Thus it is possible to have a cubic lattice that lacks the full symmetry of a solid cube. From considerations such as these, Johann Hessel showed in 1830 that there were only 32 possible combinations of rotation axes, mirror planes and centers of symmetry. Such combinations of symmetry elements Hessel called "point groups," as they all pass through a point at the center of the object. As the point-group symmetry a crystal possesses can be determined by consideration of its external form, all crystals could be categorized into 32 "crystal classes." Cubic crystals fall into five point groups, only one of which has full cubic symmetry.

Combining the 14 Bravais lattices with Hessel's 32 point groups produced a total of 72 three-dimensional lattices. However, it soon transpired that these were not all the possible crystal structures. In 1879, Leopold Sohncke, Professor of Physics at the Technische Hochschule of Munich, recognized the existence of two additional types of three-dimensional symmetry element: Screw axes and glide planes. Points related by a screw axes can be interconverted by a rotation associated with a translation along the rotation axis; two-, three-, four-, and six-fold screw axes are possible. Points related by a glide plane can be interconverted by a reflection associated with a translation parallel to the reflection plane. In 1891, Evgraph Fedorov and Artur Schoenflies independently combined Hessel's point groups with Sohncke's translational symmetry elements and determined that the total number of three-dimensional lattices—for which Schoenflies introduced the term "space groups"—was 230. At this point, the theory of crystal symmetry was complete; every crystal had to belong to one or other of Schoenflies' space groups. However, since the existence of screw axes and glide planes could not necessarily be determined from the external form of the crystal, it was not possible to assign any crystal to a particular space group.

A paper by Friedrich, Knipping, and Laue was presented to the Bavarian Academy of Sciences on June 8, 1912, by Sommerfeld, who was a Fellow of the Academy. On July 6, 1912, Laue presented a detailed attempt to explain the zincblende diffraction spectrum. Röntgen seconded the motion to accept these works for publication in the proceedings of the Academy (*Sitzungberichte der Bayerische Akademie der Wissenschaften*).

The two papers appeared in late August. However, many physicists had heard about X-ray diffraction before then. Laue discussed the work at a meeting of the Berlin Physical Society on June 8. On his way back to Munich, he gave a talk at Würzburg. A physicist from Göttingen present at the Würzburg lecture obtained copies of Laue's slides to take back with him. Laue also sent copies of one of the photographs to a number of "eminent colleagues."[70] Ewald, the unwitting catalyst for Laue's idea, attended a talk Sommerfeld gave in Göttingen in June 1912. That evening, he derived the theory of the reciprocal lattice and the sphere of reflection, two very important concepts in the post-First World War development of X-ray crystallography.[71]

The first paper consisted of a "theoretical part" by Laue and an "experimental part" by Friedrich and Knipping.[72] In the former, the theory of the three-dimensional diffraction grating was derived. Laue assumed that the radiation "emitted" by an atom in the crystal is of a definite wavelength. The "wavelets" originating from atoms along a row of the crystal lattice will only result in a diffraction maximum or minimum—that is, give a spot on the photographic plate—in a direction for which the wavelets from neighboring atoms are in phase—that is, reinforce on another by coinciding crest-to-crest and trough-to-trough. Because a crystal is a three-dimensional lattice, however, this condition must be satisfied not only for neighboring atoms along a row of the lattice, but rather along rows corresponding to all three axes of the crystal. Laue therefore derived three equations, containing terms for the secondary X-ray wavelength, angle of secondary X-ray emission, and distance between atoms along each crystallographic axis, which must all be solved for each spot on the diffraction pattern.

So far, so good. However, in order to explain the diffraction patterns obtained for zincblende by Friedrich and Knipping in terms of Laue's theory—the experimental part of the paper—it was necessary to make the ad hoc assumption that the diffracted radiation consists of components with five different wavelengths. Because there is, as noted above, a relationship between the inter-atomic distance in the crystal and the wavelength of the diffracted radiation, and neither of these values could be independently measured, one could only be expressed in terms of the other. Thus, Friedrich, Knipping, and Laue concluded that the wavelengths present in the diffracted radiation were in the range 0.038–0.15 a, where a is the axial length of the zincblende lattice (in a cubic lattice, all three axes are equivalent).

The lack of a clear-cut relationship between theory and observation must have been a bit of a disappointment. On the bright side, though, the results contained strong evidence for the wave nature of X-rays. The sharpness of the spots, and the fact that the secondary radiation was highly penetrating, were properties of waves. Further, if the incident radiation were corpuscular, only atoms on rows parallel to the incident beam could scatter coherently and the scattering of neighboring rows would occur without any phase relationship, so the photographic plate would exhibit uniform circles. One important point was left unresolved, though: "We will for the present leave undecided whether the periodic [secondary] radiation is formed in the crystal by fluorescence or whether it is already present in the primary radiation itself, together with the [*Bremsstrahlung*] pulses, and is simply separated by the crystal." In the latter case, the scattering of X-rays by the zincblende crystal would be essentially identical to optical diffraction by a grating. In the former, the zinc atoms of the crystal would form a lattice of points generating secondary X-rays, which would then interfere to produce a diffraction pattern.

In the second paper,[73] Laue analyzed the four-fold symmetrical diffraction patterns of zincblende. Because of the symmetry, each quadrant of the pattern was identical, containing 12 spots. Laue found that each of these non-symmetry-related spots could be explained if the secondary radiation contained five different homogeneous components, whose wavelengths were in the ratios $4:6:7:11:15$. From the density of the crystal and the atomic weights of zinc and sulfur, and assuming that there was one molecule of ZnS per unit, he calculated that the value for a, corresponding to the dimensions of the cubic unit cell, must be 3.38×10^{-8} cm. Knowing a, one can then calculate the wavelengths; these were in the range $1.3–5.2 \times 10^{-9}$ cm, which was consistent with Sommerfeld's estimate based on diffraction by slits.

Willie Bragg heard about these exciting new findings when, no doubt relieved to escape the frustrations of the Cavendish, he joined his family for a holiday in Cloughton on the Yorkshire coast in August 1912. WHB had had advance notice of Laue's work from the Norwegian physicist Lars Vegard, who had spent some time in his department at Leeds and was now working in Würzburg. As described above, Laue gave a talk about X-ray diffraction in Würzburg in early June; on the 26th of that month, Vegard wrote to WHB about "new curious properties of X-rays" enclosing a photograph that he had obtained from Laue.[74]

Not unnaturally, WHB was not ready to accept the apparent evidence for the wave nature of X-rays. His son was also "an ardent supporter of my father's corpuscular theory."[75] Together, they tried to come up with alternative interpretations of Laue's findings that were consistent with the neutral-pair concept. They decided that the putative X-ray particles might be being channelled through "avenues" that lay between rows of atoms in the crystal lattice. When the family returned to Leeds, Bragg used his father's laboratory for an experiment to test the avenue hypothesis. A thin X-ray beam was directed at a crystal forming the aperture of a light-tight box containing a photographic plate. The angle between the box and the X-ray beam was varied, in the hope that spots would be formed on the plate when the X-ray beam was parallel to an avenue. No evidence in favor of the hypothesis was obtained.

On his return to Cambridge in October, Bragg borrowed a copy of Laue's paper from Richard Whiddington, a Fellow of St John's College, and continued to puzzle over the interpretation of the zincblende diffraction pattern.[76] It would have been fitting if the solution had occurred to him in the Trinity College garden where Isaac Newton had once sought inspiration. However, it was while walking along the "Backs" (parklands on the far side of the River Cam from the main College buildings), opposite St John's College, that Bragg realized that X-ray diffraction could be regarded in the same way as the diffraction of light by a line grating, with the sheets of atoms in the three-dimensional crystal corresponding to the lines of the two-dimensional grating.[77] If this were the case, then there would be a simple relationship between the angle at which diffracted waves reinforced one another, and therefore give detectable diffraction maxima and mimina, and the spacing between atomic planes in the crystal. What had tipped Bragg off were Laue's findings that the spots of the diffraction pattern became less circular as the plate was moved away from the crystal, and that the spots changed in intensity when the crystal was tilted away from the symmetry axis.[78]

According to his wife's memoir, Bragg "rushed back and put it to Cecil, who though an engineer himself, could grasp it."[79] However, when Bragg applied the concept of reflection from atomic planes to Laue's zincblende data, he found that it did not work! His bitter disappointment did not last long. On further reflection, a less glamorous but even more important idea occurred to him. At one of the meetings of the little discussion group to which Bragg belonged, Gossling had read a paper on a theory of crystal structure developed by William Pope, Professor of Chemistry at Cambridge, and William Barlow, the last of the gentleman-scientists. In this paper, Pope and Barlow had mentioned that the most efficient way to pack spheres of identical size into a cubic crystalline structure was not the primitive type, in which lattice points occcur only at the corners of the cube, but rather the face-centered type, in which lattice points occur both at cube corners and the centers of the six cube faces (Figure 2.1).[h]

[h] It is not clear which paper Gossling presented, but a major review of Pope's and Barlow's ideas had been published five years earlier. [Barlow, W. and Pope, W. J. (1907). The relation between crystalline form and the chemical constitution of simple inorganic substances. *Journal of the Chemical Society* **91**, 1150–214.]

Bragg immediately realized that the two cubic lattices would result in different diffraction patterns, because the spacing between atoms would be different—in the face-centered lattice, for example, the inter-atomic spacing along face diagonals would be half that in the primitive cubic lattice. When he analyzed the zincblende pattern as a face-centered cube, he was able to explain all the spots found without making any ad hoc assumptions.

How was it that a 22-year-old student found the solution that had eluded the experienced physicists of Munich, Berlin, Göttingen, and—for that matter—Leeds? Most physicists of the time would have been familiar with the idea that crystals were space lattices and aware of some of its physical implications, but it is safe to say that most would never have heard of the different types of cubic lattice. According to Ewald, "crystals were liable to be treated as museum pieces and freaks of nature" rather than "typical representatives of solid matter."[80] Laue, however, was working in a city with a strong crystallographical tradition. Sohncke's cigarbox models of his "point systems" were in the "museum room" of Sommerfeld's Institute for Theoretical Physics. Paul von Groth, Professor of Mineralogy at the University of Munich, was in 1912 halfway through writing his five-volume bible of crystal structure, *Chemische Krystallographie*. Laue had done his doctorate—with Max Planck—on "Theory of Interference Phenomena in Plane Parallel Plates," which sounds exactly like Bragg's concept of X-ray diffraction. But he was unfamiliar with the intricacies of crystal structure: "During my first stay in Göttingen I had made a half-hearted attempt to attend a mineralogy course but had given up very soon. From books I then learned the rudiments of crystallography, that is to say, crystal classes, that was all."[81]

As Bragg himself realized, his successful interpretation of Laue's data was largely due to a "concatenation of fortunate circumstances."[82] As a recent graduate, the lectures of Thomson on the relationship between cathode rays and X-rays, and those of Wilson on optical diffraction, were still fresh in his mind. The concept of the face-centered lattice—"the decisive factor, as far as I was concerned"[83]—was not something Bragg had encountered in his physics teaching, but it had dropped into his lap from another direction. Even so, his great discovery would never have happened had Bragg not been his father's son—WHBs privileged knowledge of the Munich experiment and his keen desire that Laue's interpretation be proved wrong provided the prepared mind of his son with the impetus it needed. If the discovery of X-ray diffraction could only have happened in Munich, its interpretation could only have happened in Cambridge.

According to the historian John Heilbron, the correct interpretation of Laue's findings was also arrived at in Manchester by two research students of Rutherford's: Henry Moseley and Charles Galton Darwin, grandson of the famous naturalist. Their analysis was presented at the Manchester physics colloquium on November 1, 1912. WHB, who was present, told Moseley and Darwin that his son had arrived at a similar conclusion. Three days later, Moseley wrote to his mother: "After much hard work Darwin and I found out the real meaning of the [German] experiments and of this I gave the first public explanation on

Friday. I knew privately however that Bragg and his son had worked out an explanation a few days before us, and their explanation although approached from a different point of view turns out to be really the same as ours. We are therefore leaving the subject to them."[84]

If Moseley and Darwin were able to explain Laue's diffraction pattern of zincblende almost simultaneously with Bragg, it would certainly take some of the luster off the latter's achievement. However, the Manchester explanation was incorrect, as Moseley acknowledged in a November 18 letter to WHB: "I see from Tutton's letter to Nature and from Pope and Barlow's papers that the ZnS crystal is not nearly as simple as we [Moseley and Darwin] thought . . . I am at present too muddled over the geometry of the Pope and Barlow crystal to go into the question. Perhaps your son's paper at Cambridge will make this all clear."[85]

Rather than Moseley and Darwin agreeing not to publish a significant finding in order to let the Braggs get the credit for it, the opposite appears to be the case. According to a letter Bragg wrote to John Desmond Bernal in 1942, "Rutherford asked my father to hold back his results on X-ray spectra until Moseley's paper was ready, in order to encourage a young researcher, which my father did, I think with almost too much generosity."[86]

By the time Moseley had written to WHB, Bragg had already presented his interpretation of the Laue phenomenon to the Cambridge Philosophical Society. Entitled "The Diffraction of Short Electromagnetic Waves by a Crystal," the paper was read on November 11, 1912,[i] and subsequently published in the Society's *Proceedings*.[87] Bragg used the term "short electromagnetic waves" rather than "X-rays" because his father and he had not given up the particle hypothesis and felt the radiation being scattered by crystals might not be the (neutral-pair) characteristic X-rays but rather the *Bremsstrahlung* caused by the stopping of cathode rays in the X-ray tube.[88]

Bragg started by demolishing Laue's interpretation. By making his arbitrary assumption about the presence of five specific wavelengths, Laue had claimed that he could account for all the spots in terms of the cube length, a, and a set of integers, h_1, h_2, h_3, which represented vectors along the three crystallographic axes. However, Laue was vague on how close an approximation to his equations was required in order to obtain a detectable spot. As Bragg pointed out, there were various combinations of h_1, h_2, and h_3 that agreed "very closely" with one of the five wavelengths, and yet no spots were observed at positions corresponding to these values. An experiment in which the Munich group had tilted the crystal at 3° from the cube axis was also incompatible with Laue's analysis. The tilted crystal gave a distorted diffraction pattern but with the same spots as those given by the untilted specimen. If diffraction resulted from specific wavelengths in the incident beam, tilting the specimen would abolish the conditions for diffraction.

[i] Bragg's presentation to the Cambridge Philosophical Society has also been dated to November 2 (RI MS WLB33D/124) and November 10 (RI MS WLB33D/125). The date given here is from the published paper.

Next, the concept of internal reflection of X-rays was introduced. Bragg noted that a crystal consisting of sets of parallel planes will produce an interference maximum when the diffracted radiation has wavelength $2d \cos \Theta$, where Θ is the angle between the incident beam and a line normal to the plane, and d is the shortest distance between adjacent planes. This, in part, explains the variable intensity of the spots, because, when the wavelength is too small, the "successive pulses . . . begin to neutralize each other"; when the wavelength is too large, the energy of the reflected beam is small. Spot intensity is likely also a function of the density of atoms in the plane.

Bragg's view of diffraction cleared up the question about the relationship between the primary and secondary X-ray radiation that Laue had left open. Even if the primary beam consisted of completely heterogeneous, "white" X-radiation, at any given value of Θ only wavelengths that corresponded to values of d present in the crystal would result in diffraction: "Considered thus, the crystal actually 'manufactures' light of definite wavelengths, much as, according to [Arthur] Schuster, a diffraction grating does."

Laue had assumed a primitive cubic lattice. However, according to Pope and Barlow, a face-centered lattice gave closest packing of spheres of equal size in a cubic arrangement. Pope's and Barlow's "valence volume" theory of crystal structure stated that ions of the same charge would have the same volume. Since zinc and sulfur are both divalent, they could therefore be more snugly packed into a face-centered lattice than a primitive cubic one. This argument was another lucky break for Bragg; the valence volume theory was wrong, but it led Bragg to a correct conclusion—that zincblende is based on a face-centered lattice.

Bragg then applied his ideas of "reflection"[j] and the face-centered lattice to the diffraction pattern of zincblende. An infinitely extending lattice will contain an infinite number of different planes, which can be described by assigning coordinates to representative atoms that lie on them. However, the density of lattice points in the planes with higher coordinate values will be lower (Figure 0.2). If the atoms of a crystal lie on lattice points, low lattice-point density will correspond to low atomic density, and therefore such planes should diffract X-rays only weakly. Bragg showed that he could explain every spot in the zincbende pattern as originating from a crystal plane with low coordinate values, and that no combinations of such low values failed to result in a spot (Figure 2.2).

Finally, Bragg emphasized that his interpretation was "fundamentally the same as that employed by Laue." He showed that spots on the zincblende diffraction pattern occur for all integral values of the Laue coefficients h_1, h_2, h_3 that correspond to wavelengths within a certain range. The wavelength range was expressed in units of a/λ, since neither a nor λ could be independently determined; a could be estimated from the density of the crystal and the atomic

[j] Technically the scattering of X-rays by crystals is diffraction rather than reflection, but the latter term was preferred by Bragg and will be used because of its historical significance.

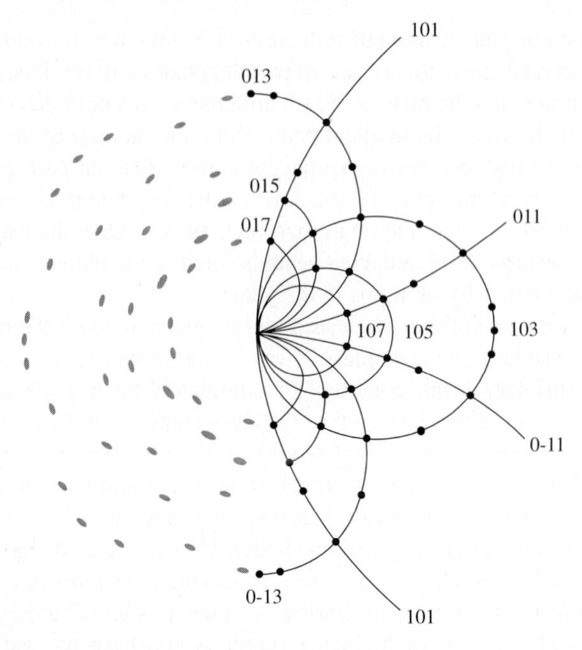

Fig. 2.2 Bragg's interpretation of the X-ray diffraction pattern of zincblende. Every spot on the diffraction pattern lies at the intersection of two ellipses. These ellipses represent the directions of radiation scattered by a zone of the crystal (a set of planes all parallel to a particular axis). The intersection of two ellipses therefore represents the direction of X-rays scattered by a single type of atomic plane. Adapted, with permission, from Figure 4 of Bragg, W. L. (1914). The diffraction of short electromagnetic waves by a crystal. *Proceedings of the Cambridge Philosophical Society* **17**, 43–57. Published by Cambridge University Press.

weights of zinc and sulfur, but only if the numbers of molecules of ZnS per unit cell were known. As Bragg pointed out, the Pope–Barlow structure for zincblende had four zinc atoms grouped in tetrahedra at each lattice point of a face-centered cube.

Laue and Bragg both viewed crystals as three-dimensional diffraction gratings. They derived relationships between the spacing of the crystal lattice and its resulting X-ray diffraction pattern that were fundamentally the same. However, Laue's treatment was unnecessarily complicated, as it contained terms for the spacing between lattice points along all three crystallographic axes. Bragg's treatment, in contrast, envisaged X-rays being reflected by sheets of atoms in the crystal, and therefore the condition for diffraction depended only upon the spacing between the planes. Laue viewed the crystal as a three-dimensional array of points; Bragg as a one-dimensional array of planes.

The demonstration of X-ray diffraction by crystals had shown that X-rays are transverse waves and that crystals are three-dimensional atomic lattices. However, the relationship between the primary and secondary X-ray beams was left unclear and Laue's interpretation of the zincblende diffraction pattern was erroneous. Bragg's work now clarified the first issue and corrected the latter. The primary X-ray beam consists of "white" radiation (as described below, this is only partly true), the secondary X-rays are homogeneous beams of particular wavelengths selected from the white radiation by the crystal. The spots on the diffraction pattern correspond to specific planes of atoms within the crystal, and the plane responsible for a particular spot can easily be determined from its location relative to the origin of the diffraction pattern. The wavelength of a secondary beam and the spacing between the planes from which it arises are related, and, as neither could be independently quantified, it was not possible to determine the dimensions of the crystal lattice. However, it was possible to distinguish between different types of lattice.

In the discussion following Bragg's presentation to the Cambridge Philosophical Society, C. T. R. Wilson suggested that "reflection" of X-rays could be obtained from external faces of crystals as well as internal planes, as long as they were smooth enough.[89] Bragg decided to try this with the mineral mica, the fracturing of which was known to produce extremely smooth faces. Using a strip about 1 mm thick, he was able to demonstrate reflection of X-rays from the mica surface.[k] As Bragg later recalled, "I remember so well taking my terribly crude picture of reflection [from mica] to show to J.J. Thomson. He betrayed his excitement in a characteristically J. J. way by thrusting his spectacles up on his forehead, ruffling his hair violently, and making a peculiar mixture of grin and chuckle. It was a great moment."[90] He wrote to WHB: "I have just got a lovely series of reflections of the rays in mica plates, with only a few minutes exposure! Huge joy, I think the mirror work is a possibility."[91] This letter is undated, but on December 5, WHB wrote to Rutherford, who had replaced Schuster as Professor of Physics at Manchester in 1907: "My boy has been getting beautiful X ray reflections from mica sheets, just as simple as the reflection of light in a mirror."[92] By examining the properties of the rays reflected by mica, WHB convinced himself that they were indeed X-rays, an observation that led him to develop the X-ray spectrometer (see below).

The "mirror work" that Bragg referred to involved bending the mica into a semi-circle so that the reflected X-rays could be brought to a focus—a spectacular demonstration of Bragg's concept that was reported in a letter to *Nature* on December 12, 1912, while the earlier paper was still in press.[93] Another fruitful suggestion came from Pope, to whom Bragg was introduced by Arthur Hutchinson, University demonstrator in mineralogy at Cambridge.[94] Pope proposed that Bragg study the alkaline halides, and obtained for him suitable crystals of rock salt (NaCl) and sylvine (KCl) from Germany.[95]

[k] In principle this is identical to Friedrich's and Knipping's first experiment on copper sulfate. This failed to demonstrate reflection because the angle of incidence used was very high. [Ewald, P. P. (1962). The immediate sequels to Laue's discovery. In *Fifty Years of X-Ray Diffraction* (P. P. Ewald, ed.), pp. 57–80. International Union of Crystallography, Utrecht.]

Bragg had more than enough avenues of research to follow, but was frustrated by the poor facilities available to him at the Cavendish Laboratory. "I had to manage with bits of cardboard and drawing pins, and a very poor [X-ray] tube worked by an induction coil."[96] In his eagerness to try out his ideas on as many crystals as possible, Bragg burned out the 10-shilling platinum contact of the induction coil. This greatly displeased the head mechanic, Fred Lincoln, who had joined the Cavendish in 1893 and whose "fierce eye and even fiercer moustache . . . induced a very proper respect in the young research worker applying to him for apparatus and stores."[97] Lincoln "regarded himself as the appointed executive of J.J.'s parsimony"[98] and made Bragg wait a month for a replacement.[1]

During the winter of 1912–13, Bragg was also learning crystallography. Pope, "my kind counsellor,"[99] sent him to London to meet William Barlow. Although "purely a geometrician," Barlow was "an inexhaustible mine of suggestions."[100] From him, Bragg began to augment his meager store of knowledge about crystallography. Like Pope, Barlow must have been only too happy to help. Bragg's analysis of zincblende had vindicated Pope's and Barlow's conception that in crystals "the component atoms are homogeneously arranged to form a close-packed assemblage." True, Bragg's work would soon lead to the demise of their valence-volume theory, which stated that atoms of the same valency are of equal volume. However, this was a small price to pay—Pope and Barlow had realized that the anions and cations of binary compounds must be of at least slightly different sizes in order to account for the reduction in symmetry seen in many crystals of these compounds.[101]

His study of crystallography exposed the sensitive Bragg to some embarrassment over the naivety of his paper for the Cambridge Philosophical Society. In that paper, he had developed his own nomenclature for describing different crystal planes. He now found that a more convenient system had been devised by William Miller in 1839 and widely used ever since. The Miller indices, (hkl), of a crystal plane are defined as the reciprocals of the distance along the unit cell that the plane cuts the a, b, and c axes, respectively. Negative values indicate planes that cut the axes on the negative side of the origin of coordinates. Thus, the six faces of the unit cell have Miller indices (100), $(\bar{1}00)$, (010), $(0\bar{1}0)$, (001), and $(00\bar{1})$. One of the planes connecting opposite corners of the unit cell would be (111), and so on.

WHB was initially much less interested than his son in the possibility of using X-rays to determine the structures of crystals. He was preoccupied with what the apparent diffraction of X-radiation implied for his neutral-pair theory, which envisaged X-rays as particles rather than waves. Laue, a partisan of the wave theory, had noted both similarities and differences between the primary

[1] David Schoenberg wrote: "I remember one of my friends, who was trying to separate the isotopes of lithium, wasted nearly a year because he did not dare to ask Lincoln for a fresh supply of metal and it turned out in the end that Lincoln had made a mistake and given him sodium instead of lithium." [Schoenberg, D. (1987). Teaching and research in the Cavendish: 1929–35. In *The Making of Physicists* (Williamson, D., ed.), pp. 101–112. Adam Hilger, Bristol.]

and secondary X-ray beams. Bragg believed that the secondary beam was a reflection of part of the spectrum of the primary beam, and therefore if the latter was a wave the former must be too; however, he had claimed that the primary radiation might contain neutral particles in addition to "short electromagnetic waves." WHB decided to characterize the reflected radiation. To that end, he designed an instrument by which the reflected rays could be directed into an ionization chamber, the production of ions being a fundamental property of X- and γ-radiation. WHBs X-ray spectrometer contained a platform on which the crystal could be rotated with respect to the X-ray beam and an ionization chamber that could be rotated around the crystal. The ionization chamber contained a gas that was ionized by X-rays and an electrometer so that the amount of radiation detected could be quantified (Figure 2.3).

Father and son joined forces during the Christmas holiday of 1912–13, studying the radiation reflected by crystals of rock salt.[102] Although WHBs laboratory was better equipped than the Cavendish, it was still tricky work. As Bragg wrote in 1961, "You must find it hard to realize in these days what brutes X-ray tubes then were." [103] Running one was as much an art as a science. Traces of gas in the evacuated glass tube were necessary to supply electrons for the cathode ray. Gas was emitted from the metal parts of the tube as it heated up and removed by the sputtering of the cathode. As the gas level varied, so did

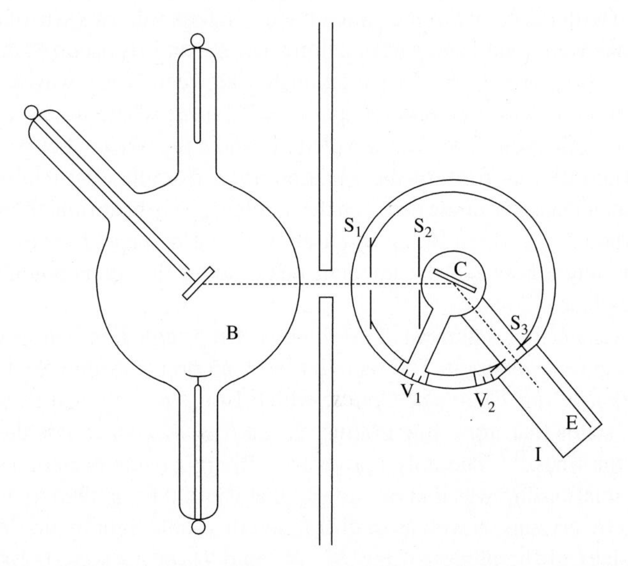

Fig. 2.3 The X-ray spectrometer. B: X-ray bulb; S_1, S_2, S_3: slits; C: crystal; V_1, V_2: verniers; I: ionization chamber; E: electrode. The table on which the crystal sits and the ionization chamber can both be rotated. V_1 is used to measure the angle between the crystal face and the X-ray beam, V_2 to measure the angles at which reflection of X-rays occur. Adapted from Bragg, W. L. (1914). X-rays and crystals. *Journal of the Röntgen Society* **10**, 70–82.

the voltage that had to be applied and the hardness (penetrating power) of the resulting X-rays. If the voltage became too high, the anticathode would melt; if the voltage became too low, the insulation would fail.[104] When his tube got too hard, Bragg had to hold a match to a little palladium tube, thereby allowing some gas to diffuse into the tube. Measurement of the diffracted radiation was almost as bad: "The ionization was measured with a Wilson tilted gold-leaf electrometer. I well remember the fiddly job of cutting strips of gold leaf and fixing them to the plate with a bit of lick. The regular sweep of the crystal through the rotating angle was achieved by having a capstan with spokes which moved the crystal, and pulling the spokes with ones finger in time to the beat of a metronome."[105]

The results the Braggs obtained from this first period of collaboration were published in a joint paper in April 1913.[106] Using the prototype X-ray spectrometer, it was possible to make accurate measurements of the angles at which the diffracted X-rays emerged. When they irradiated various faces of various crystals with X-rays from a platinum anticathode and plotted the ionization current against the angle at which the detector was set, the Braggs found three peaks, which they labeled A, B, and C. These peaks always occurred with the same relative magnitudes and spacings. "There can be little doubt the three peaks are, in all cases, due to the same three sets of homogeneous rays, rays which do not change with the state of the bulb [i.e. whether the X-ray tube was producing more penetrating "hard" or less penetrating "soft" X-rays], but may well do so with the nature of the anticathode." In rock salt, two sets of the A, B, and C peaks were found and part of a third. B_1, B_2, and B_3 occurred at θ values of 11.6°, 23.6°, and 36.6°. The relationship between X-ray wavelength and diffraction angle was now restated as $n\lambda = 2d \sin\theta$, where n was an integer, the order of reflection; λ was the wavelength; and d the spacing between atomic planes. The different form of the equation arose from the fact that θ was the angle of incidence (the angle between the incident X-ray beam and the reflecting plane), whereas Θ, which Bragg had used in his earlier paper, was the glancing angle (the angle between the incident beam and a line perpendicular to the reflecting plane).

This was the first statement of the famous "Bragg equation." However, there was nothing novel about it. The second (1909) edition of Arthur Schuster's *An Introduction to the Theory of Optics*, which Bragg had studied as an undergraduate, states that, for a line grating, $2e \sin\theta = n\lambda$, where e is the spacing between the lines.[107] The only reason why Bragg's name became associated with this relationship was that he showed that it could be applied to diffraction of X-rays by crystals as well as to diffraction of visible light by gratings.

The sines of the angles of the B_1, B_2, and B_3 peaks were 0.200, 0.401, and 0.597—very near a ratio of 1 : 2 : 3. "There can be little doubt as to the interpretation of these results. The three peaks A, B, and C represent three sets of homogeneous rays. Rays of a definite quality are reflected from a crystal when, and only when, the crystal is set at the right angle." The idea of orders of diffraction was already familiar from optical diffraction, and arose naturally from the Bragg equation. For any crystal plane, the condition for reflection would be satisfied at various values of n, corresponding to multiples of $\sin\theta$. If reflection occurred from a set of planes at a certain value of $\sin\theta$, it would also occur at $2\sin\theta$, $3\sin\theta$, etc., so long as radiation of the appropriate wavelength were present in the X-ray beam and θ was less than 90°. It was clear that the $A_1B_1C_1$ and $A_2B_2C_2$ sets of peaks were the first- and second-order spectra, respectively, of rock salt.

Despite their statement that "These results do not really affect the use of the corpuscular theory of X-rays," the Braggs were now getting very close to the view of X-radiation held by wave proponents. The primary beam was now seen to consist of a heterogeneous range of wavelengths (Sommerfeld's *Bremsstrahlung*), superimposed upon which were a few homogeneous (single-wavelength) components that were characteristic of the metal of the anticathode.

These homogeneous components would prove to be an invaluable tool in the development of X-ray analysis as they made possible absolute measurements of crystal dimensions.[m] Indeed, the Braggs were now able to make the first measurements of atomic spacings in crystals and of X-ray wavelength. Bragg's work at Cambridge had now shown that rock salt is a face-centered cube. From the density of the crystal and atomic weights of its constituent atoms, the Braggs calculated that the length of the cube, a, was 4.45×10^{-8} cm. From this, the wavelength of the homogeneous X-ray components could be determined. However, the Braggs were not sure how a, the length of the unit cube, was related to d, the spacing between reflecting planes. If these were the same thing, the wavelength of the B peak was 1.78×10^{-8} cm. If, however, d were actually $a/2$, taking into account the planes of atoms on the cube faces as well as those at the cube corners, then the B wavelength would be halved.

Moseley and Darwin now obtained compensation for losing out to the Braggs on the interpretation of the Laue phenomenon. Using a spectrometer six times more sensitive than WHBs, they showed that two of the three peaks found by the Braggs were in fact doublets. The X-ray spectrum from a

[m] In a Laue photograph, multiple spots occur only because of the range of wavelengths present in the incident X-ray beam. Thus, the "Bragg equation" is satisfied for atomic planes with different values of d only because the incident beam consists of different values of λ. By definition, therefore, the spots on a Laue photograph result from secondary beams of different wavelengths. Finding values of d, which is necessary to determine the structure of the crystal, is essentially impossible, because the values of λ are not known. With the X-ray spectrometer, however, the ionization chamber could be turned so that the Bragg equation was satisfied for the same set of atomic planes at all wavelengths present in the incident radiation. Instead of the spot on the Laue photograph, one obtains a spectrum of X-radiation reflected by that set of planes. If the incident beam were truly "white," the spectrum would be uninformative. However, the presence of the homogeneous components meant that values of θ at which a given peak—platinum "B," say—appear in the reflected X-ray spectrum all corresponded to the same wavelength of secondary radiation. Values of d calculated by this technique are therefore in the same units. What those units are depends upon the values of λ, but once the wavelength of platinum "B" radiation was known, absolute measurements of lattice spacings could easily be made. Even without that information, relative dimensions could be determined. Because of the presence of the characteristic radiation, and the ability of the spectrometer to detect it, one of the variables in the Bragg equation, λ, could be kept constant.

platinum anticathode consisted of five fine lines, which Moseley and Darwin renamed α, β, γ, δ and ϵ. By studying the spectra of X-rays emitted by a large number of elements, Moseley showed that there were up to six spectral lines per element, and that these occurred at higher frequency with increasing atomic weight of the element. Plotting the square root of frequency against an integer that Moseley called N, and which represented the position of the element in the periodic table, gave two series of straight lines; one, corresponding to Barkla's K rays, went up to $N = 50$ (tin); the other, corresponding to Barkla's L rays, went up to $N = 79$ (gold). The close linear relationship between N and X-ray frequency showed that the former must be, as the Braggs put it, "more than a mere ordinal: it must represent some fundamental attribute of the atom." [108] Influenced by the view of the atom arising from Rutherford's studies on radioactivity—massive, positively charged particles surrounded by an equal number of electrons—Moseley stated: "This integer N, the atomic number of the element, is identified with the number of positive units of electricity contained in the atomic nucleus." [109] Among other things, atomic number provided a theoretical basis for periodic classifications of the elements that had been pioneered by Dmitri Mendeleev around 1870. It also soon became clear that the ability of an atom to reflect X-rays was related to its atomic number and not, as the Braggs had previously believed, its atomic weight.

Early in 1913, Bragg was approached about a job at the new University of British Columbia, which had vacancies for a professor and an assistant professor. WHB felt that he should insist on the higher position, even though it would mean "little time for research." As he told Rutherford, "I really think the boy would do well, he is not at all one-sided." [110] However, it is doubtful that even the prospect of his own department would have drawn Bragg away from Cambridge and Leeds at this point; as he later wrote: "It was a wonderful time; we were like prospectors who had discovered a new gold-field in which nuggets were to be found just beneath the surface." [111] In any event, Thomson and others advised Bragg against taking the job in Vancouver.[112]

The reaction to Bragg's interpretation of the zincblende diffraction pattern was more skeptical in Germany in general and Munich in particular. When Laue's paper was republished in *Annalen der Physik* in 1913, he appended three notes dated March 1913. In these, he did not discuss face-centered cubic lattices and explicitly dismissed the idea that the homogeneity of the diffracted radiation is due to selection by the crystal.[113] In May, Ewald returned to Munich to find Friedrich "in despair about Laue's obstinacy regarding the 'characteristic radiation of the crystal' . . . " [114] This stubbornness gave Bragg a clear field.

Back at the Cavendish in early 1913, Bragg used Laue photographs to analyze the alkaline halides, including the rock salt and sylvine crystals provided by Hutchinson. In an undated letter from this period, Bragg wrote to his father: "Such an exciting photo today, with rock salt! I have worked it out, and it is almost perfectly characteristic of the point system which has points at the cube corners alone, not at the centers of the faces . . . I am sure this is because Na and Cl have more the same molecular [sic] weight than Zn and S . . . I will try

KCl next, which has two nearer atoms."[115] The diffraction pattern of sylvine confirmed Bragg's suspicion; he wrote to WHB: "My last photograph, taken with KCl, has turned out toppingly. It is perfectly characteristic of the point system with points at the cube corners alone."[116]

These studies on the alkaline halides led to a paper that provided, for the first time, a complete analysis of crystalline structures by X-ray methods and more than made up for the limitations of his *Proceedings of the Cambridge Philosophical Society* paper. Bragg was the sole author, although his father was acknowledged for collecting the spectrometer data used.[117]

Combining his new grasp of crystallography with X-ray optics, Bragg noted that all planes in a zone (a set of planes whose intersections are all parallel, like the sides of a pencil) will produce spots lying on an ellipse passing through the central point of the pattern, the axis of the ellipse being the zone axis (the direction of the intersections). By identifying which ellipse belongs to which zone, Miller indices can be assigned to spots lying on intersections of ellipses.

To simplify the diffraction patterns, Bragg used stereographic projection to transform the ellipses into circles. For KCl irradiated on a cube face, the pattern obtained contains circles derived from the $(0kl)$ and $(h0l)$ zones, where $k = \pm 1$ and $l = 0, 1, 2, 3, 4$, or 5. The intersections of these planes represent spots with indices of the form $(hk1)$, where h and k can be ± 1, ± 2, ± 3, ± 4, or ± 5. For example, the spot occurring where the $(h21)$ ring intersects the $(2k1)$ ring arises from planes with Miller index (221). Spots are present at all intersections within a certain range of values of h and k, indicating that KCl forms a primitive cubic lattice (Figure 2.4(a)). The range of spots obtained corresponded to glancing angles between 12° and 20°.

KBr, KI, and ZnS had similar diffraction patterns, which differed from that of KCl in that spots were not present at all intersections of the circles (Figure 2.4(b)). However, all planes with all indices odd form a complete series (all are present within certain values of h and k), as do those planes in which all indices are even (although in the latter case these are further from the center of the pattern).

The obvious explanation was that KCl is based on a primitive cubic lattice, while KBr, KI, and ZnS are based on face-centered ones. The diffraction behavior of rock salt, however, was intermediate between KCl on the one hand and ZnS on the other. Some spots present in KCl were absent in NaCl, while some spots absent in ZnS were present in NaCl (Figure 2.4(c)). Bragg thought that the confounding factor in these analyses might be the weights of the constituent atoms. As Barkla had realized, X-ray diffraction, like X-ray absorption, might be a function of atomic weight. Also, it was now well known that the absorption of X-rays showed an approximate proportionality to the atomic weight (or number) of the absorbing element.

Taking the relative atomic weights of the metal and halogen atoms into account provided a plausible explanation for the diffraction behaviors of the alkaline halides. In KBr, the halogen (80) is so much heavier than the metal (39) that the diffraction essentially occurs only from the former. In KCl, the two

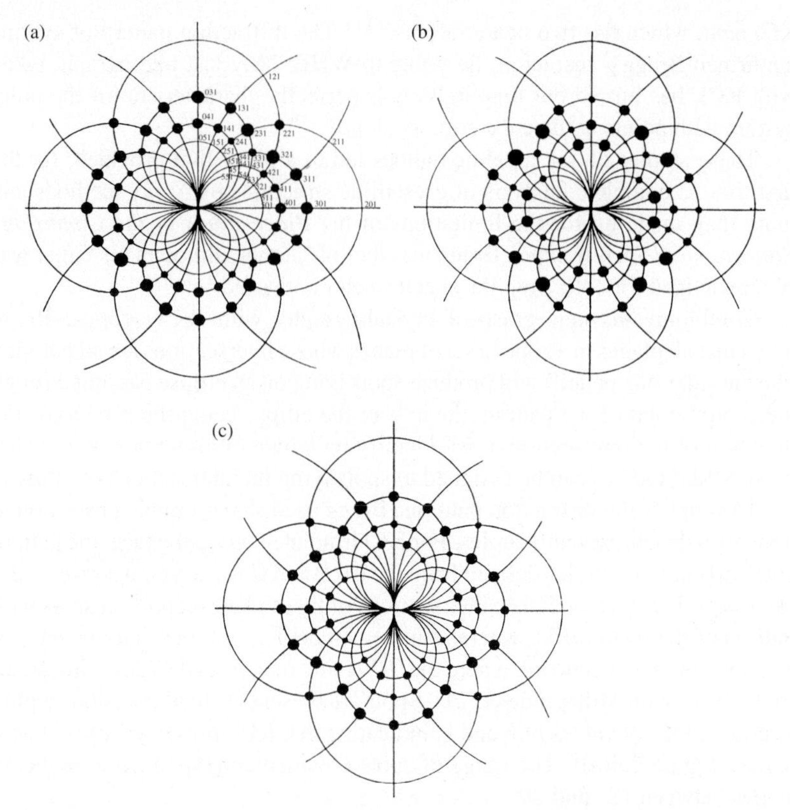

Fig. 2.4 Diffraction patterns of alkaline halide salts. (a) Stereographic projection and interpretation of potassium chloride diffraction pattern. All spots occur at intersections of circles (actually ellipses) corresponding to X-rays diffracted by particular "zones" (sets of atomic planes that are all parallel to the same axis). Within a range of values of h, k, and l, spots occur at all intersections. (b) Stereographic projection of potassium bromide diffraction pattern. Compared to potassium chloride, there are systematic absences in the spots. (c) Stereographic projection of sodium chloride (rock salt) diffraction pattern. Reproduced, with permission, from Figures 3, 4, and 9 of Bragg, W. L. (1914). The structure of some crystals as indicated by their diffraction of X-rays. *Proceedings of the Royal Society of London A* **89**, 248–277. Published by the Royal Society.

atoms are similar in weight (39 and 35.5), so both diffract. NaCl is intermediate, as the atomic weight of sodium is 23. Bragg therefore argued that "the atoms of alkaline metal and halogen have precisely the same arrangement in all these cases." This arrangement is a chessboard pattern of metal and halogen atoms; in KBr and KI, the metal is light enough to be effectively invisible, and the crystal diffracts as a face-centered lattice; in KCl, the two types of atom "become identical," and the crystal diffracts like a primitive cubic lattice (Figure 2.5).[n]

[n] Strictly speaking, all the substances studied are on face-centered lattices as the asymmetric unit is the entire "molecule" rather than single atoms.

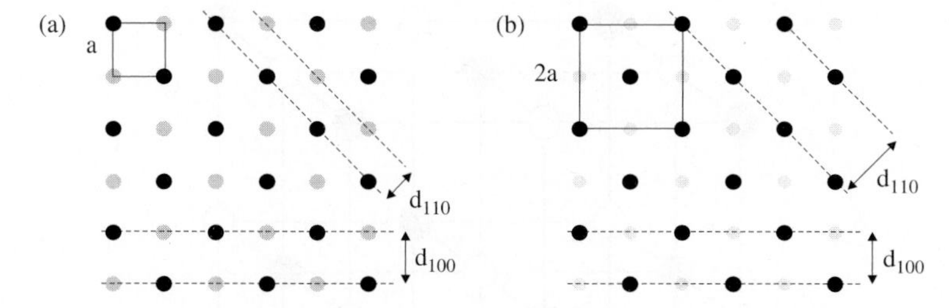

Fig. 2.5 Lattice spacings of primitive and face-centered cubic lattices. (a) A cubic crystal with two atoms contributing equally to scattering of X-rays (e.g. KCl) will diffract as a primitive cubic lattice of unit-cell length *a*. (b) A cubic crystal with two atoms contributing unequally to scattering of X-rays (e.g. KBr) will diffract as a face-centered cubic lattice of unit-cell length 2*a*. Some lattice spacings (e.g. d_{100}) will be the same as in A, others (e.g. d_{110}) will be different

This provided a rationale for the systematic absence of spots in the diffraction patterns of KBr, KI, and ZnS. Introduction into a primitive cubic lattice of lattice points at face centers will have different effects on planes with odd and even indices. In the case of odd-index planes, the new lattice points will lie on these planes, merely increasing their density. In the case of even-index planes, the new points will lie halfway between the planes, halving the *d*-spacing of these planes and thereby doubling the wavelength.

All this was based on Laue photographs. However, corroboration could be obtained from spectrometer measurements. Bragg noted that the first-order ($n = 1$) *B* peaks from the (100), (110), and (111) planes of rock salt occurred at angles of 11.4, 16, and 9.8°, respectively. From the Bragg equation, the ratio $d(100) : d(110) : d(111)$ is therefore $1 : 0.718 : 1.16$—very close to the ratio $1 : 1/\sqrt{2} : 2/\sqrt{3}$ characteristic of a face-centered lattice.

In his previous paper, Bragg had left open the question of whether the lattice points of the zincblende crystal were occupied by atoms, molecules or (as proposed by Pope and Barlow) groups of atoms. An important corollary of the present analysis was that the lattice points could not correspond to molecules because then the presence and absence of spots would not differ among the alkaline halides—each compound would diffract like a primitive cubic lattice no matter the relative weights of metal and halogen. The conclusion that atoms lie at the points of the face-centered cubic lattice was supported by spectrometer measurements which showed that the number of molecules associated with each "diffracting center" in KCl was half that in NaCl, ZnS, CaF_2 (fluorspar), $CaCO_3$ (calcite), and FeS_2 (iron pyrites). Bragg therefore proposed not only that there were no molecules of NaCl at the lattice points of the rock salt crystal, but also that there were no NaCl molecules at all! "in sodium chloride the sodium atom has six neighbouring chlorine atoms equally close with which it might pair off to form a molecule of NaCl" (Figure 2.6).

This finding of a one-to-one correspondence between lattice points and atoms cleared the way for measurement of the dimensions of the unit cell and

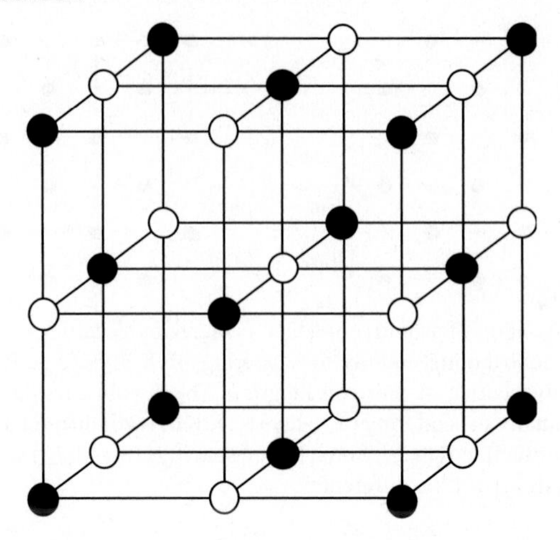

Fig. 2.6 Bragg's 1914 structure of sodium chloride. Each sodium atom (closed circle) is equidistantly surrounded by six chlorine atoms (open circles), and *vice versa.*

the wavelength of the homogeneous components of the primary X-ray beam. In rock salt, for example, the cube faces were 2.81×10^{-8} cm apart; from this it could be calculated that the wavelength of the platinum B peak was 1.1×10^{-8} cm.

When an X-ray beam is directed at a specific crystal face, the angles at which the homogeneous peaks occur are proportional to the spacings between planes parallel to that face. The relative sizes of the homogeneous peaks, however, provided information about the nature of the crystal lattice. The normal situation was that the first-, second-, third-, fourth-, and fifth-order spectra have approximate relative intensities $1:0.2:0.07:0.03:0.01$. This was the case for the (100) spectra of rock salt, which arise from planes containing both Na and Cl atoms. For the (111) reflections, however, the second-order spectrum was more intense than the first-order. According to Bragg, this is because (111) planes consisting entirely of Na atoms alternate with (111) planes consisting entirely of Cl atoms (similar to the alternation of the two sets of (110) planes shown in Figure 2.5). For a given wavelength of X-rays, waves reflected from the chlorine planes will be in phase at an angle of incidence at which the Bragg equation is satisfied and n, the order of reflection, equals 1. If the more weakly reflecting sodium planes are halfway between the chlorine planes, the waves reflected from the two types of planes will be 180° out of phase, and so the first-order (111) reflection is less intense than if the sodium planes had not been present (Figure 2.7(a)). At an angle of incidence where the Bragg equation is again satisfied and n equals 2, the X-rays reflected from the sodium-containing and chlorine-containing planes are exactly *in* phase, and so the second-order (111) peak is *more* intense than it would have been were the sodium planes not present (Figure 2.7(b)). This idea that parallel planes containing different atoms could cause a degree of interference that is dependent upon the relative

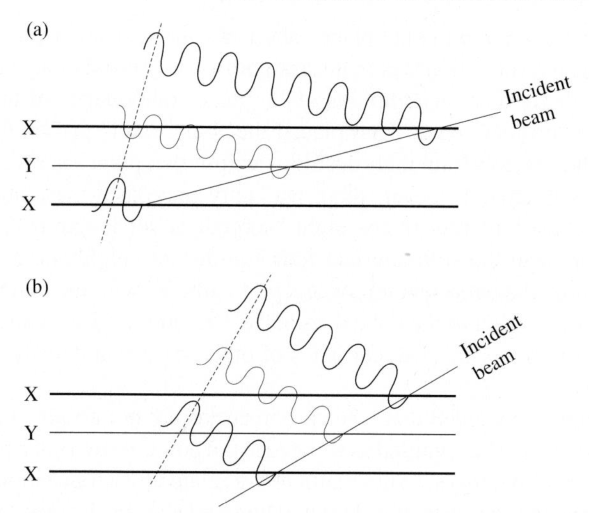

Fig. 2.7 Effect of equally spaced planes consisting of different atoms on intensity of diffracted X-rays. (a) First-order reflection: waves from planes X and Y are completely out of phase (dashed line). (b) Second-order reflection: waves from planes X and Y are completely in phase (dashed line)

scattering powers of the atoms present was one of Bragg's greatest insights—it would become the key to all analyses of crystal structure by X-rays.

Back-to-back with the 1914 paper on the alkaline halides was another blockbuster—a joint paper with WHB on the structure of diamond.[118] Bragg had obtained a diamond of suitable size from Hutchinson and unsuccessfully tried to analyze its structure by the Laue method before turning the project over to his father.[119] WHBs analysis involved a rhodium anticathode, which had two advantages over the platinum ones used previously: A higher proportion of the X-rays emitted are homogeneous; and these homogeneous components are of lower wavelength, which allows more orders of reflection to be studied.

The X-ray spectra of diamond exhibited several striking features: The first-, third-, fourth-, and fifth-order (111) reflections were present, but the second-order was completely missing. From the angles of incidence at which reflection occurred, the distance between the (111) planes, d_{111}, could be calculated from the Bragg equation to be 2.03×10^{-8} cm. Diamond was known to belong to the cubic crystal system. Assuming that it, like the other minerals studied so far, was face-centered, the length of the cube axis, $2a$, could be calculated (for a face-centered cube, d_{111} is $2a/\sqrt{3}$). This in turn allowed the Braggs to calculate that there are eight carbon atoms per unit cell, as opposed to four lattice points: "We therefore have four carbon atoms which we are to assign to the elementary cube in such a way that we do not interfere with the characteristics of the face-centred lattice."

The spectrometer measurements made this easy to do. In NaCl, the presence of weakly diffracting Na-containing (111) planes between the strongly diffracting Cl-containing (111) planes abolishes the first-order reflection. For diamond, the interpretation was more straightforward, as only one type of atom is present—there must be additional (111) planes one-quarter of the distance between the "normal" ones. That placed the four "extra" carbon atoms on a plane, but did not constrain their positions within that plane. In order to maintain cubic symmetry, however, there was only one place these atoms could go—at the centers of four of the eight "sub-cubes" of length a. This meant that every atom in the structure had four equidistant neighbors, as would be expected from "the persistent tetravalency of carbon," with the bonds between them all lying parallel to the cube diagonals. The proposed structure was confirmed by examining the characteristics of other spectra and analysis of Laue photographs.

The Braggs concluded that "the carbon atoms are not arranged on a space lattice, but they may be regarded as situated at the points of two inter-penetrating face-centred space lattices." These lattices are related by a translation along the cube diagonal of one-fourth of its length. Diamond was the first crystal structure that did not correspond to one of the 14 Bravais lattices. (The alkaline halide structures consist of inter-penetrating primitive cubic lattices of cations and anions, but together these form the face-centered lattice predicted by Bravais.)

Bragg's 1913 paper on the alkaline halides has a fair claim to be regarded as his *magnum opus*, and probably ensured his share of the Nobel Prize. At the time, though, it was overshadowed by the diamond paper. "The structure of diamond was widely acclaimed with satisfaction by the chemists . . . My structure of rocksalt had a very different reception." [120]

The reason for these different reactions is not hard to find. For chemists, the diamond structure represented proof of the tetravalency of the carbon atom, first proposed 40 years earlier by Jacobus van's Hoff and by Joseph LeBel, and the theoretical basis of stereochemistry ever since. In addition, this paper provided the first measurement of a bond length in an organic compound—the distance between adjacent carbon atoms of 1.52 Å (the modern value is 1.54 Å). As Bragg wrote in 1965, "I think it was the diamond structure which first brought home to the scientific world the importance and power of the new method." [121]

If the diamond structure validated one of the great achievements of nineteenth-century chemistry, the sodium chloride structure undermined another—the idea of the molecule. When chemists looked at Bragg's structure, with every sodium atom surrounded equidistantly by six chlorines, they saw a salt crystal containing no NaCl molecules. Bragg recalled "the Professor of chemistry at Leeds [Arthur Smithells] begging me to find that one sodium [was] just a tiny bit nearer to one chlorine than it was to the others!" [122] The Professor of Inorganic Chemistry at the Municipal University of Amsterdam "was against Bragg's model because no separate molecules could be discerned in it, whereas these were 'the basis of the whole of chemistry.' "[123]

This was all a replay of a debate that had occurred 30 years earlier. In 1883, Barlow had described five ways of packing spherical atoms, including the primitive cubic, face-centered cubic, body-centered cubic, and hexagonal systems. He suggested that the body-centered lattice was likely to occur in crystals composed of equal numbers of two different atoms, including "sodic" chloride.[124] In response, Leopold Sohncke, one of the founders of space-group theory, wrote: "The atom of Cl seems consequently to be in equally close connection with eight atoms of Na; it has exactly the same relation to these eight atoms. It appears, therefore, as *octovalent*, certainly not as univalent; for it would be entirely arbitrary to suppose any *two* neighboring atoms of NaCl in an especially close connection and to take this couple for the chemical molecule of NaCl. By this example we see *that from Mr. Barlow's point of view both the notion of chemical valency and of chemical molecule completely lose their present import for the crystallised state*." [125] (Emphases in original.)

Later in life, Bragg came up with the following analogy for the chemists' view of rock salt: "... it is as if, having found that the numbers of ladies and gentlemen at dinner parties is [sic] generally equal, we had falsely concluded that they were all necessarily married couples, instead of realizing that it was because each lady liked to have a gentleman on either side, and vice versa." [126]

As time went by, chemists abandoned their instinctive dislike of Bragg's structure for the alkaline halides. The evidence was unambiguous, and X-ray methods could provide great insights into all substance of interest to chemistry. The last volley was fired in 1927 by Henry Armstrong, Emeritus Professor of Chemistry at Imperial College London. In a letter to *Nature* entitled "Poor Common Salt!," Armstrong wrote: "... Prof. W.L. Bragg asserts that 'In sodium chloride there appear to be no molecules represented by NaCl. The equality in number of sodium and chlorine atoms is arrived at by a chess-board pattern of these atoms; it is a result of geometry and not of a pairing-off of the atoms.' This statement is more than 'repugnant to common sense'. It is absurd to the *n*...th degree, not chemical cricket... It were time that chemists took charge of chemistry once more and protected neophytes against the worship of false gods; at least taught them to ask for something more than chess-board evidence." [127] Then almost 80 years old, Armstrong was a scientific gadfly who had previously written fairy tales satirizing the ionic dissociation theory of Arrhenius.[128] However, Bragg appears to have taken "Poor Common Salt!" seriously; throughout his career he cited Armstrong's letter as evidence of the way his alkaline halide structure was received.

The Long Vacation of 1913 provided the first opportunity for an extended collaboration between father and son: "We worked furiously in 1913 and 1914, going back in the evenings to the deserted university to get more measurements." [129] The structures of zincblende, fluorspar, iron pyrites, and calcite were all worked out in the summer of 1913.[130] Zincblende proved to be very similar to diamond, with the zinc atoms occupying the points of a face-centered cubic lattice and the sulfur atoms the centers of four of the eight sub-cubes, four zinc atoms forming a tetrahedron around every sulfur. Fluorspar (CaF_2), having twice as many fluorine atoms as calcium atoms, was easily explained by placing the former at the centers of all eight sub-cubes. This meant that the fluorine-containing (111) planes had twice the atomic density of the calcium-containing ones. Since it was found from the relative intensities of the different orders of (111) spectra that the scattering contributions of the two types of planes were exactly equal, and since the atomic weight of fluorine (19) is almost exactly half the atomic weight of calcium (40), this allowed Bragg to state that the diffracting power of an atom is proportional to its atomic weight.

However, it was the structure of iron pyrites, FeS_2, which "provided the greatest thrill."[131] The density of the crystal and measurement of the principal reflections suggested a face-centered cubic lattice. Since S is about half the weight of Fe, a face-centered lattice should give spectra like those of fluorspar. However, this was far from being the case: The first-order (100) spectrum was strong, the second- and third-order were absent and the fourth- and fifth-order were in normal proportions to the first. This was "somewhat similar" to the orders of reflection of the (111) planes of fluorspar, suggesting that the S atoms of pyrites are displaced from the cube centers. However, Bragg could find nowhere else to place them that was consistent both with the "queer succession of spectra" and with the cubic symmetry of the crystal.

The solution came from a paper of Barlow's. Each sub-cube of a cubic crystal normally has 4 three-fold rotation axes that intersect at its center. However, cubic symmetry can still occur if each sub-cube only has 1 three-fold axis—so long as these axes are chosen such that they do not intersect. (As noted above, cubic crystals can lack some of the symmetry of a solid cube.) Bragg realized that the sulfur atoms in pyrites could lie anywhere along the non-intersecting three-fold axes running through the sub-cubes, each of which has an iron atom at one end and a vacant sub-cube corner at the other. The sulfurs had to lie on the axes, as otherwise each would be multiplied three-fold by the symmetry operation of the axis and this would result in a number of atoms per unit cell that was inconsistent with the empirical formula.

As the sulfur atoms have to lie on the three-fold axes, their positions are defined in the two dimensions perpendicular to that axis. The only variable in the positions of such an atom is its distance along the axis. Atomic coordinates not defined by the symmetry of the crystal became known as "parameters." The structure of iron pyrites therefore has only one parameter. The value of this was easily determined from the relative intensities of the (100) spectra. Placing the sulfur at a distance of four-fifths of the length of the axis from the iron-containing end would explain the observed absence of the second- and third-order spectra (Figure 2.8).

This was the first time that a structure with a parameter had been solved. Bragg wrote: "I got so excited about it when I worked it out in the drawing room of our house at Leeds that I tried to explain it to an aunt who happened to be sitting in a corner. The result was completely unsuccessful!"[132]

The paper describing this work, entitled "The Analysis of Crystals by the X-Ray Spectrometer," was submitted in November 1913.[133] Bragg was the sole author, although he was scrupulous enough to mention that "many" of the measurements had been made by his father. For the first time, he used

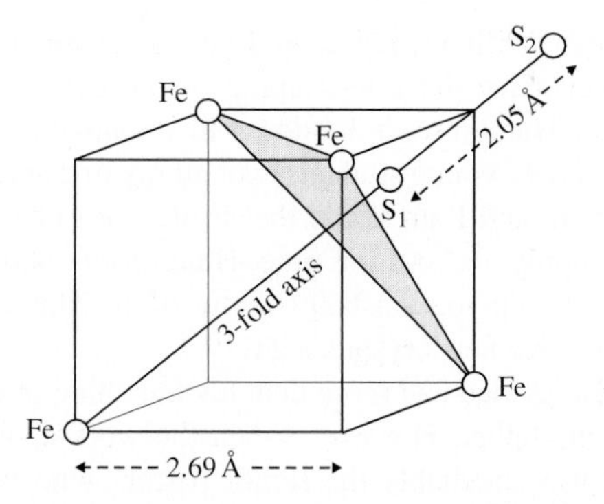

Fig. 2.8 Bragg's 1914 structure of iron pyrites. Shown is one of the 8 "sub-cubes" of the face-centered cubic unit cell. S_1 is the sulfur atom in the sub-cube, which lies on the three-fold rotation axis, four-fifths of the distance between the iron (Fe) atom at one and and the vacant sub-cube corner at the other. S_2 is the sulfur atom in an adjacent sub-cube. The shaded area is a (111) plane, which is perpendicular to the three-fold rotation axis. Adapted from Figure 1a of Bragg, W. L. (1920). The arrangement of atoms in crystals. *Philosophical Magazine* **40**, 169–89. (http://www.tandf.co.uk)

the name "W. Lawrence Bragg"—an attempt to avoid confusion between the two William Braggs.

Bragg wrote that, using the spectrometer method, "we can obtain enough equations to solve the structure of any crystal, *however complicated*, although the solution is not always easy to find" (emphasis added). Considering the embryonic state of X-ray analysis, this was a remarkably bullish statement. However, he had elaborated a general law for deriving the distances between different atomic planes from spectrometer measurements. For planes of atom A with atomic weight m_1 and separated by distance D, which have between them planes of atom B with atomic weight m_2, displaced from the A planes by distance D/n, the relative intensities of the orders of spectra will be as follows: First, 1 $[m_1^2 + m_2^2 + 2m_1m_2\cos(2\pi/n)]$; second, 0.2 $[m_1^2 + m_2^2 + 2m_1m_2\cos(4\pi/n)]$; third, 0.07 $[m_1^2 + m_2^2 + 2m_1m_2\cos(6\pi/n)]$; fourth, 0.03 $[m_1^2 + m_2^2 + 2m_1m_2\cos(8\pi/n)]$; and fifth, 0.01 $[m_1^2 + m_2^2 + 2m_1m_2\cos(10\pi/n)]$.[o] To perform a complete analysis of crystal structure, similar calculations could be performed for three or more nonparallel planes. This first attempt to relate atomic weight and location to reflection intensity, although eventually requiring substantial modification, marks the beginning of quantitative X-ray analysis.

In the acknowledgments section of this paper, Bragg thanked Hutchinson for supplying him with crystals. A quarter-century later, he recalled: "the professor

[o] This law assumed that "The intensity of the spectrum is taken to be proportional to the square of the amplitude of the resultant reflected wavelet"—an unproven proposition in 1913 that proved to be usually the case.

[of mineralogy] was [William] Lewis and he had given strict orders that no mineral should ever leave the safe-keeping of the collection at Cambridge. I shall never forget Hutchinson's kindness in organizing a black market in minerals to help a callow young student. I got all my first specimens and all my first advice from him, and I am afraid that Professor Lewis never discovered the source of my supply."[134] At one time, Hutchinson who, as Demonstrator in Mineralogy was ex officio Assistant Curator of the Mineralogical Museum, was sending Bragg three new crystals a day![135]

To this point, Bragg had had more than his fair share of credit for the joint publications with his father. However, when the work came to be presented at conferences, it was inevitably the senior partner who was invited. WHB discussed their joint work at the annual meeting of the British Association for the Advancement of Science in Birmingham in September 1913, and at the Second Solvay Conference on Physics in Brussels the following month. The theme of the Solvay Conference was, appropriately enough, "The Structure of Matter." WHBs talk was enthusiastically received, and presentations by Laue, Pope, and Barlow also referred to Bragg's work.[136] Presumably at WHBs instigation, a group of conferees that included Sommerfeld, Laue, Einstein, Lorentz, and Rutherford sent Bragg a postcard that read: "The most heartfelt congratulations on your fine scientific success and many greetings from Brussels."[137]

Most 23-year-old scientists would be delighted to receive this kind of acknowledgement from the leading authorities of their discipline. Few young scientists, however, have to labor in the giant shadow of a famous father. Bragg's great fear was that his discovery of the physical basis of X-ray diffraction would be attributed by the scientific community to WHB. Their ongoing collaboration increased the risk of this happening, but, both for personal and professional reasons, he could hardly strike out on his own. As he wrote in his autobiography, "A young researcher is as jealous of his first scientific discovery as a kitten is with its first mouse, and I was exceedingly proud of having got out the first crystal structures. But inevitably the results with the spectrometer, especially the solution of the diamond structure, were far more striking and far easier to follow than the elaborate analysis of Laue photographs, and it was my father who announced the new results at the British Association, the Solvay Conference, lectures up and down the country and in America while I remained at home."[138]

As Bragg well knew, WHB was giving him more than his fair share of the credit for their joint work, to the point of refusing authorship on papers to which he had made major experimental contributions. Bragg also knew that without WHBs X-ray spectrometer his ideas about reflection from crystal planes would not have progressed far. Pride in his son's precocious achievements, in addition to his innate fair-mindedness and personal modesty, made WHB keen to acknowledge the collaborative nature of their work. As he wrote to Gwen concerning a lecture he had given at the American Academy of Arts and Sciences in Cambridge, Massachusetts, in November 1914: "Of course I always tell them where Bill's work came in: This time I was particularly . . . keen that they should understand Bill's achievement so at the end I asked to be allowed to say another word or two and I described his position again. One old boy who had come in from Worcester got up and said that did not at all diminish their debt to the lecturer because 'Quid facit per filium facit per se' 'What a man does through his son he does himself' . . ."[139]

Had he been emotionally better adjusted, Bragg may have taken the long view that his solo papers would ensure his scientific legacy. The reality was, however, that he could be overly sensitive to perceived slights and was probably already subject to the black moods of depression that would plague him all his life. Had Bragg had a better relationship with his father, they could have discussed the situation openly and "issued a joint communique on their relative contributions," as Bragg's son Stephen put it.[140] But, as Gwendy wrote in her biography of WHB, "Father and son never managed to discuss their scientific relationships thoroughly, WHB being very reserved and WL inclined to bottle up his feelings."[141] It appears they were never close: Jenkin has written: "I suspect that he [WHB] was somewhat distant and aloof [from his children],"[142] and, as described above, Bragg blamed WHB in part for his poor experience at the University of Adelaide. Gwendy wrote to David Phillips in 1979: "Poor Willy, it must have been very galling—and WHB too innocent, somehow, to understand—he was so proud of his son, and trying to give him praise; just excitement about the work could have blinded him to what was happening."[143] Much damage had been done by January 1915, when WHB wrote in the preface to *X-Rays and Crystal Structure*: "I am anxious to make one point clear, viz., that my son is responsible for the 'reflection' idea which has made it possible to advance, as well as for much the greater portion of the work of unravelling crystal structures to which the advance has led."[144]

Alleviating Bragg's anxiety about his prospects for scientific glory were his friendships with Cecil Hopkinson and with other Trinity College physicists. In his biographical memoir of Bragg, David Phillips wrote: "A.L. Goodhart remembered dining regularly in Trinity with Bragg, E.D. Adrian, F.W. Aston and G.P. Thomson, probably during 1913, and listening fascinated to the chaffing of the scientists: Braggs were good at experiments, Thomsons were not!"[145] Edgar Adrian, a physiology graduate, became a Fellow of Trinity in 1913. George Thomson, the son of J. J., was to become one of Bragg's closest friends.

Another reason for optimism was that Bragg's college was appreciative of his success. In an undated letter probably written in early 1914, Bragg wrote to WHB to tell him that he had been offered a lectureship in Natural Sciences at Trinity "from 1st October next," with the promise of a College fellowship also. The lectureship was worth £450 a year, with additional payment for the fellowship. Bragg was "very excited about it all" and thought it was "a very fine chance," but would not reply until he heard from his father.[146] The paternal verdict must have been favorable, for Bragg accepted the offer and was made a fellow of Trinity on August 11, 1914.[147]

Up to this point, all the crystals whose structures had been determined were cubic, which is the crystal system of highest symmetry. During the 1913–14 academic year, Bragg in Cambridge and WHB in Leeds were exploring crystals of lower symmetry, including quartz (SiO_2, hexagonal), wurtzite (ZnS, hexagonal) and calcite ($CaCO_3$, hexagonal).[p] In July 1914: Bragg wrote to his father: "I have been writing up the aragonite but am a bit puzzled about the structure . . . I would love to thoroughly examine aragonite by your method of revolving the crystal. Every crystal should have such a series of measurements made on it."[148]

Aragonite, an orthorhombic form of calcium carbonate, would have to wait. On August 4, Great Britain declared war on Germany. That day, Edward Appleton, a recent graduate who had been working with Bragg on the crystal structure of copper, tracked him down in the Cavendish with the news that he had decided to enlist. Bragg had made the same decision. The First World War would put his scientific career on hold for almost 5 years.

[p] By 1915, 33 crystals had been studied by X-ray methods: 18 which had been characterized "with some completeness," nine for which the "marshalling" of atoms was known, but not the precise atomic locations; and six for which only the number of molecules per unit cell had been determined. These were mostly cubic, hexagonal, or rhombohedral, the crystal systems of highest symmetry, but the group of partly characterized crystals included three members of the orthorhombic system. [Bragg, W. H. and Bragg, W. L. (1915). *X-Rays and Crystal Structure*, pp. 173–4. G. Bell and Sons, Ltd., London.]

3
Our show is going famously: World War One

Since the autumn of 1909, Bragg had been a "reserve trooper" in King Edward's Horse. This was a reserve cavalry regiment whose members, men from the Dominions, practised marksmanship, riding and care of their mounts, and had a camp in the summer. Bragg was discharged in November 1913, having completed his term of service. On the discharge certificate, he was described as 5 ft. $9\frac{1}{2}$ in., with brown eyes and dark hair. His conduct and character were "very good."[149] On his coming to Cambridge, Bob Bragg also became a member of King Edward's Horse, as did their cousin Stenie Squires, the son of Aunt Lizzie. When war broke, Bob, who was still a member of the regiment, joined its camp, eventually applying for a commission and being posted to the Royal Field Artillery in Leeds. Bragg applied for a commission immediately and trained in Cambridge while awaiting a posting. This decision was made by a group of Cambridge Dons, who decreed that his experience in horsemanship and mathematics qualified him for the Horse Artillery.

On August 24, 1914, Bragg was sent to a Territorial battery, the Leicestershire Royal Horse Artillery, as a 2nd lieutenant.[150] For the first year, the battery was quartered in Diss, Norfolk, "an attractive town, centred round a small pretty lake." Bragg was initially billeted with the maker of Gossling's Pig Powders, whose wife was reputed the finest cook in Norfolk. Subsequently he was billeted with Mrs Taylor, a "grand old lady" who occupied the Manor House. In February 1915, Bragg wrote to his mother: "Besides myself there is only a parrot in the house, with no feathers on, which has been with them for 20 years . . . He is always calling for a dog called Duke and I expected to meet him until Mrs Taylor told me Duke had departed thirteen years ago, it was rather funny."[151]

Bragg kept his horses in the stables, where they were looked after by his groom.[152] He also had a "servant," Cobley. It was probably Cobley to whom Gwendy referred in the following anecdote: "Long years after the 1st world war he met the man who had been his batman in the Leicestershire Horse Artillery [sic] . . . who told him 'When I first had to look after you, I thought you were the worst officer in the British Army, but after a month I'd have done anything for you. Poor Willy, he never could remember to give the right order at the right time to start with."[153] Bragg agreed that he was not a natural soldier. As he admitted in a February 1915 letter to his mother: "I make an awful bad officer but I suppose it can't be helped."[154]

The billets may have been undemanding, but it was a frustrating time for Bragg, who was "very much a fish out of water" among the hunting men of the battery. He wrote to his mother: "I get very sick sometimes because I am slack with the men, and get moments of awful despondency, and then again I feel I am much better than anyone else. The former feeling generally preponderates."[155] No doubt Bragg's state of mind was not improved by the realization that, while he was cooling his heels in rural Norfolk, older scientists were free to exploit an actively developing field of research that he had initiated. "Crystals seem a long way away at present," he wrote to his father in December 1914.[156] He whiled away the time by writing, with WHB, a book on X-ray analysis. This would be published in 1915 as *X-Rays and Crystal Structure*.[157]

Bragg's spirits were raised considerably in May 1915, when he learned that he and WHB had been awarded Columbia University's Barnard Gold Medal, presented every 5 years, on the recommendation of the National Academy of Sciences, for the most outstanding discovery or application of science in the physical or astronomical sciences. The NAS report, which was signed by Robert Woodward, President of the Carnegie Institution of Washington, read in part: "By means of a rare combination of experimental skill and theoretical insight they have been able to show that X-rays are essentially light rays of excessively short wave length and that crystalline structure is definitely molecular [!]. They have determined the order of wavelength of X-rays and they have also determined the order of the intervals that separate the molecules in certain crystalline forms. In a noteworthy series of luminous publications and in an equally noteworthy series of masterly public expositions they have supplied the initial methods and furnished many of the preliminary results which must lead to a still more productive era in the advancement of molecular physics and hence in the advancement of the entirety of physical science."[158] The medal was worth $200, but of far greater value to Bragg was the stature of the previous winners—Lord Rayleigh and William Ramsay (jointly), Wilhelm Röntgen, Henri Becquerel, and Ernest Rutherford—and the fact that he and his father had been honored jointly. He wrote to his mother: "I am so awfully braced about that Gold Medal, what a score for us it is."[159]

For the Bragg family, the phoney war period ended in the summer of 1915. Bob sailed for Alexandria in June. The following month, WHB became a founding member of the Board of Inventions and Research of the Admiralty. He had been appointed Quain Professor of Physics at University College London, but had not yet taken up the position.[160]

In August, Bragg was ordered to the War Office, where he was informed of French efforts to find the locations of German artillery pieces using sound waves. Colonel Winterbottom of the Royal Engineers had obtained permission to set up an experimental section under "an officer with scientific training." For Bragg, this was infinitely better than vegetating among the horsey set in Diss—he came out of the interview "walking on air." He and Harold Robinson, who was from Rutherford's department at Manchester, were sent over to France to report to Colonel Jack at General Headquarters in St Omer. Jack, as officer in charge of Field Survey sections, was also responsible for surveying and map-making. Bragg and Robinson were sent to a French sector in the Vosges, which was so quiet that no gun fired during the 2 or 3 weeks they were there. The two men lived in a ski chalet and messed with the French officers.[161]

In early September, Gwen and WHB learned that Bob had died in the Gallipoli landings. Bragg was informed of his brother's death by a military padre when he returned to St Omer from the Vosges. A few weeks after he landed at Gallipoli, a dugout in which Bob and another officer were censoring letters was hit by a Turkish shell. One of Bob's legs was blown off and the other severely damaged. He was invalided to a hospital ship but died the next day.[162]

It is hard to tell how Bragg was affected by Bob's death. The only letter that refers to this is undated, but written to WHB from the Strand Palace Hotel in London, where Bragg was presumably staying on arriving in England on compassionate leave. It read: "I got Mother's letter telling me about Bob just before I was leaving for England yesterday. I will come up to Rough [?] House as soon as I have reported here and can get away. I will be able to arrive either tonight or tomorrow morning and will telegraph to you what time. Give my very dearest love to Mother."[163] In his autobiography, Bragg wrote: "When we got back to St. Omer to report to Colonel Jack, I heard from my parents that Bob had died of wounds at Gallipoli, very soon after the landing."[164] Like many veterans of war, he rarely discussed his experiences. Jenkin wrote of this period: "WLB seems to have remained largely untouched by the hell around him."[165] The brothers had been close in Australia. However, Bragg was envious of Bob's skill at games and more-outgoing personality,[166] as well as the respect in which he was held by his military superiors.[167] According to Gwendy, he also felt that his mother preferred her younger son.[168] It is possible that his grief at Bob's death was mixed with guilt.

To make things worse, Bragg learned in November that his best friend had also become a casualty. Wounded in the head in France, Cecil Hopkinson was sent back to England in serious condition.

Two factors helped Bragg come to terms with Bob's loss. The first was that he had a job to do. Robinson and Bragg were sent to Kemmel Hill, southwest of Ypres, to set up the first sound-ranging station. The microphones were in a line about a mile behind the front, with the recording station at La Clytte. In October, Bragg spent a couple of weeks in Paris, where he visited the Institut Marey to consult with Lucien Bull, the Irishman who had developed the prototype French sound-ranging system. Bull's laboratory was in "a low white building in a very shaggy garden which smells just like the old Observatory." The research staff consisted of two assistants, one with an artificial leg, both of whom seemed to be mainly concerned with polishing the apparatus. The laboratory was also inhabited by two refugees, four dogs, seven cats, three goats, "n fowl and a lot of pigeons."[169]

From Paris, he wrote to his mother: "Tell Gwendy that on the Metro trains each guard makes a different kind of noise to tell the engine driver he is ready, and there is a guard to each carriage on the train, like our tube. One has a whistle, another a little trumpet and another a tooter . . . You know I am always thinking of you and hoping you are feeling a little less sad about Bob."[170] He also mentioned that he had received a letter from "Elaine." This was Elaine Barrow of Leeds, Bragg's first girlfriend.

The second thing that helped Bragg overcome Bob's death was that, on November 17, he received a letter from his parents containing wonderful news—he and WHB had been jointly awarded the 1915 Nobel Prize for physics. He wrote back to his mother: "You can imagine how I felt, really I am the most lucky fellow in the world I think. It is so awfully nice to be coupled with Dad in this way. You are a lucky mother to have married into so distinguished a family. I got many congratulations today, everyone had seen it in the papers which get to us the next morning here . . . I can't realize it a bit. Will Dad go over to Stockholm to get the booty? What awful fun it all is. I was sorry to see that a German [Richard Willstätter] had got the chemistry prize, that was tactless."[171] Four days later, he wrote: "Today the Curé, who had seen my photo in the paper, came in and offered me a bottle of wine with his best bow as a little present to felicitate the occasion. Generals humbly ask my opinion about things, it is great fun."[172]

Nobel Prizes in physics, chemistry, physiology or medicine, literature, and peace had been first awarded in 1901, in accordance with the will of the Swedish industrialist Alfred Nobel, who had died 5 years earlier.[173] The Nobel Foundation, which was set up to administer the awards, delegated selection of winners for the physics and chemistry prizes to the Royal Swedish Academy of Sciences, which established two committees for that purpose. Nominations are solicited from eminent scientists (including previous winners) in September of each year with a deadline for submission of February 1 in the following year. On the basis of the nominations received, the Nobel Committees for Physics and Chemistry make recommendations to the Academy each November. The prizes are awarded on December 10—the anniversary of Nobel's death—in Stockholm. One provision of Nobel's will—that the Prizes should recognize work performed in the previous year—was recognized to be impractical by the Foundation and relaxed. As a result, the first Nobel Prize in physics was awarded in 1901 to Wilhelm Röntgen for his discovery of X-rays, which, as described above, had occurred in 1895.

The Nobel Prizes are not the oldest awards for scientific research—for example, the Copley Medal of the Royal Society was established in 1731. However, the Nobels almost immediately became the most prestigious, probably because the large sums of money involved provoked intense public interest.

When war broke out in the summer of 1914, the Nobel Committees for Chemistry and Physics had already made their selections: Theodore Richards of Harvard University and Max von Laue, respectively. On behalf of the selecting bodies, including the Academy of Sciences, the Nobel Foundation successfully petitioned the Swedish government to defer the awards until the following year. Another deferral was requested in 1915, but this time permission was denied. The Academy of Sciences then awarded prizes for both years: The physics Prize for 1914 to Max von Laue, as previously agreed; that for 1915 to the Braggs "as a reward of your works upon the intimate structure of crystals as investigated with the help of the Röntgen ray."[174] However, the Karolinska Institute, which selects winners for the Nobel Prize in Physiology or Medicine, exercised its privilege to reserve the 1915 prize on the grounds that no nomination of sufficient quality had been received. Had the Academy of Sciences taken the same position as the Karolinska Institute, it is quite possible that one or both of the Braggs would have ended up being overlooked.

There being no realistic prospect of the war ending by December, the award ceremony for the 1914 and 1915 Nobel Prizes was postponed until June 1, 1916. When that estimate of the war's duration proved overly optimistic, the ceremony was postponed indefinitely, and the prize money—146,900 Swedish crowns—diplomas and medals were given to the British embassy in Stockholm.[175]

At 25, Bragg was the youngest person to have won a Nobel Prize; at the time of writing, he still is. The magnitude of this achievement is perhaps best shown by the fact that Bragg received the Nobel Prize in Physics before Albert Einstein, who had already interpreted the photoelectric effect and developed the special theory of relativity, and before Max Planck, who had already done the work on black-body radiation that initiated the quantum revolution. Although it by no means resolved all his problems with his father, Bragg's recognition by the Nobel Foundation finally alleviated the secret fear he had been bearing since 1912—that WHB would receive the credit for his great insight into X-ray diffraction.[176]

Indeed, Bragg could easily have missed the Nobel Prize. In 1914, the only person to nominate a Bragg was the biochemist Emil Warburg, who proposed to divide the physics Prize between WHB and Laue. The following year, four nominations named one or both Braggs. Svante Arrhenius proposed to give the Nobel Prize for Physics solely to Henry Moseley or to divide it between Moseley, WHB, and Charles Darwin. Oddly, Arrhenius made the same nominations for the chemistry Prize. Henry Bumstead of Yale University proposed to give the physics Prize to Max von Laue or to divide it between Laue, WHB, and Bragg. Stefan Meyer suggested sharing this prize between WHB and Laue; Theodore Richards also proposed a two-way split, but between WHB and Bragg.[177]

As the deliberations of the Nobel Committees are not made public—indeed, they are not even recorded—it is impossible to say what influenced the decisions regarding the 1914 and 1915 physics Prizes. The decision-making process was complicated by Moseley's death at Gallipoli in August 1915—after the nomination deadline but before the physics and chemistry committees submitted their selections to the Academy of Sciences. It may be that the death of Moseley disqualified Arrhenius' nominations and that the decision to give the 1914 prize to Laue meant that Bumstead's and Meyer's 1915 nominations became moot. This left Richards' suggestion of a father-and-son Prize.

In retrospect, probably the ideal outcome would have been to award one Nobel Prize for Physics for the discovery of X-ray diffraction by crystals, and another for the insights this phenomenon provided into the nature of X-rays and the consequent discovery of the atomic numbers. The former Prize could have been shared by Laue and Bragg, the latter by WHB and Moseley. However, the death of Moseley made this impossible, and under the circumstances the Nobel Committee for Physics made a good—even visionary—decision.

Bragg had returned to Kemmel Hill from Paris "convinced our show will work."[178] Others, however, were not convinced. As J. F. Wright wrote to Bragg in 1952: "the outfit I joined at La Clyte [sic] which consisted of your goodself, Mr. Robinson and, I believe, Mr. Bouquet [Bosanquet], Sergeants Albrect [Albrecht] and Clark, a large lorry and two drivers, a set up which was disowned by everybody. When I was told I was joining you, on loan, and asked who it was I was joining, I was told; 'Some barmy scientists with a lot of gadgets trying to find some German guns. This will soon fizzle out and you will be back here again in a week or two, and we want all wire, insulators and poles back in good condition.' "[179]

This skepticism seemed to be justified. The apparatus designed by Bull consisted of six microphones spread along a "base" about 8300 meters long and 3700 meters behind the front line. The microphones were attached by wires to a single galvanometer at a headquarter station behind the base. Well in front of the base were one or two forward observation posts equipped with an telephone and a switch to activate the recording mechanism at headquarters. When enemy guns fired, the observer had a few seconds to start the recording. He then phoned headquarters with an estimate of the guns' approximate direction and target. As Bragg noted, "The difference between good and bad observers was colossal, and in fact the section depended entirely upon their judgment for its material." The six galvanometer "strings," one for each microphone, cast shadows on a film that had time markings every 1/100 of a second. "It was fascinating to watch the strings kick one after the other when a gun was firing, and then a few seconds later to hear the gun itself at headquarters."[180] The film could be developed and fixed in a few seconds. It was then handed to a "computer" who plotted the position of the gun.

There were a number of serious limitations to this approach. Sound-ranging could not be used if the wind was westerly, as this carried the sound of the enemy guns away from the microphones. According to a French version of Murphy's Law that Bragg recounted as "Principe de l'embêtement maximum" ("Principle of maximum cussedness"), the wind on the western front was generally unfavorable. Another problem was the complexity of the wiring system that connected the microphones, forward observation posts and headquarters—a total of about 40 miles of wire that had to be of low resistance and free of earths and other faults. This meant that a great deal of time had to be spent inspecting and maintaining the wire. A third drawback to sound-ranging was that the microphones used were much more sensitive to the shell-wave, the sound made by a projectile traveling faster than the speed of sound, than to the gun report, which arrived a few seconds later. The microphones also picked up extraneous sounds such as "rifle fire, traffic noises, and dogs barking."[181]

In November 1915, Bragg wrote to his mother with a description of La Clytte: "Where we are, there is a gentle little rise in front of bare ploughed fields, and with big farmhouses with a square of trees round them looking rather bleak. Beyond that is a valley dotted with wee cottages, very broad and shallow and a little brook running down it. It is practically untouched and people still live there and you see cattle in the fields. Beyond it is another rise topped with gaunt skeletons of houses, and with shell bursting along it pretty well all day. Beyond that again the trenches. I have only peeped at them a few times."[182]

An undated letter to Gwen, probably from about the same time, described the sound-ranger's daily routine: "Well, generally in the course of the morning something goes wrong, and we go out to put it right. It takes about five minutes to wriggle into our little Singer cars with the hood up, they are so tiny. They get over the ground in the most marvellous way. The roads are either bumpy pavé or else slime made of clay and occasional unfathomable holes, some of them made by shell. We scoot out and find the damage. About half past one we come back to lunch, which is always steak and potatoes and beer, except when one of us has a parcel from home. The beer comes out of a large blue coffee-pot, which we all love dearly. Most of the afternoon we work like blazes in our bedroom amongst piles of instruments and things. Dinner is at eight and is again steak potatoes and beer, though sometimes we go a bust and buy a leg of mutton. We have veges too and and once had macaroni cheese. We live jolly well really. After dinner we generally peg away again for a bit and knock off at about 11.45 when we feel jolly ready for bed, I can tell you."[183]

While on leave in England, Bragg would often take his sister Gwendy rowing or to shows: "She had rather a serious time with no other young family."[184] From France, Bragg wrote her amusing letters. One from Paris read in part: "There are ladies who go about selling balloons, they are without exception extraordinarily fat, I suppose because all the thin ones have been wafted away long ago."[185] In July 1917, he wrote: "Some of the small girls in this town are very pretty. They never seem to have anything to do, the children, but sit in buttercuppy fields beside the river fishing, and chatting to each other. All the small boys wear little blue soldier's caps, it's rather smart. If they are alone they are very polite to me and say 'B'jou, M'sieu' but if they are with a cow they feel grown up and simply say 'B'jou'. Mummy will explain."[186] It is perhaps not too fanciful to see these letters to his young sister as early examples of Bragg's gift for communicating with children, later to stand him in such good stead at the Royal Institution.

In April 1916, WHB became Resident Director of Research at the Admiralty experimental station in Hawkcraig, where he experienced a great deal of difficulty in mediating between the civilian scientists and naval personnel. In 1917, he moved to a new facility at Parkeston Quay, Harwich, where he led a team that developed the underwater microphone and other anti-submarine devices.[187]

Bragg was experiencing his own problems with the "brass-hats." On May 1, 1916, Bragg wrote to his mother: "I am having a dreadful time too, doing everything wrong and altogether sure that anybody else whatever could do it all better than I do."[188] A week later, Bragg lamented a flap over some apparatus he had asked WHB to have made: "I have had a fearful cussing over it all because it is costing ever so much while everyone is waiting for it."[189]

His superiors' attitude towards Bragg had improved by June 20, when he wrote to Gwen: "Did you see that I had scored a mention in despatches and a second pip?"[190] Perhaps the reason was that a solution to the shell-wave problem had been found. The first clue came from an unlikely source. The farmhouse in which Bragg was billeted had a privy in a small room leading off the kitchen. "When one sat down, one closed the only aperture between the hermetically sealed farmhouse and the outer world." If a large Allied gun fired over the house, the shell-wave was much louder than the gun-wave, but the pressure change associated with the latter "caused one to rise slightly but perceptibly." One of the soldiers in Bragg's section was Corporal William Tucker, who, as a lecturer at University College London, had done research on the cooling of platinum wire by air currents. Tucker had two small mouse holes in the wall beside his bed, and noticed a draught of cold air whenever the gun-wave arrived. He devised a microphone consisting of a thin, electrically heated wire stretched over a small hole in a container—"empty rum jars were very convenient."[191] The decrease in the electrical resistance of the wire as the gun-wave struck was recorded by a galvanometer. When wire of the right type arrived from England, a trial was made at Kemmel Hill. Bragg wrote: "I shall never forget the occasion when we first tried out the new device . . . The shell waves produced hardly any effect because their vibration was so rapid, but the gun reports made large clean 'breaks' on the ciné film used due to the deflection of the wire."[192]

By August 1916, Bragg was in a jubilant mood, writing to his father: "I am doing great things just now, I can tell you! I will have great yarns with you about all these things when I next come back on leave."[193] That autumn, Captain Joseph Gray joined Bragg's Experimental Sound Ranging Section. There were still problems because of the effects of wind and temperature and the large number of extraneous sounds detected. Gray initially used precise meteorological records and data from batteries whose positions were known to develop a set of empirical rules. Later, he developed the "Wind Section" a semi-circle of microphones 8 km in radius at the center of which a bomb was detonated. Results from this were telephoned to all Sound Ranging Sections in the area.[194] Gray's other innovation was the "regular base," a system of spacing microphones at precise distances apart in a semi-circle facing the enemy front line. This meant that the "kicks" of the six galvanometer strings which registered the arrival of the sound at the microphones formed a regular pattern on the ciné film, making it possible to calculate the positions of individual guns even when many were firing.[195]

The successes of Tucker and Gray meant that Bragg's experimental section now had to become a sound-ranging school. On November 5, 1916, Bragg wrote to his mother: "We are still in the middle of alterations to our camp, I have such a tremendous big place now and 56 men with nine officers . . . We have four officers learning, six gunners, and four instrument makers and n. linemen, all standing by waiting to be taught things."[196] To get men for new sound-ranging sections, Bragg went to base camp at Rouen and told the new recruits "B.Sc.s, one pace forward." He later recalled: "I remember after one such selection the sergeant coming to me and saying 'Could we not perhaps pass Jackson, sir, he says he comes from the Royal Observatory at Greenwich and is the man who found another satellite of Jupiter'. I think he had been failed for faulty arithmetic, but on the basis of this claim we let him through. He was John Jackson who later became the Astronomer Royal at the Cape!"[197]

From 1917 on, Bragg's assistant at the sound-ranging school was Reginald James, who had been in his Part II class at Cambridge. James had then become a member of Ernest Shackleton's ill-fated Antarctic expedition of 1914–16. After their ship "Endurance" was trapped in and then crushed by pack ice, Shackleton's men spent 6 months on the ice before sailing to Elephant Island in small boats. Most of the men, including James, were left there for 4 months until Shackleton returned with a rescue party from the inhabited island of South Georgia.[q] Bragg considered James "a tower of strength"[198] and an "A1 physics man."[199]

Another sound-ranger was Charles Darwin, grandson of the famous naturalist and a fellow Trinity man. On completing the Mathematical Tripos in 1910, Darwin had gone to Rutherford's department in Manchester as Schuster Reader in Mathematical Physics. He struck Bragg as "not the keen painstaking type, but rather the thinker with big ideas."[200] Darwin's "big ideas" would play a major role in the post-war development of Bragg's career. He would also become a close friend.

Bragg was eventually pulled back to GHQ in a supervisory role. Each section had a mechanic with a lathe to facilitate development of equipment improvements. The section heads met every two months to exchange information, "followed by a dinner which was rather an orgy."[201]

In January 1917, Bragg wrote to his father: "Our show is going famously, my only fear is lest the war should end before it has reached its full stage of perfection!"[202] An intercepted German order circulated to all Field Survey Companies on June 23 showed how effective Bragg's "show" had become: "Group Order. In consequence of the excellent sound ranging of the English, I forbid any battery to fire when the whole sector is quiet, especially in an east wind. Should there be occasion to fire, the adjoining battery must always be called upon, either directly or through the Group, to fire a few rounds."[203] The

[q] In his biographical memoir of Reginald James [Bragg, W. L. (1965). Reginald William James, 1891–1964. *Biographical Memoirs of Fellows of the Royal Society of London* **11**, 115–25], Bragg claimed that James' measurements of longitude using the occultation of stars by the dark limb of the moon allowed boats to be launched from the ice at the point of closest approach to Elephant Island. However, James' own account makes it clear that such observations were only possible during the period "Endurance" was ice-bound. I am grateful to Dr Durward Cruickshank for pointing this out to me.

recommended distracting fire did not help the Germans much—by this time positions of half a dozen batteries firing simultaneously could be distinguished. By the end of the war, there were about 40 sound-ranging sections covering the Western front completely. The technique could not be used in westerly winds, but was excellent in the fog that disabled the Flash Spotter unit headed by Harold Hemming.

However, there were problems on the home front. In the summer of 1916, Cecil Hopkinson suffered a serious relapse. Bragg wrote to his mother: "I can't bear the idea of losing him either for you know how different he was to everybody else when we had such good times together."[204] However, in early 1917 Hopkinson died; "so the war robbed me of the two people who meant most to me."[205] In early 1918, Bragg broke up with Elaine Barrow. There had seemingly been an understanding of marriage, if not a formal engagement. However, something had happened, for which Bragg blamed himself, writing to his mother on February 6: "I can tell you when I realized what an absolutely rotten thing I had done I went nearly off my head with shame."[206] Bragg went into a period of depression that was virtually incapacitating, on February 8 writing to Gwen: "At the present mo I'm very much in disgrace here because I was so ashamed I couldn't think of any work and I neglected all my jobs hopelessly... It was so hopeless because I nearly go off my head thinking about things when really they aren't so bad as all that, or rather they are bad but not so important. It's just the most awful weakness and I get so sick about it... I try so hard to pretend to you to be frightfully good and decided and nice and I'm not at all."[207] Later that month, having received an understanding letter from Elaine, Bragg had regained his equilibrium.[208] They continued to correspond, although apparently only as friends.

In his letters to his mother, Bragg was obviously minimizing the danger of his situation. In May 1918, he wrote: "Just to show you how lightly we take the war here I may mention that we have installed a [five?]-hole golf course on the common here. It's a most exciting course because the common is all over bumps of mole hills, and covered with gorse. One loses balls galore and nearly all one's time is spent looking for them. Also if a small French boy sees your ball land anywhere he runs and picks it up and politely returns it to you."[209] However, although the sound-rangers were well behind the front line, there was still danger from shellfire. According to his son Stephen, Bragg kept as a souvenir of the First World War a pair of binoculars that had been torn in half by shrapnel while hanging near him.[210] A former sound-ranger wrote to Bragg in 1965: "I well remember the German breakthrough, with seconds warning we were ordered to fall in and retreat, no time to dress properly, we looked a real untidy lot, and the remarks from troops going up the line are not repeatable, they did not know we were not to be exposed to the risk of capture."[211]

Bragg's work was being recognized by civilian and military authorities. In April 1918, he was awarded the Order of the British Empire. In June he wrote to his mother: "When I got back [from leave] I had an awful blow. I found I had been a major for some time. Think how I could have swanked around in town!"[212] He was also awarded the Military Cross.

As the Allies went on the offensive and the stalemate along the Western front was broken in the summer of 1918, the sound-rangers were put out of business. A typical wind section, now involving about 48,000 yards of wire covering 35 square miles, was far too elaborate to be used in the fluidity of the German retreat.[213] However, the sound-rangers did get a chance to examine the results of their activities. In September 1918, Bragg gleefully noted that "we must have given the Bosche beans by the way his battery positions had been knocked about."[214] He also interrogated German prisoners.[215] As the Allies advanced, Bragg had the chance to experience the liberation of occupied territory. On October 13, 1918, he wrote to his mother: "You would love to see the joy of the civilians who have got amongst friends again . . . The first one I saw was an old lady riding back in front of one of our lorries. She had a smile that nearly met round the back of her head and was bowing to right and left, just like the queen, while all the men shouted 'Hello Mother' to her . . . The people say the German men were generally not so bad but their officers were the absolute outside edge."[216] Following the armistice on November 11, Bragg collected material for a sound-ranging handbook at Colonel Jack's H. Q. at Campigneulles, near Montreuil-sur-mer, in the Pas de Calais.[217]

As the war wound down, Bragg's thoughts increasingly turned to the resumption of his scientific career. In August 1918, he wrote to his father: "Has any new work on our job come out lately? What do you mean to go on with? I don't know quite what line to take up. Anyhow, I expect I will be so busy trying to learn my job at Cambridge that I will have no time for experimental work."[218] However, his return to Cambridge was by no means certain. Just before the war, there had been talk of his being offered a position in his father's department at Leeds. Now, of course, the stakes had been raised considerably. Although still very junior, with only a handful of publications, very little experience of teaching and none of administration, the Nobel Prize meant that he could reasonably aspire to senior positions. In October 1918, Bragg was offered a position with the General Electric Company in the United States at a salary equivalent to £1000—far higher than his Cambridge lectureship—but he was reluctant to take anything other than a university job and even more reluctant to leave Britain.[219]

By December, Bragg and his father had agreed that it was better for him to leave Cambridge. Chairs at Leeds, Manchester, and Birmingham were being discussed.[220] In January 1919, Bragg was in a quandary: "I never felt more in want of advice and a greater difficulty in deciding what to do, than I do at present." The advantages of Cambridge were little teaching, no administration, presence of friends and loyalty to Trinity. The disadvantages of taking the job at "B." (presumably Birmingham) University were "a lot of new responsibilities," "older people who would be fed up at working under a new and very inexperienced man" and that "It would be very awkward if I went to B. and in a year or two was offered some better job at Cambridge." The "better job at

Cambridge" he had in mind could have been the Cavendish Professorship of Experimental Physics, which J. J. Thomson had held since 1884. Of the posts available, Bragg preferred Birmingham: "My reasons for B. are that I think it is the most sporting thing to do. I do want a place [of] my own finally, and I do not want to settle down as a college don." However, "I am just a bit doubtful of my powers of tackling the B. University job right away as I know so little physics! I've had no experience lecturing."[221]

At the beginning of 1919, Bragg went on a sightseeing trip to Germany with Harold Hemming and a Canadian named Beatty. They stayed the first night at a "most palatial" hotel in Brussels, "a gay place, lit up far more than London." They had a "sumptuous repast," went to the theatre, saw some of the "attractions" and at 2 a.m. went to see the Hotel de Ville by moonlight, "it looked just gorgeous." From Brussels they drove to Cologne via Waterloo, Namur, Huy, and Aix la Chapelle (now Aachen). Bragg wrote to his mother: "Cologne is a very fine place. The cathedral is immense . . . All the part of Germany we saw was quite untouched by the war and looked very prosperous, awfully good train and tram services everywhere and everything so well-arranged and neat. It is a wonderful country. *But* they are an awful looking crowd . . . The German girls are the very plainest I have ever seen in all my life. As Hemming says, there is not one that would not stop a clock . . . One day the major of one of the squadrons, who was working in conjunction with our show, took me up in his own private aeroplane and we flew along the Rhine to Bonn and back, it was the greatest treat I have ever had . . . I could not resist buying a Bosche helmet when I was in Bonn, an officer's one with a silver star on it, just as a souvenir . . . I expect I will get demobilized in a week or two now."[222]

Indeed, on January 24, 1919, Lt. W. L. Bragg, O.B.E., M.C., was "disembodied" from the British Army.[223] The "big job" not having materialized, he returned to Cambridge.[r]

[r] Bragg was not decommissioned until November, 1921. He was then allowed to retain his war-time rank of Major (RI MS WLB37A/4/8).

4

A system of simple and elegant architecture: Manchester, 1919–30

When Bragg returned to Cambridge, he took up the fellowship and lectureship he had been awarded just before the outbreak of war. As part of his duties of teaching and supervising students, he demonstrated in Searle's "famous" Part I physics practical class. Searle "despised research" but "took infinite pains over the practical class." He was "a terrific tyrant"; Bragg remembered Searle shouting at him: "Bragg, come here, and see what this fool has done. This is the kind of thing you have got to look out for."

It was during this "brief interlude" that he attended a "thé dansant" and met Alice Grace Jenny Hopkinson, Cecil's first cousin. He had heard a great deal about Alice, including that she was an extremely pretty girl. However, they had not met before the war, as Alice lived in Manchester, where her father, Albert, was a doctor. Now 19, she was studying history at Newnham College Cambridge.[224]

Bragg and Alice met through *his* cousin, Vaughan Squires, a medical student at Cambridge. In her memoir, Alice described her future husband as "a dark man in uniform, with decorations of military O.B.E. and M.C. on his tunic . . . He was not at all what I had expected and I remember telling him so at once. Somehow I had had the idea that he would be a small man with spectacles, shy and vague."[225]

Alice learned more about Bragg when they met again at a dance hosted by the Professor of Engineering. On this occasion, he informed her that he would be moving to Manchester. Despite his sense of loyalty to Trinity, which had supplemented his army pay until he became a major, Bragg had resolved to leave when a suitable senior post became available. This occurred when J. J. Thomson resigned the Cavendish Professorship to devote himself full-time to the position of Master of Trinity College. Ernest Rutherford was appointed in Thomson's place, vacating the Langworthy Professorship of Physics at Manchester University. Whether or not Bragg was disappointed at not being approached for the Cavendish chair, he quickly set his sights on Rutherford's old job. He had something of an "in" at Manchester, as the Vice-Chancellor, Sir Henry Miers, was also Professor of Crystallography[s] and Horace Lamb, Professor of Mathematics, had been WHBs predecessor in Adelaide.[226] Despite his youth and inexperience, Bragg was offered the professorship, at a stipend of £1000 per annum.[227]

When Alice questioned his decision to leave Trinity for "wet and ugly" Manchester, Bragg mentioned that he thought it a mistake to spend one's whole career in Cambridge—although he admitted that the Cavendish Professorship "would be any physicist's dream." It appeared that Alice would have little occasion to return to Manchester, as her parents had just moved to Cambridge, Albert having developed a serious illness due to overwork.

In May, Bragg proposed; Alice turned him down on the grounds that she did not know him well enough. In addition, she still had another year to go at Newnham and wanted to complete her studies—although, as a woman, she could not be awarded a Cambridge degree. Aunt Evelyn, who was very annoyed that Alice had rejected a man who was not only a Nobel Laureate but also her late son's best friend, told WHB and Gwen that Bragg had had a lucky escape, as Alice was "unstable and a sad flirt."[228] On the other hand, another Hopkinson aunt, Monica Wills, "profoundly disapproved of scientists, knowing nothing about them, and was convinced that they were all atheists."[229]

Bragg and Alice continued to meet during the Long vacation. However, she sometimes felt it was "an uneasy friendship," as "W.L.B. was very serious minded in those days."[230] "He was at great pains to tell me that he was a 'lone wolf' . . . ."[231]

In September 1919, Bragg took up his new position. Rutherford had created a strong department at Manchester, which included Hans Geiger and Niels Bohr. By 1914, "The visible spearhead of British physical research had clearly gone to Manchester."[232] To this center for nuclear physics Bragg brought his fellow sound-ranger Reg James to join another sound-ranger, Harold Robinson, who had already returned to his old job at Manchester. Robinson was appointed Assistant Director of the laboratory. The other lecturing staff consisted of David Florance, Evan Evans, and Norman Tunstall, who had kept the department going during the war.

The quality of research at Manchester might have been better than that at Cambridge, but the physical surroundings were no match for the graceful Cavendish Laboratory: Rudolf Peierls recorded that the Physics Department "consisted of two buildings, of which the newer one had much of the inside covered with brown-glazed tiles, suggestive of a public lavatory."[233] Bragg lived with Douglas Drew, a former sound-ranger and now lecturer in Classics; Gwen sent Charlotte to look after the two bachelors.

Bragg's new job quickly degenerated into a fiasco. As he later wrote, "I made every mistake it was possible to make in planning the courses and the examinations."[234] In addition to his inexperience, Bragg had forgotten most of his physics and found studying it very boring after the excitement of the war. The

[s] Miers would later become godfather to Bragg's son David.

students, who included many ex-servicemen, were "a tough crowd."[235] During lectures given by the new men, Tunstall recalled, "panels of the benches were kicked into matchwood."[236] Bragg boxed the ears of a student who set off a firework in class.

He was also intimidated by having to replace Rutherford, a brilliant lecturer and charismatic leader, which his sister thought was "a bit like a chamber orchestra (however excellent) succeeding a full orchestra in the Albert Hall."[237] Peierls recalled of this period: "He [Rutherford] was visiting Manchester to give a colloquium talk, and he was introduced by Lawrence Bragg, who spoke at length about his feeling of inadequacy in succeeding Rutherford as Professor at Manchester. People on the Faculty Board, he said, had been accustomed to turn for wisdom on difficult decisions to the Professor of Physics, and were disappointed to see it was only him sitting there. This left the audience with some feeling of embarrassment, which was happily defused when Rutherford got up and said 'Professor Bragg has expressed some doubts about his ability to fill my chair, but' and here he pointed to his own bulk and to the beginning of middle-age spread of Bragg, 'I think he is well on the way to doing so.' "[238]

There is no doubt that Rutherford and Bragg were very different personalities. At Rutherford's first faculty meeting at Manchester, there was a discussion about reallocating space in the physics building to the Chemistry Department. When asked for his opinion, the new Professor of Physics banged his fist on the table, roared "By thunder!" and embarked on a tirade that ended with him chasing the Professor of Chemistry back to his office.[239] Rutherford recommended that Bragg take the same approach, advising him about the Senate of the University, "Don't let those chaps bully you, Bragg. You give them hell."[240] It quickly became clear Bragg was temperamentally incapable of following this advice. Soon after his appointment, he was summoned by the Bursar of the University and informed that the headmistress of a local girls' school had complained that one of its pupils had been refused acceptance to the honors physics class. The Bursar insisted that Bragg "appear on the mat" to explain himself to the school's headmistress.[241] One cannot imagine the volatile Rutherford submitting, meekly or otherwise, to this kind of dressing down.

Before even leaving France, Bragg had worried that his appointment to a physics chair might be resisted by older subordinates. This fear was justified. According to Alice, Evans and Tunstall "deeply resented a young man of 29 with no experience of lecturing or running a department being made their professor."[242] Bragg soon started to receive poison-pen letters, clearly written by someone with inside knowledge, accusing him of incompetence. According to his sister Gwendy, the low point came when Bragg was going into a committee meeting and overheard a colleague say "Bragg ought to resign."[243] He could not sleep, felt unable to confide in his parents or anyone else, and suffered "what was really a nervous breakdown."[244]

Bragg had no interest in or gift for administration. In Manchester, he learned to delegate administrative duties to James, who Alice described as "in every way WLB's right hand."[245] Other key subordinates were the former sound-ranger

Ernest Scott Dickson, who joined the department in 1920, and Rutherford's old assistant, William Kay, who was "a prince of laboratory stewards . . . brilliant at devising experiments and demonstrating them in class."[246] Mair Jones, Bragg's secretary throughout all his time in Manchester, also played an important role in the department. She "became formidable" if she thought that his good nature was being taken advantage of. Alice wrote: "Once I remember he set off on a dark, foggy morning, in the brown trousers of one suit and the blue jacket of another; Mair Jones quietly rang me up, and asked if I could bring into the lab. one or other of the missing partners."[247]

Bragg's state of mind improved when he discovered that poison-pen letters were also being sent to James. After surviving Shackleton's expedition, James was not about to lose any sleep over gossip. The anonymous criticism then extended to Rutherford and WHB. Eventually it was discovered that the author was Evans' wife, distraught after the death of their child, and Evans left the department.

By 1921, the difficult transition was largely over. Bragg was learning the ropes and most of the ex-servicemen had graduated. Relations with the students improved to the point that the Professor played center-forward on the departmental hockey team.[248] Bragg had survived his baptism of fire.

Meanwhile, Bragg was attempting to establish a research program in Manchester. Soon after arriving there, he wrote to his father, who had now taken up his post at University College London: "I have been wondering what you were intending to go on with. I do hope you will never keep from doing any bit of work, Dad, because you think that may be the line I am going on. I hope you will always just chase the idea you are keen on and never think whether our ideas may overlap. I am sure that is the right way to set about it."[249] Some of Bragg's "scientific friends"—apparently including Darwin—advised him to abandon the X-ray analysis of crystals, on the grounds that all crystalline structures would soon be solved.[250] Bragg seems to have given some credence to this argument, as his initial research efforts at Manchester used X-rays not to solve the structures of crystals, but rather to determine the structures of atoms.

Although the war had seriously disrupted scientific activities in Europe, a number of important advances had occurred in the five years since Bragg had been involved in research. It was now clear that it was the electrons, not the nuclei, of atoms that diffracted X-rays. Atomic models proposed by Gilbert Lewis in 1916 and Irving Langmuir in 1919 were based on the idea that the nucleus is surrounded by "shells" containing 2, 8, 8, 18, and 18 electrons. A number of crystal structures had been solved, including quartz, with four parameters.

Enough structures were now available for Bragg to assert in a 1920 review article that crystals were composed of "inelastic spheres in contact," and it was therefore possible to estimate atomic dimensions. Using the rule that "two atoms may not be placed closer together than a distance equal to the sum of the radii of the spheres representing them," Bragg was able to determine atomic diameters for a number of elements. Particularly useful were series of isomorphous (identically shaped) crystals which shared a common anion or cation. For example, from the difference between neighboring atoms in MeS and MeO compounds (where Me is a metal), Bragg was able to calculate the increase in atomic diameter in going from oxygen to sulfur. By this means, Bragg was able to calculate the diameters of a considerable number of metallic and non-metallic elements. Plotting the atomic diameters against atomic number showed that for a particular row of the periodic table, the alkali metals are of greatest diameter, then the alkaline earths, reaching a minimum with the electronegative elements. Between the inert gas of atomic number n and the alkali metal of atomic number $n + 1$, there is a large increase in diameter.[251]

Bragg's idea that he and his father should follow their own research directions without any attempt at coordination was not working out. In an undated letter probably from 1920, he wrote to WHB: "I am so sorry I was so stupid when you said you wanted to try rock-salt. It was awfully selfish and I want to take it all back."[252] In another undated letter to his father, Bragg wrote: "I have been thinking over the research on the Debye [powder diffraction] method and I am sorry I was so cantankerous about it when I talked it over with you."[253] Eventually it was decided that WHB would concentrate on organic crystals and his son on inorganic ones.

Bragg's first major post-war project was inspired by two papers published in 1914 by his friend Charles Darwin, now a Fellow of Christ's College Cambridge.[254] In analyzing the intensity of reflection of X-rays by crystals, Darwin distinguished between two types of crystals. The first were perfect crystals, such as diamond, in which the intensity of the reflected wave is proportional to its amplitude, F. The second were imperfect crystals, such as rock salt, consisting of slightly misaligned "blocks," in which the intensity of the reflected wave is proportional to the square of its amplitude. Paradoxically, a perfect crystal would reflect less well, as reflection could occur only at a very specific angle, whereas in an imperfect crystal different blocks would reflect at slightly different angles.

Bragg realized that Darwin's work provided a theoretical framework for making absolute, rather than relative, measurements of X-ray diffraction. To do this, he used an X-ray spectrometer that had been built in his father's laboratory in Leeds. He was greatly helped by another gift—a new type of X-ray tube donated by William Coolidge of the General Electric Company in Schenectady, New York. The Coolidge tube gave a much more stable output because it used a much higher vacuum and provided electrons for the cathode rays by thermal emission from an incandescent wire.[255] Using his old friend rock salt, Bragg worked on determining absolute intensities of reflection with James and another former sound-ranger, Charles Bosanquet of Oxford University, who came to Manchester during the university holidays. The first of what would become known as the "BJB" papers was published in March 1921.[256]

This paper aimed at determining a relationship between the amount of X-radiation reflected by a crystal and the number of electrons in its unit cell. To determine the amount of X-radiation reflected, it was necessary to use a homogeneous incident beam. To do this, Bragg and James first reflected an X-ray beam from a crystal to make it homogeneous, then reflected it from a second crystal to make the quantitative measurement. The "reflecting power" of the second crystal face was measured using a technique WHB had developed in 1914, in which the amount of radiation reflected by a crystal was measured while it was being rotated at a constant rate.[257] WHBs "integrated intensity" was defined as $E\omega/I$, where E is the amount of energy reflected when the crystal is rotated through the glancing angle at ω radians per second and I is the amount of energy detected when the incident beam is admitted to the detector for one second. The integrated intensity of reflection consisted of contributions from all the blocks of an imperfect crystal and had the advantages, as Bragg later noted, of being dimensionless and independent of "all the purely geometrical features of the experiment."[258] At the most favorable angle for reflection, $\frac{1}{25}$th of the incident beam was found to be reflected by the (100) face of rock salt.

The theoretical part, performed by Bosanquet, included a relationship between reflecting power and number of electrons based on a formula derived by J. J. Thomson for the amount of radiation scattered by an electron and including terms for polarization and temperature formulated by Hendrik Lorentz of Leiden and Peter Debye of Göttingen, respectively. Lorentz had shown that the intensity of diffracted X-radiation decreases with increasing order of reflection (n in the Bragg equation), providing an explanation for Bragg's 1912 observation that the diffraction pattern of zincblende contains only spots corresponding to planes with low Miller indices. Debye had shown that diffraction intensity decreases with temperature (particularly for higher orders of reflection) because thermal motion of the atoms causes destructive interference of the diffracted waves—an effect that Arnold Sommerfeld had told Max von Laue in 1912 might prevent diffraction from being observed at room temperature.

Bosanquet's relationship was:

$$E\omega/I = (N^2\lambda^3/2\mu \sin 2\theta) \cdot F^2(e^4/m^2c^4) \cdot ([1 + \cos^2 2\theta]/2) \cdot e(-\mathrm{B} \sin^2 \theta)$$

where N is the number of atoms per unit volume, μ is the absorption coefficient, e and m are the charge and mass of the electron, respectively, and c is the velocity of light. F, the amplitude of the reflected beam, depends upon the angle of incidence and the positions of electrons within the atom, and tends to Z, the number of electrons, at low angles. All other quantities having been measured, F could be calculated for a range of values of θ. This was done for Na and Cl. If the theory were correct, the value for F at $\theta = 0$ should be the numbers of electrons in the ions—18 for Cl^- and 10 for Na^+. At the lowest angle studied, $\sin\theta = 0.1$, F was 11.67 for Cl and 6.9 for Na, so the agreement with theory was reasonable.

To study the effect of F upon the distribution of electrons within the atom, three models were considered: Uniform distribution of electrons within a sphere; spherical shells containing 2 and 8 electrons (for the sodium ion) or 2, 8, and 8 electrons (for the chloride ion); and spherical shells of electrons oscillating along a line joining them to the center of the atom. The third model gave the best agreement with the observed curves of F against θ.

In the second BJB paper, the theory was refined to correct for the effects of extinction.[259] The relationship between reflecting power and number of electrons derived in the first paper included an absorption-coefficient term, μ. This could be measured simply by directing an X-ray beam through the crystal of interest and measuring how much the beam was attenuated. However, Bragg and colleagues pointed out that "When X-rays pass through a crystal in such a direction that the crystal reflects the rays, the absorption of the transmitted beam is greater than that for other directions." This is because (most) reflected X-rays have to travel through part of the crystal in order to reach the ionization chamber of the spectrometer, and during this passage there is a chance that they may be reflected again by the same crystal planes. If this does occur, the resulting doubly reflected beam will be parallel to the incident beam but exactly out of phase with it; the strength of the incident beam will therefore be decreased by interference. For this reason, the effective absorption coefficient is higher at the reflecting angle than it is at other angles—in the case of rock salt, 52% higher. Therefore, the effective absorption coefficient, μ, will be equal to $\mu_0 + \epsilon$, where μ_0 is the uncorrected absorption coefficient and ϵ is a new term, which Bragg and co-workers referred to as the extinction coefficient. The effect of extinction depends upon the average size of homogeneous elements in the crystal. For Darwin's (hypothetical) perfectly imperfect crystal, this effect could be ignored; for a highly perfect crystal, it would be quite significant. Extinction also depends upon order of reflection, being much more significant for low-order reflections. For example, for rock salt, $\mu_0 = 10.7$. For the first-order (100) reflection, $\epsilon = 5.6$; for the second-order, $\epsilon = 1.96$; and for the third-order, $\epsilon = 0.02$. Making allowance for extinction, curves of F_{Cl} and F_{Na} against $\sin\theta$ extrapolated to the expected values of 18 and 10, respectively, at $\sin\theta = 0$.

The stated rationale for making accurate absolute measurements of reflection was "in order that they may serve as a basis for an analysis of the arrangement of the electrons in the atom." Quantum mechanics, not X-ray diffraction, would determine the issue of atomic structure. But the "BJB" approach would turn out to be of great importance in another area, the structural analysis of complex crystals. Relative measurements of reflection might be enough to calculate the distances between atomic planes in simple unit cells, as Bragg had shown in the seminal case of iron pyrites, but would be inadequate for the complex unit cells of many minerals and organic molecules. Whether Bragg, James, and Bosanquet realized it at the time, what would soon be required in X-ray analysis was a means of determining electron density at different points in the unit cell; for this, absolute measurements of reflection were absolutely necessary.

The third and final "BJB" paper was published in September, 1922.[260] It had been known since the very early days of X-ray diffraction that the intensity of reflected radiation falls off with increasing order of reflection. WHB had proposed in 1915 that this effect is due to the arrangement of scattering matter in the atom. Using absolute measurements, Bragg and co-workers now placed different numbers of electrons at different distances from the nucleus, calculated the resulting theoretical intensities, and compared these with intensities actually measured. The best agreement was found for sodium if 7 electrons were located at a radius of 0.29 Å and 3 at 0.76 Å, and for chlorine if 10 electrons were located at a radius of 0.25 Å, 5 at 0.86 Å and 3 at 1.46 Å. The general conclusion was that neither sodium nor chlorine could have an outer "shell" containing eight electrons in circular orbitals.

In the early post-First World War period, Bragg remained close to his Cambridge friends and former sound-rangers. He went on sailing holidays with George Thomson, who owned the yacht "Fortuna," and Arthur Goodhart, both of whom were now fellows at Corpus Christi College. Now an avid outdoorsman, Bragg also spent holidays climbing in the Lake District and on the Isle of Skye, although "I never became good enough to lead." He was impressed with Skye, which he visited with Francis Aston, Darwin, and (Edward?) Milne. "The rock is so crystalline and rough that one can stick to quite steep slopes. The scenery was wonderful too, with the contrast of the jagged black main Coolin range with the smoke-red Blaven."[261]

In February 1921, Bragg received the welcome news that he had been elected a fellow of the Royal Society[262]—a recognition that was perhaps overdue. He had lost touch with Alice Hopkinson—depressed by his problems in Manchester, he seem to have abandoned his suit—but she now sent him a letter of congratulation. Bragg took this opportunity to suggest another meeting, and Alice invited him for tea at Newnham on April 24.[263] The first train Bragg could take from Manchester arrived only 15 minutes before their date; taking no chances, he went up to London the night before. His careful planning was rewarded; he and Alice became engaged that day. Symbolically, Bragg popped the question in Free School Lane, the street in which the Cavendish Laboratory was located.[264]

On his way back to Manchester, Bragg stopped off in London to tell his parents about the engagement. Only WHB was home when he arrived. Gwen and Gwendy realized something was afoot when they went in for dinner and saw a bottle of champagne on the table. Told by WHB that her son was engaged, Gwen asked "Who to?"

Soon thereafter, Alice visited her future in-laws. It was the first time she had met Gwendy—"a most attractive teenager" who "positively squeaked with excitement." Seeing Alice off at the station, Gwen offered her some cryptic advice: "You must make the running, my dear, and hold his hand as I have always had to do with Dad."

Back at Newnham, Alice was stopped in the corridor by a fellow student: "Is it true you are engaged to W.L. Bragg? Look, I'm just reading one of his books and you don't know the first thing about science. What a waste."[265] If Alice had not already realized that her married life would be very different, she got the message when she was sent by her parents to announce her engagement to

Horace Lamb, now retired from Manchester and living in Cambridge: "Lady L. was in a bath-chair in the garden and poked me with a stick and said I was much too young and gay to become a Professor's wife."[266]

In the summer of 1921, Alice and Bragg went with WHB, Gwen, and Gwendy to St Briac, Brittany, and then to Ullswater with her family. They were married on December 20, 1921, at Great St Mary's Church in Cambridge. Alice had five bridesmaids, including her sister Enid and Gwendy, and the best man was Bragg's cousin, Vaughan Squires, who had introduced the bride and groom to one another. The wedding was enlivened by the unexpected appearance of Alice's Aunt Emma—one of many eccentric Cunliffe-Owen relatives.

The Braggs honeymooned at Wrington in Somerset and Valescure on the French Riviera. On their way to the Mediterranean coast, they stopped in Paris and had lunch, "with footmen in white gloves," at the home of the Duke and Duchess de Broglie; Bragg had met Maurice de Broglie, who was the leading French authority on X-ray diffraction, during the war. Alice also met Lucien Bull, the father of sound-ranging, who took her and her new husband to the theater.[267] In Valescure, they walked through the hills, eventually reaching the highest ridge of the Esterel range, where Alice saw snow-capped peaks for the first time, Brittany having been her furthest previous expedition. To make the Riviera feel more Christmas-like for Alice, who was spending her first Christmas apart from her parents, Bragg dug up a juniper bush and decorated it with ornaments he bought in Nice. On the way back to England, they stayed at Avignon.[268]

Willie and Alice Bragg had an exceptionally happy relationship. David Phillips wrote: "In all respects it was an eventful and fruitful marriage which remained a romance to the end."[269] By all accounts, Bragg was a highly devoted husband; Alice's letters to him when they were separated by war in 1941, and had already been married for twenty years, positively sizzle with passion. However, there were some problems in the early years of the marriage. In an undated letter, Alice wrote to Bragg: "I made such a bad beginning when we were married and let you down so."[270] The problems seem to have mostly been family related. Bragg had used some of his Nobel Prize money to buy a house for them—130 Lapwing Lane in Didsbury. Alice did not have a chance to see it because she was studying for her Tripos. Also, housing was scarce and Bragg was afraid that if he did not act quickly he may lose the opportunity. When they arrived back from their honeymoon, Alice found that "Lapwings had long left the neighbourhood, but there were beautiful trees lining the road." The house was "in perfect order, and very charming." Far less charming was the housekeeper Gwen had provided—Bragg's former nursemaid, Charlotte. Alice wrote: "She disliked me on sight, and during the four or five days she remained in the house she hardly spoke to me, but went about muttering 'Ah, poor Mr. Villy, God help him.' "[271]

Once the servant problems had been sorted out—a cook and a maid were soon engaged—there remained the uncomfortable reality that Alice was back in a town she had been very happy to leave. "Manchester was a dreadfully dirty and ugly place with a vile climate, foggy and drizzly." Before going to university, Alice had vowed to her sister that she would never live in Manchester again.[272] Also, the life of a professor's wife was very different from that of a Cambridge undergraduate. According to Alice: "Twice a week in the afternoon I had to be 'At home' to callers from 3 o'clock onwards; a stream of ladies came, leaving cards in a bowl in the hall, and staying about a quarter of an hour. Mrs Mate [the maid] served an elegant tea, with our wedding cake cut in pieces on a silver dish. They were the wives of professors, doctors, Manchester business men, all, or so it appeared to me, very old, indeed many were friends of my parents."[273] As Bragg sadly recorded, "We went to formal parties of the middle-aged, where Alice as the new bride wore her wedding dress and was taken in by the host."[274] The dinner parties hosted by the Braggs were less of an ordeal, as the guests were sometimes closer to Alice's age, or interesting people such as Charles Prestwich Scott, editor of the *Manchester Guardian*. Much to Alice's bemusement, a variety of eccentric scientists arrived as house-guests, including Linus Pauling, who "already had that dedicated air"; Arnold Sommerfeld, "scratching mysterious lattices on the table cloth with a fork"; and John Desmond Bernal, who "took a little getting used to."[275]

Bragg, still struggling to get his research program established, felt torn between his responsibilities as professor and husband. Like many scientists, he would concentrate on his current research problem to the point of being oblivious to the outside world. As a result, he worried that he was "failing her [Alice] wretchedly as a companion."

As the Braggs and Hopkinsons had very different family dynamics, both Bragg and Alice had difficulties in dealing with their in-laws. Although Bragg seems to have liked Albert Hopkinson, characterizing him as "a really good man," his feelings about Alice's mother, the former Olga Cunliffe-Owen, seem to have been more mixed: He characterized her as "utterly outspoken" with "aristocratic instincts" and "endless moral courage."[276] According to Bragg's daughter Patience: "I think he was probably uneasy with his parents because he had married my mother who was a very different kettle of fish to the Bragg family. They always discussed for ages what they all wanted to do and then did something nobody wanted to do because they all thought somebody else wanted to do it. My mother would go for it and state what she wanted. She had a healthy streak of ruthlessness."[277] Alice liked WHB, but found him emotionally distant. Of Gwen, she wrote: "I found her ambivalent, utterly warm and kind, but I was never quite sure if she was expressing what she really felt."[278] Gwendy wrote: "she came to value GB, though GB must have been a maddening mother-in-law especially to start with. I did have such a time pouring oil myself!"[279]

Bragg's mixed feelings at that time are shown by a letter he wrote to WHB on March 28, 1922: "This has been such a happy term. Alice has had some worries, about household things and so on, because it is such a new job to her and she has been used to such an exceedingly gay time in Cambridge. She has been such an angel, though, and we have just had a heavenly time. Manchester *is* a grimy place to bring a darling pretty person like that to, who is so fond of gay people and pretty clothes, and it makes me wild sometimes to think that it is necessary for us to live here. The *people* are so ghastly . . . "[280]

At Easter, 1922, Bragg and Alice went on holiday to Malvern, where she encouraged him to take up watercolor painting again—he had apparently not done so since Australia. That summer, they sailed on the Norfolk Broads, where "Alice entered into the spirit of it so thoroughly and enjoyed it all so, however wet and cold it was."[281] By now Bragg was feeling more settled, writing to his father in August 1922: "I have got the lab here in order now, with first rate courses (so I fondly imagine) and I think I will have time for research in future." However, he was still suffering from feelings of inadequacy and envy: "You are so good at all this crystal work that I often feel the most awful fraud, that I have crept into a reputation towed in your wake! . . . Like a silly ass, I often felt low when you told me about all the exciting things you were doing, because I wasn't getting on."[282]

In September 1922, Bragg and Alice went to Sweden for Bragg to deliver his Nobel lecture.[283] He had been invited to a special ceremony in June, 1920, at which a number of wartime Nobel Prizes had been officially awarded, but had, for some reason, declined the invitation.[284] The introduction was given by Gustaf Granqvist, Professor of Physics at Uppsala University and chair of the Nobel Committee for Physics. Bragg must have been gratified to hear Granqvist deliver an *ex cathedra* endorsement of his alkaline halide structures: "a metal atom in the crystals of the alkaloid [sic] salts is situated at one and the same distance from the six haloid atoms nearest to it, and *vice versa*—a relationship that was found to prevail, *mutatis mutandis*, in all the crystals examined. This means the exceedingly important discovery, both for molecular physics and chemistry, that the crystals consist of atomic lattices and not, as has been always imagined, of molecular ones."[285]

The Swedes remarked upon the youth of Bragg and particularly his wife, and the fact that Bragg gave his Nobel lecture in a lounge suit rather than morning dress. Because the official award for the wartime prizes had been made in 1920, the Braggs were not received at the Palace. Bragg was not a particularly thin-skinned person, but was occasionally capable of deep resentment over seemingly trivial matters. He took his lack of a royal audience as a slight, and bore a grudge until restitution was achieved in 1965.

Arne Westgren, a young metallurgist, was delegated to show Bragg and Alice around Stockholm, the beginning of a long and scientifically important friendship between the two men. One of the eminent scientists they met was Svante Arrhenius. Much later, Bragg wrote of this encounter: "I remember when my wife and I went to Sweden in early days and were entertained by the famous Arrhenius, my wife asked a Swedish friend [Westgren?] what Arrhenius had done. He replied 'When he was a young man he made a very famous theory; since then he has gone around the world accepting honourary degrees.' "[286]

When he wrote this, Bragg no doubt appreciated the irony that the words of the "Swedish friend" could now be applied to him.

During the 1922 trip, Bragg gave lectures in Denmark, where he and Alice stayed with the Bohrs in Copenhagen. They also visited Norway, where Bragg gave a "bad popular lecture" in Oslo and had a holiday in the country above Lake Myosen "in lovely sunny still weather with the grass brilliant green and the trees yellow, red and purple." They then travelled across the country to Bergen on the scenic mountain railway and sailed from there to Newcastle.[287]

Bragg never knew why WHB did not go to Sweden to formally receive his Nobel Prize.[t] It was only after Bragg's death that his sister realized that their father may have stayed away "so that Willy could get the acclaim."[288]

The Scandinavian trip, Bragg wrote, "did a great deal to send up the value of my shares as a husband." Alice had discovered that she loved to travel and was impressed by the reception a leading scientist receives in foreign countries. This could be some compensation for living in dreary Manchester and socializing with middle-aged women. According to their daughter Patience: "My mother loved travelling and so he loved giving her the opportunity to do so. He was determined that he wouldn't be a dull old stick for her, because she wasn't particularly interested in science . . . I think my father felt that the onus was on him all the time to find splendid ways of entertaining my mother and taking her places."[289]

In 1923, however, Alice was pregnant and the traveling was more modest. The Bragg Easter and summer holidays were both spent in Cornwall, at St Ives and Morwenstow, respectively. They drove to Morwenstow in their first car, a two-seater Wolseley. A journey of this distance was a great adventure in those early days of motor travel, and there must have been some second thoughts when the rear brake cables vibrated whenever the car went over a bump, causing "terrific jars and jolts." On the way back, Bragg and Alice detoured to the Wolseley factory, where mechanics discovered that three of the five bolts holding a back wheel had sheared off. A son, Stephen Lawrence, was born on November 17, 1923, a difficult birth that was attended by the former partner of Alice's father's.[290]

After his digressions into atomic structure and absolute intensity measurements, Bragg was now getting back into the business of crystal structure analysis. His target was the calcium carbonate, aragonite, which he had previously had a go at in 1914.[291] Then it had been too tough a nut to crack. Indeed, in his 1816 textbook of mineralogy, Parker Cleaveland had written: "The analysis of no mineral has ever so much exercised the talents, exhausted the resources, and disappointed the expectations of the most distinguished chemists in Europe, as that of the Arragonite [sic]."[292] Now, however, Bragg had some new tricks

[t] WHB was invited to give his Nobel lecture at the Nobel Prize ceremony of 1920 (RI MS WHB11A/23). He declined, citing pressure of work (RI MS WHB11A/250).

up his sleeve. In August 1922, he wrote to WHB: "I was trying to work out aragonite and I think I see how it goes now... As far as I can see, aragonite is rather like calcite, with the calcium atoms in hexagonal instead of cubic close packing. The CO_3 groups lie between two sets of three calcium atoms in both."[293]

By now it had become clear to Bragg that packing of atoms was a very important consideration in crystal structure. In the simplest situation—a crystal composed of spherical atoms of equal size—closest packing is achieved by staggering neighboring rows of atoms in each layer by half a diameter, forming a hexagonal two-dimensional array. The atoms of the second layer are placed in the interstices between three atoms of the first layer. There are two ways to place the atoms of the third layer, both of which result in equal density: Directly above the atoms of the first layer, or directly above unoccupied interstices of the first layer. The former is referred to as hexagonal close packing, and had been found to occur in several crystals of the hexagonal system; the latter is referred to as cubic close packing, and was known to occur in face-centered lattices, such as the alkaline halides. In both cases, each atom has six nearest neighbors (Figure 4.1).

A paper describing the aragonite structure, authored by Bragg alone, was submitted for publication in November 1922.[294] Aragonite was known to belong to the orthorhombic crystal system, which means that the *a*-, *b*-, and *c*-axes of its unit cell are unequal in length but at right angles to one another. However, it was considered to be pseudohexagonal, as faces that would be at a 60° angle to one another in the case of true hexagonal symmetry are at 63.48° in aragonite. From the reflections corresponding to the principal axes of the crystal, Bragg was able to calculate the volume of the unit cell. Using the known density of aragonite, he was then able to calculate that the unit cell contains four molecules of $CaCO_3$.

Bragg's working hypothesis was that aragonite would be a hexagonal array if the carbonate ions were considered as single lattice points. It deviates from hexagonal symmetry because the carbonate ions are themselves asymmetrical and therefore can adopt different orientations. The trick was to find

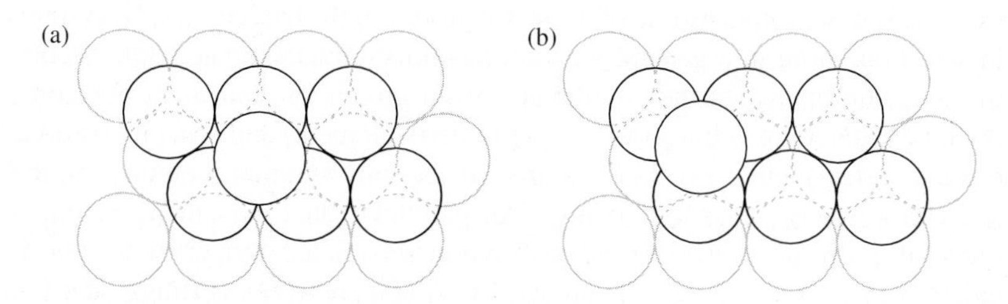

Fig. 4.1 Close packing of spheres. (a) Hexagonal close packing—the third layer of spheres is placed directly above the first layer. (b) Cubic close packing—the third layer is placed directly above a set of interstices of the first layer

arrangements of the carbonates that reduced the symmetry from hexagonal to orthorhombic.

To do this, Bragg used space-group theory for the first time. As described above, the complete set of 230 space groups had been derived by the end of the nineteenth century. As also noted above, each space group can be uniquely defined by its Bravais lattice and its combination of symmetry elements. After the First World War, several X-ray crystallographers, including Kathleen Yardley and William Astbury in WHBs laboratory, had the idea of using space-group theory as a systematic means of solving crystal structures. To do this, they compiled tables listing the positions of atoms in the unit cell consistent with the combination of symmetry elements present in each space group, thus defining sets of symmetry-related atoms characteristic of that space group.

The position of each atom in a three-dimensional lattice can be defined by three coordinates, x, y, and z, corresponding to fractions of the lengths of the a-, b-, and c-axes of the unit cell, respectively. If the values of x, y, and z are such that the atom lies at a center of symmetry, like the Na and Cl atoms in rock salt, it has no parameters (coordinates that are not defined by the space group and therefore have to be determined experimentally). If it lies on a rotation axis, like the sulfur atoms of iron pyrites, two of its coordinates are defined by symmetry and it will have one parameter. If it lies on a mirror plane, one of its coordinates is defined and it thus has two parameters. Finally, if the position of an atom does not coincide with any symmetry element, which is referred to as a general position, it has three parameters.

In most cases, the operations of symmetry elements will mean that an atom at position (x, y, z) has counterparts at a number of "equivalent positions." The number of equivalent positions for a particular atom will thus depend on two factors: The number of symmetry elements present in the space group, and whether the atom lies on one or more of these symmetry elements. Atoms in general positions will, of course, have more symmetry-related partners. However, the exact number of equivalent positions depends upon the symmetry of the crystal. One extreme is represented by a cubic crystal of highest symmetry, such as the alkaline halides, in which the large number of symmetry elements means that an atom in a general position is multiplied 192 times. The other extreme is represented by one of the space groups of the triclinic crystal system, in which an atom in a general position has no symmetry-related equivalents. Space group tables list the coordinates of all groups of equivalent positions, including general positions (not on a symmetry element) and special positions (on a symmetry element). For example, if the only symmetry element in the space group were a four-fold rotation axis parallel to the c axis, then an atom in (general) position (x, y, z) would have symmetry-related partners with coordinates $(-y, x, z)$, $(-x, -y, z)$, and $(y, -x, z)$ (Figure 4.2). As Bragg later put it, the symmetry elements of the space group are "like the mirrors of a kaleidoscope, which turn an irregular collection of objects into an attractive repeated pattern."[295]

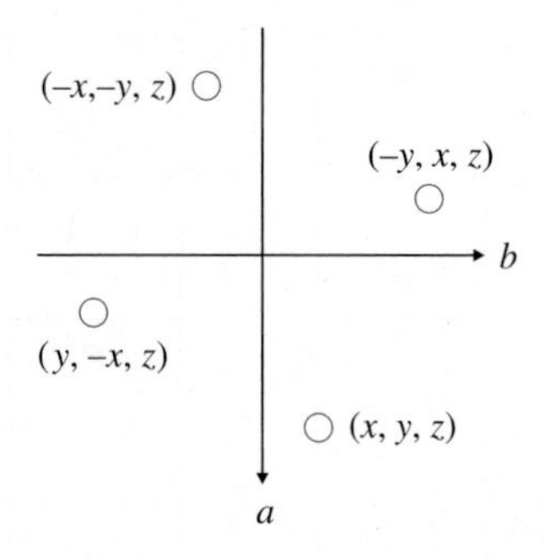

Fig. 4.2 Equivalent positions. A four-fold rotation axis perpendicular to the *ab* plane generates three symmetry-related counterparts of an atom in general position (x, y, z). The four atoms are crystallographically equivalent

It is important to note that space group tables do not specify *whether* an atom occurs at particular values of x, y, and z; only that if an atom *does* occur at particular values of x, y, and z then a certain number of atoms will occur at other specified locations. It should also be emphasized that there is no limitation on the number of atoms of a particular symmetry-related type; in complex crystals such as those of proteins, every atom is in a general position. Finally, although many crystal are composed of molecules rather than atoms, the considerations described above are true for every atom of the molecule.

The use of space-group theory was described by the American crystallographer Ralph Wyckoff in his 1922 book *The Analytical Expression of the Results of the Theory of Space-Groups*. The point group (set of symmetry elements passing through a single point) can be determined from the external form of the crystal. In the case of calcite, the two space groups consistent with that point group have different combinations of equivalent positions. As Wyckoff pointed out, experience had shown that the number of atoms in the unit cell of a crystal was usually much lower than the number of equivalent positions, and therefore many atoms lay on symmetry elements. Consideration of the number of atoms present in the unit cell and the number of equivalent positions available to place them led to a small number of possible structures for calcite. These possibilities could then be distinguished by measurement of specific X-ray reflections.[296]

For aragonite, Bragg used an approach similar to the one Wyckoff had described for calcite. From space-group tables, it was clear that four space groups were consistent with his general structure of aragonite, and each of these would cause the intensities of specific reflections to be halved. Only one of these space groups corresponded to the halvings actually observed with aragonite (Figure 4.3).

This gave Bragg a partial structure. In this structure, the calcium atoms lay on mirror planes; one coordinate was therefore determined, the other two unknown. The carbon atoms also lay on mirror planes, with two

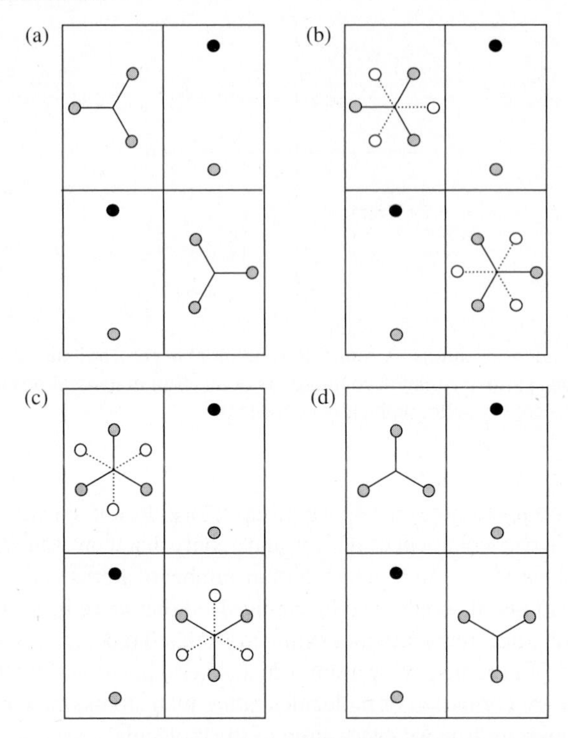

Fig. 4.3 Possible space groups of aragonite. Projections on the (001) plane. Individual circles represent calcium ions, linked circles represent carbonate ions. Heavier shading represents uppermost atoms. A: Q_h^{11}; B: Q_h^{6}; C: Q_h^{16}; D: Q_h^{13}. Adapted, with permission, from Figure 4 of Bragg, W. L. (1923). The structure of aragonite *Proceedings of the Royal Society of London A* **105**, 16–39. Published by the Royal Society

unknown coordinates. The three oxygen atoms associated with each carbon were of two types. The "unique" oxygens were on mirror planes, with two unknown coordinates. The remaining oxygens were in general positions and therefore had three unknown coordinates, but were mirror images of one another. The total number of unknown atomic coordinates—parameters—was therefore nine.

To determine values for these parameters, Bragg used two-dimensional projections of the proposed structure on the three planes formed by the crystallographic axes (*ab*, *ac*, and *bc*, corresponding to the (001), (010), and (100) faces of the unit cell) to show which reflections would be altered by varying the values of the parameters. When a structure that satisfied these criteria had been arrived at, he compared the intensities of all orders of all reflections observed with the theoretical amplitudes of reflection that he calculated based on the postulated structure of the crystal. Because aragonite is an "ideally perfect" crystal, the intensity of the reflected beam is proportional to its amplitude, rather than the square of the amplitude for an "ideally imperfect" crystal. However, Bragg's comparison of observed intensities with calculated amplitudes was far from exact: He did not correct for the decrease in intensity which occurs with increased reflection angle, and he assumed that the scattering powers of calcium, carbon, and oxygen were proportional to their atomic weights rather than their atomic numbers. Nonetheless, he was able to show that large observed intensities were associated with large calculated amplitudes, in agreement with the proposed structure.

The final structure of aragonite was a distorted hexagonal close-packed lattice. In contrast to calcite, in which the oxygens lie between two calcium atoms, the carbonate groups of aragonite are rotated 30° so that the oxygens lie between three calcium atoms. Subsequently, Bragg was able to use this structural difference to explain the different refractive indices of the two calcium carbonates.[297] Meanwhile, Alice celebrated her husband's success by embroidering the aragonite pattern on her nightdress.[298]

The structure of aragonite broke the sound barrier of X-ray analysis. As Bragg wrote in 1938: "For ten years (up to 1923) it was considered practically impossible to analyse crystals in which the atoms were in 'general' positions, or crystals of lower symmetry than cubic, tetragonal, or hexagonal."[299] Bragg had not only solved a structure with more parameters than had been possible to solve before, he had also invented a general method for the solution of crystals with many parameters. Although it would be refined in several aspects, the general thrust of the trial-and-error method was already clear: Comparison of the observed intensities of *all* orders of *all* reflections with the theoretical amplitudes calculated for the proposed structure would be the acid test of X-ray analysis.

However, Bragg still seemed to be feeling some ambivalence about the X-ray analysis of crystals. This was probably related to his wish to be thought of as a physicist rather than a crystallographer. In his early years in Manchester, as we have seen, he used X-rays to study the structure of the atom; in the mid-1920s, he studied crystals by non-X-ray means. The latter included studies on the relationship between the structure of a crystal and its refractive index (the ratio of the speed of light in a vacuum to the speed of light through the crystal)[300] and an attempt to explain the values of the angle between the axes of rhombohedral crystals in terms of lattice energy.[301]

In the summer of 1924, Bragg and Alice combined an invitation to lecture for 6 weeks at the University of Michigan, Ann Arbor, with the British Association meeting in Toronto. Baby Stephen was left in Cambridge with a nurse and his Hopkinson grandparents. In Michigan, each week Bragg gave four lectures on crystal analysis, which he found easy, and four on X-ray spectra, which was "not really in my line." It was too heavy a teaching load but he "struggled through." He and Alice loved Ann Arbor: They bathed and canoed in the river, played golf and tennis, and explored the surrounding countryside. An outing to a baseball game in Detroit brought out Alice's mischievous streak; she supplied a British view of baseball for the student newspaper "in very racy terms," which prompted a visit by an outraged representative of the faculty wives.

After the British Association meeting, the Braggs went to Bigwin Island in Lake of Bays, north of Toronto, where they sailed small boats and took long walks around the island, much to the puzzlement of the American guests. This was by way of a warm-up for their next stop—a ten-day canoe trip through Algonquin Provincial Park. A guide, Jack, paddled a canoe with most of the gear. On portages, Jack and Bragg made three trips: First with the canoes, then with the tents and blankets, finally with the food packs. Alice made one trip, carrying the eggs and fishing rods! Bragg marvelled at the isolation of the Ontario wilderness—during the ten days they saw only two other parties—and the abundant wildlife. At night, they listened to the haunting cry of the loon and the howl of wolves. "It was an unforgettable experience."[302]

Bragg found himself much in demand in North America. After their Canadian backcountry holiday, he and Alice visited Schenectady, Philadelphia, Princeton, Baltimore, Boston, New Haven, and New York before returning to England. In New York, they went for lunch with representatives of the Rockefeller Foundation, who assured Bragg of their support if needed.[303] This pledge would be redeemed 14 years later.

In April 1925, Alice attended her first scientific conference, the Second Solvay Conference on Chemistry in Brussels. She presumably did not find it too onerous, as Madame Solvay arranged excursions for the wives and reserved a box at the opera for their use. That summer, the Braggs stayed with Alice's parents and her Aunt Monica Wills at a rented laird's house at Aviemore in the Scottish Highlands. They went out shooting but found the deer to be "quite unnecessarily wary." They also learned fly-fishing. They then went on a holiday arranged by another aunt of Alice's, Mabel Hopkinson, to Portsalon in Donegal.[304]

In Britain, Bragg and his father had a virtual monopoly on X-ray diffractometry. Elsewhere, the reception of their work was much cooler. In 1954, Bragg wrote: "I still remember the scepticism with which our results were received. Our German colleagues, in particular, were very certain that one could not trust the answer when there were so many variables."[305] In 1925, Jean Wyart was in Paris. "I had read the works of Laue and of the Braggs; I was enraged to see that none of my friends [at the *Ecole Normale Supérieure*] believed in the reality of the arrangement of atoms such as the Braggs proposed it. The Braggs' structure, they thought, was only a clever hypothesis to explain the X-ray diffraction, just as if two sets of atoms existed: The atom of the chemists and the atom of the Braggs. The majority of the chemists don't think much of such research and one of them asked me if I would really enjoy playing at cup-and-ball with atoms."[306] When James was in Peter Debye's lab in Leipzig in the early 1930s, he was asked: "Tell me, how does Bragg discover things? He doesn't know anything."[307]

Paul Ewald, who had inspired Laue to attempt the diffraction of X-rays by crystals, engineered an Anglo-German *rapprochement* in September 1925, when he organized a conference on the relationship between crystal structure and X-ray intensity. This was held at Ewald's mother's house in Holzhausen, Bavaria, and was attended by a small group of X-ray physicists, including Bragg, James, Darwin, Max von Laue, Wyckoff, and Debye. Holzhausen, a village of about 200 people, did not have much in the way of conference facilities. Ewald rented the local inn to accommodate the visitors and set up a blackboard borrowed from a school in his mother's studio. Apart from discussing physics, conference participants bathed in the Ammersee and were entertained by a play staged by Ewald's children. Bragg, who had forgotten to bring a bathing suit, was forced to borrow one from Ewald's chambermaid. A more significant lapse of memory afflicted Darwin, the "champion" of the British group. "To our consternation we found that Darwin had typically omitted to refresh his memory of his own theories and was unable to explain them when he got up to make his contribution!"[308] Bragg's own priority is shown by the comment he made at the end of one session: "I will not be satisfied until I can determine a structure with 19 parameters."[309] It is not clear to which crystal he was referring, but in fact he was not satisfied by solving a structure with 19 parameters—although he would be rather pleased about his role in solving one with 2500.

The discussions of the Holzhausen conference were summarized in a 1926 paper by Bragg, Darwin, and James.[310] Bragg still clung to the hope that X-rays could be used to study the structure of the atom, writing "direct information as to the positions of the electrons in the atom . . . must be regarded as the main objective, but a knowledge of the scattering powers of the atoms is also essential to the application of X-rays to discover the structure of complex crystals." The factors affecting the intensity of X-rays reflected by crystals were discussed, including those related to lattice structure, such as the amplitudes and interactions of waves scattered by each atom in different directions; factors related to crystal perfection, such as extinction; and purely physical factors, such as temperature. The significance of crystal perfection was that the intensity of reflection was proportional to the amplitude of the diffracted wavelet for a perfect crystal and to the square of amplitude for a perfectly imperfect crystal; the state of perfection of the crystal was therefore essential information in comparisons of absolute intensity measurements with (theoretical) structural amplitudes. Rock salt, fluorspar, and barytes ($BaSO_4$) appeared to be highly imperfect, whereas calcite and diamond were better described by the perfect-crystal model.

It had by now occurred to Bragg that the quantitative measurement of X-ray reflection represented a powerful tool for crystal structure analysis. Previously, a proposed structure was tested by calculating the extent to which a particular reflection would be increased or decreased and comparing this prediction to the relative intensities of measured reflections. Now, however, it was possible to test a structure by calculating the *absolute* intensity of reflections and comparing these to the measured values. This was a far more rigorous test of a structure, particularly in the case of complex crystals. The first structure solved using quantitative techniques was barytes, with 11 parameters, solved by James and W. A. Wood in 1925.

These ideas were put to the test in Bragg's efforts, together with a student, Guy Brown, to solve the structure of the gemstone chrysoberyl ($BeAl_2O_4$), also known as alexandrite. Measurements of unit cell dimensions, together with the known density, showed that chrysoberyl has four molecules of $BeAl_2O_4$ per unit cell, or 28 atoms, making it the most complex crystal Bragg had yet studied. An important clue came from the structure of a crystallographically related crystal, corundum (Al_2O_3). Bragg had published a partial analysis of corundum in one of his papers on refractive index, but the complete structure had been worked out by Linus Pauling. In this crystal, the oxygen atoms are very nearly in hexagonal close packing, with the Al atoms placed symmetrically between six oxygen atoms. The structure of corundum, like those of beryllium oxide (BeO) and spinel ($MgAl_2O_4$), is principally determined by the closest distance which neighboring oxygen atoms can achieve, approximately 2.7 Å, with the smaller metal ions occupying some of the interstices between oxygen atoms. It seemed likely that, although chrysoberyl is orthorhombic, its structure, like that of aragonite, would be based on almost hexagonal close packing of O, as this permits the most "economical use of the available space."

Once Bragg and Brown had determined the space group, and therefore the symmetry elements, it became clear there were only two possible arrangements of the eight aluminum atoms in the unit cell. An aluminum atom in a general position, and therefore with three parameters, would be multiplied eight-fold by the operation of the symmetry elements; alternatively, four aluminum atoms could lie on reflection planes, with two parameters, and four on centers of symmetry, with none. These two possibilities having quite different implications for the intensities of specific reflections, it was easy to show that the latter arrangement was the case. It was also possible to approximately determine the x and y parameters of the Al atoms on reflection planes (z being dictated by symmetry). The oxygen atoms were of three types: One quarter on a reflection plane, defined by two parameters; one quarter on a reflection plane but not equivalent to the first set, and therefore also with two parameters; and half in general positions and equivalent, and therefore with three parameters. The beryllium atoms could lie on a reflection plane (two parameters) or the center of symmetry (none); however, their effect on X-ray reflections was too small for these alternatives to be distinguished.

The method of determining the nine or eleven parameters was to assume an "idealized" structure in which the Os are in exact hexagonal close packing and the Als are exactly at the same distance from 6 Os; the Als on reflection planes were then supposed to be in the approximate positions found by the preliminary analysis. The structure was tested by calculating theoretical amplitudes and comparing these with the intensities measured both by photographic and spectrometer methods. The photographs used were of the rotation type, in which a large number of spots is obtained by turning the crystal, in this case by 30°, through the X-ray beam. The intensity of the resulting spots could not be quantified, but were ranked in five categories from "very weak" to "very strong"; the most significant findings from the rotation photographs were the absence of spots that should appear if a particular structure were correct.

The intensities measured by the X-ray spectrometer were compared with the structure factors (theoretical amplitudes) calculated for the corresponding crystal planes. In the proposed structure, for example, the second-order reflection from the (011) crystal planes would result from destructive interference between aluminum-containing and oxygen-containing planes such that the amplitude of the resultant wavelet was equivalent to a structure factor of $Al_2 - O_2$. The sixth-order reflection from the (011) planes resulted from *constructive* interference between the aluminum-containing and oxygen-containing planes such that the resultant wavelet was equivalent to a structure factor of $Al_2 + O_4$. The structure factors could not be calculated simply by adding or subtracting the atomic structure factors of these elements because the sixth-order reflection occurs at a higher value of θ than the second-order and the amplitude of the diffracted wave decreases with angle of incidence; however, the appropriate correction could be made using curves of atomic structure factor against θ that had been published by Douglas Hartree, Professor of Applied Mathematics at Manchester. Once the structure factor was calculated, it was squared before comparison with the intensity of that reflection measured by the spectrometer, as the intensity is proportional to the square of the amplitude for an imperfect crystal such as chrysoberyl.

This analysis confirmed the proposed structure and the values assigned to the nine parameters. The hexagonal lattice of oxygen atoms that was the framework of the crystalline structure was slightly distorted by some of the aluminum atoms. The beryllium atoms were assumed to be surrounded by oxygen atoms arranged in a tetrahedron, by analogy with the magnesium atoms in spinel.[311]

This paper on chrysoberyl was the first reference in Bragg's writings to Linus Pauling, who was to become his great rival. Pauling obtained a degree in chemistry at Oregon Agricultural College (now Oregon State University) before moving to the California Institute of Technology (Caltech) in 1922 to work on a Ph.D. project supervised by the crystallographer Roscoe Dickinson. In the summer of 1922, Pauling prepared for his postgraduate research by reading the Braggs' book *X-Rays and Crystal Structure*.[312] He later claimed that he "didn't learn a great deal from it."[313] However, Bragg's paper on ionic radii "stimulated me to begin collecting experimental values of interatomic distances and to attempt to analyse them, in searching for basic principles about chemical bonding."[314] After obtaining his Ph.D. in 1926, Pauling spent 18 months doing post-doctoral research in Europe before returning to Caltech as Assistant Professor of Theoretical Chemistry and Mathematical Physics.

In the mid-1920s, Bragg made a strategic decision to study the silicate group of minerals. The silicates were chosen for analysis because they "were of the right order of complexity." However, there is no doubt that their complex yet symmetrical patterns appealed to the artistic temperament of

Bragg. As he wrote of the silicates in 1961, "the order and simplicity of their structural schemes revealed by X-ray analysis is a generalization that gives the deepest aesthetic satisfaction."[315] Aesthetics played an important role in Bragg's scientific approach.

The first silicate studied in Manchester was beryl, $Be_3Al_2Si_6O_{18}$. A structure for beryl was published by Bragg and his student Joseph West—yet another former sound-ranger—in 1926.[316] Using now-standard methods, Bragg and West identified the space group of beryl and showed that its unit cell contains two molecules of $Be_3Al_2Si_6O_{18}$. Because of beryl's high symmetry, any atom in a general position is multiplied 24-fold by the operation of symmetry elements. This meant that none of the six beryllium atoms, four aluminum atoms, or 12 silicon atoms could be in a general position; rather, they must lie on symmetry elements. Of the 36 oxygen atoms, at most 24 could be in a general position. Such considerations greatly limited the possible structures. If an atom is not on a mirror plane, for example, its center must be at least one radius from the plane, otherwise the atom would overlap with its symmetry-related counterpart. Because there were only four Al atoms, these could only be either all on the six-fold rotation axis or two on each of the three-fold rotation axes. Likewise, the 12 Si atoms must be all on the six-fold axis or in two rings of six around this axis, the members of each ring being equivalent. Twelve of the O atoms were governed by the same considerations as the Si atoms, the remainder being in a general position. Taking all this into account, Bragg and West were able to determine an approximate structure with seven parameters: Two for the silicons, three for 24 of the oxygens and two for the remaining 12 oxygens.

In previous studies, Bragg had used structure-factor analysis to test proposed structures. Now he felt sufficiently bullish about the power of his approach to reverse the process and use the measured intensities to determine the structure: "*In the present analysis we will use the observed intensities of X-ray reflection to determine F for a number of crystal planes. From these values of F the atomic positions will be directly deduced.*" (Emphasis in original.) In actual fact, this direct deduction of the structure from observed intensities was not achieved for beryl. The above-mentioned considerations of symmetry and packing already limited the possible values of the seven parameters, and in the end Bragg and West established the most difficult parameter by old-fashioned trial-and-error-method tinkering.

Despite the complexity of its unit cell and respectable number of parameters, the high symmetry of beryl made it an easy structure to solve: "When West and I had determined the space group, I remember well that we found all the atomic positions in about a quarter of an hour, and all subsequent work only altered our first estimates slightly."[317] As shown in Figure 4.4, the atomic structure of beryl conforms very closely to the space-group symmetry.

Beryl had some interesting features. Although it was classified as a metasilicate, with an O/Si ratio of 3, there were no SiO_3 groups. Instead, the structure was "composed of SiO_4 groups, each group joined to its neighbour on either

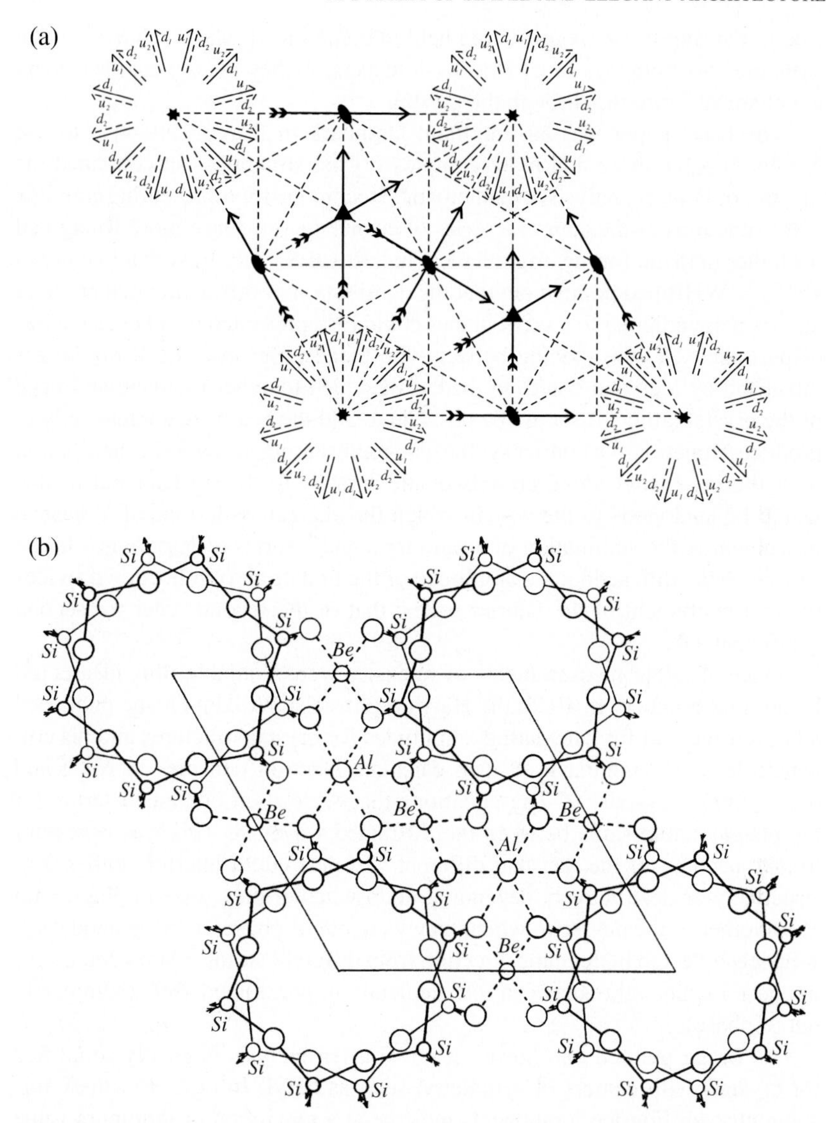

Fig. 4.4 Beryl symmetry and structure. (a) Symmetry elements of the D_{6h}^{2} space group. Projection on the (0001) plane. Perpendicular to this plane are two-fold (oval), three-fold (triangle) and six-fold (hexagon) rotation axes. Adapted, with permission, from Figure 1 of Bragg, W. L. and West, J. (1926). The structure of beryl, $Be_3Al_2Si_6O_{18}$. *Proceedings of the Royal Society of London A* **111**, 691–714. Published by the Royal Society. (b) Bragg and West's 1926 structure of beryl. Section parallel to the (0001) plane. Note the 12 symmetry-related oxygen atoms (large circles) in rings around the six-fold rotation axes, aluminum atoms on the three-fold rotation axes and beryllium atoms on the two-fold rotation axes. Reproduced, with permission, from Figure 14 of Bragg, W. L. (1930). The structure of silicates. *Zeitschrift für Kristallographie und Kristallgeometrie* **74**, 237–305

side in the ring by an oxygen atom held in common." It was a relatively open structure. No atom lay close to the six-fold axes, so these axes were the centers of "channels" running through the crystal.

The beryl paper was an important landmark in Bragg's attempts to use X-rays to solve more and more complex crystal structures—it contained his first use of Fourier analysis, a mathematical technique for representing complex wave functions as the sum of a series of cosine curves. Once more Bragg had his father to thank for the original idea. In a Royal Society Bakerian Lecture in 1915,[318] WHB had pointed out that the distribution of diffracting matter along an axis perpendicular to a set of planes could be represented by a Fourier series of periodicity d, the spacing between the planes. Because the X-ray beams diffracted by a crystal could, in theory, be added together to create an image of the crystal lattice, it should be possible to add the beams mathematically to produce a function—a Fourier synthesis—representing the periodic distribution of scattering matter along an axis of the crystal. As Bragg later put it, this would be analogous to the way in which the characteristic tone of a musical instrument is the summation of a pure tone and a series of harmonics. In the case of X-ray diffraction, the amplitude of the first order of reflection provides the first coefficient of the Fourier series, that of the second order the second coefficient, etc.

Once absolute measurements of intensity were available, this idea could be put into practice. In 1925, the Harvard physicist William Duane published a Fourier method for calculating amplitudes for crystal structures and his colleague R. J. Havighurst used it to calculate electron densities in NaCl and other simple crystals.[319] The equations that were used contained terms for the phase relationships between the diffracted wavelets. This was necessary to take account of the fact that different wavelets could interfere either constructively or destructively depending upon whether they were in phase with one another or the degree to which they were out of phase. In the general case, wavelets reflected in the same direction from different atomic planes could vary in phase by any value between 0 (completely in phase) and 180° (completely out of phase).

As Bragg pointed out, however, the Fourier analysis is greatly simplified for crystals with centers of symmetry, such as beryl. In order to satisfy this symmetry, all Fourier components must be at a maximum or minimum value at the center of symmetry. If this symmetry element is taken as the origin of coordinates, the phases of all diffracted wavelets must be 0 or 180°. In beryl, the (000l) planes,[u] which are perpendicular to the c-axis, are of three different types: Si_6O_6, O_6, and Al_2Be_3. Bragg was able to calculate an electron density distribution along the c-axis and resolve this into contributions from the three types of atoms present. Thus, he was able to determine that each O atom had about 8.95 electrons and each Si about 12.5. In a like manner, he calculated electron density distributions along two other axes of the crystal.

[u] In an hexagonal crystal like beryl, lattice planes are usually described by four Miller indices.

Bragg was frank both about the theoretical power of Fourier analysis and its practical difficulties. He wrote: "if the amplitude of reflexion in a number of orders is measured experimentally, a curve representing the electron distribution in sheets parallel to the planes can be built up by adding together the terms of the Fourier series." This electron density distribution is equivalent to the structure of the crystal. However, "The Duane method cannot be applied until the signs of the coefficient 'F' are fixed by preliminary analysis, for the observed intensities only give the squares of these quantities." For imperfect crystals, which constitute most cases, the intensity of reflected X-radiation, as measured quantitatively by the X-ray spectrometer or semi-quantitatively by photography, is proportional to the square of the wave amplitude. Even if all other factors contributing to the intensity value are known, one cannot calculate the amplitude, F, but only its modulus, $|F|$. For a non-centrosymmetrical crystal, the phase angle corresponding to each $|F|$ would be needed in order to perform a Fourier synthesis. In the more favorable case of a crystal with a center of symmetry, the phase of a reflection could only have the values of 0 or 180°: In the former case, the wavelet corresponding to that reflection would have a positive amplitude relative to the origin; in the latter case, the wavelet would have a negative amplitude relative to the origin. For centrosymmetrical crystals, it was therefore only necessary to determine the signs (positive or negative) of the amplitudes, rather than their phases.

Although reducing the possible phase values from infinity to two represents a considerable simplification of the problem, it is still the case that construction of a Fourier series and thus an electron-density distribution requires information not available from what is actually measured by X-ray diffraction—the intensities of reflections. The absence of information about signs—or, in the general case of a non-centrosymmetrical crystal, about phases—would prove the great stumbling block to the use of Fourier methods in X-ray analysis.

Bragg was only able to construct electron-density plots along specific axes of beryl because he had already determined the structure of the crystal by trial and error and therefore could calculate what the signs of the amplitudes must be. However, this limitation did not render Fourier analysis useless. Its "supreme merit," for Bragg, was in summarizing the data from all orders of reflection in a single curve. One could slide the electron-density curves for various atoms along the axis in ways consistent with symmetry until the best fit with the "observed" Fourier was obtained. At this point, therefore, Bragg envisaged Fourier analysis as complementing, not replacing, the trial-and-error approach.

In 1923, WHB had succeeded James Dewar as Fullerian Professor of Chemistry at the Royal Institution of Great Britain in London. He brought with him from University College London a group of young researchers that included William Astbury, Alexander Müller, and Kathleen Yardley (later Lonsdale). In the hitherto moribund Davy-Faraday Laboratory of the Royal Institution, they were joined by John Desmond Bernal, Reginald Gibbs, Arthur Lindo Patterson, and John Monteath Robertson.

As mentioned above, WHBs group concentrated on the X-ray analysis of organic compounds. These typically form crystals of low symmetry with most or all atoms in general positions. One factor that facilitated the X-ray analysis of organic compounds was the presence of covalent bonds, which means that the distances between atoms linked by such bonds are fixed. However, the tremendous complexity of organic structures represented a far greater challenge than anything Bragg faced in his inorganic domain.

One of the first organic substances studied by WHBs group was graphite, a form of carbon, which was shown to consist of two-dimensional sheets of atoms. This finding provided important insights into the subsequent analyses of naphthalene ($C_{10}H_8$) and anthracene ($C_{14}H_{10}$). However, the latter studies merely confirmed structures that had already been arrived at by the methods of organic chemistry. It would be a long time before X-ray crystallography was capable of solving organic molecules of unknown structure.[320]

In November 1926, Gibbs stepped over the demarcation line between Manchester and the Royal Institution by starting to work on the structure of mica. Bragg wrote to his father: "I am awfully sorry he has got on to this line, and I was honestly under the impression that we had the field to ourselves as far as your people were concerned . . . I get so hot and bothered when I think we are overlapping."[321] A few days later, he added: "If I get worried when I think we are overlapping, Dad, it's just because I want so much to make my laboratory known for some lines quite its own in order that it may have a chance to be seen in the firmament alongside the very much brighter luminary at the Royal Institution!"[322] The somewhat-ingratiating tone that Bragg often adopted in letters to his father was not characteristic of him, and should perhaps be seen as a symptom of the emotional problems he had in handling a scientific collaboration within the context of a difficult father–son relationship.

Demarcation disputes of this type cast a shadow over several visits Bragg and Alice made to the Royal Institution. Although she never claimed to understand the scientific issues involved, Alice felt that "the difficulties which were inevitable and natural have been exaggerated." In her view, the fact that Bragg and WHB worked in the same area and thought in similar ways made it difficult for each of them to disentangle his own contribution from that of the other. Alice also felt that it was because of these conflicts with his father that Bragg was so careful in ensuring that his own students received more than their fair share of credit for work done with him.[323]

The next silicate Bragg tackled was phenacite, Be_2SiO_4. A paper describing its structure was submitted for publication in November 1926.[324] Once more it was a solo effort. Because of Bragg's scrupulousness about assigning scientific credit, we may assume that he was still performing his own research projects in addition to supervising his group of workers and running the department.

Phenacite can be seen as a dry run for the most difficult silicate Bragg would tackle, diopside. The unit cell was found to contain six Be_2SiO_4 "molecules" and symmetry elements that multiplied each atom in a general position six-fold. Therefore, each of the seven atoms of Be_2SiO_4 lay in a general position, giving a total of 21 parameters to be determined. As in the case of beryl and chrysoberyl, the scattering by the beryllium atoms was too small for their positions to be determined, so Bragg concentrated on the 15 parameters describing the Si and O atoms, only tentatively assigning the beryllium positions.

The approach used to solve the structure of phenacite did not involve any experimental innovations. However, the perfection of the crystal meant that extinction was a major factor, and as a result Bragg had to rely on high-order reflections for which extinction was negligible. The assumptions that the SiO_4 group is tetrahedral and that oxygen atoms cannot approach closer than 2.7 Å greatly narrowed the options. The final structure was slightly expanded from hexagonal closest packing of oxygen atoms, with pores along the three-fold rotation axes and Si and Be atoms at the centers of O tetrahedra (Figure 4.5).

In 1927, Bragg and West wrote a review article that discussed the structures of the silicate minerals so far studied.[325] On the one hand, these tend to form large crystals with well-known crystallographic and optical properties. Also, there is a large amount of isomorphous replacement —substitution of one atom for another without effect on the lattice structure—which greatly facilitates X-ray analysis. On the other hand, the constituent molecules are very complex and the crystals of low symmetry. The guiding principle was that "The [silicate]

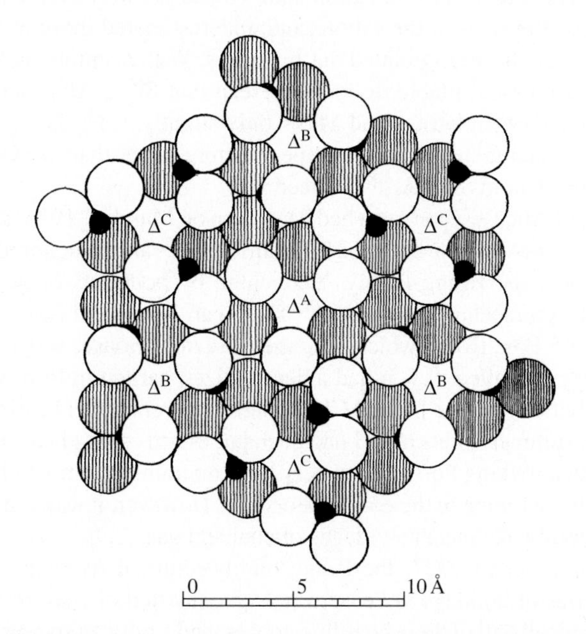

Fig. 4.5 Bragg's 1926 structure of phenacite. Section parallel to (111) plane. Large open circles are oxygen atoms above the plane of the diagram, large shaded circles are oxygen atoms below the plane of the diagram, small filled circles are silicon atoms. Triangles represent axes of three-fold rotational symmetry. Reproduced, with permission, from Figure 4a of Bragg, W. L. (1926). The structure of phenacite, Be_2SiO_4. *Proceedings of the Royal Society of London A* **113**, 642–57. Published by the Royal Society

structure may be regarded as an assemblage of oxygen atoms, cemented together by silicon and by metallic atoms."

If the O atoms are no more than 2.7 Å apart, which usually was the case, they must either be in cubic or hexagonal close packing. In cubic packing, the O atoms lie on a face-centered lattice with a unit-cell volume of 55.7 $Å^3$ containing four atoms. In hexagonal packing, the O atoms lie on two inter-penetrating hexagonal lattices with a unit cell volume of 27.8 $Å^3$ containing two atoms. Both are therefore equally closely packed. Crystals which had been shown to be in hexagonal or near-hexagonal packing included beryllium oxide, corundum, chrysoberyl, and possibly cyanite (Al_2SiO_5). Olivine [$(MgFe)_2SiO_4$], was known to be an expanded form of chrysoberyl and monticellite ($MgCaSiO_4$), an even more expanded form. In beryl and phenacite, the O atoms occupy their minimum volumes but there are channels around the six-fold and three-fold rotation axes, respectively.

The presence of the silicon and metal atoms in the oxygen lattice leads to a reduction in symmetry and an increase in unit cell dimension. The ability of one metal to replace another (or silicon) in the lattice depends mainly upon atomic volume. Bragg had erred in his 1920 attempt to determine atomic diameter by including anions containing "homopolar" (covalent) bonds, such as carbonate and sulfate. His dimensions for cation–anion distances were correct, but he had overestimated the sizes of the cations and underestimated those of the anions, as had subsequently been pointed out by others. With accurate measurements of atomic size now available, it could be seen that Si^{4+}, Al^{3+}, and Be^{2+} do not distort the oxygen lattice, and Mg^{2+} only slightly. Ca^{2+} is so large that it is the Ca—O spacing that determines the structure rather than O—O. Be^{2+} lies between four O atoms and Al^{3+} between six.

Bragg and Alice's second son had been born on March 1, 1926. Presumably for that reason, he was named David William. Perhaps also in honor of the patron saint of Wales, the Bragg family, augmented by both sets of grandparents, holidayed at Abersoch in Gwynedd.[326] That year, Bragg and Alice moved to a new house, 45 Pine Road, which, like their previous house, was in Didsbury. Stephen Bragg recalled: "... it had a decent-sized garden in front with a path around. It abutted onto the Liberal Club where I can remember hearing the clink of bowls on summer nights after I had been put to bed ... We had a big nursery upstairs with a rocking horse. I was very keen on trains, and my father laid out a model railway for me in the conservatory."[327] However, it was a difficult time as the General Strike meant shortages of coal and gas.

In the summer of 1927, the Bragg family returned to Wales, taking the first of a series of holidays in Pensarn, which consisted of "two rows of gaunt houses, practically all of them boarding houses, and a railway station." The boys played on the beach and in the pools of the river Elwy; Stephen also watched the trains. The unreliable Wolseley had now been replaced by a "magnificent" Humber car, "our pride and joy." This had been bought with a £350 gift from Alice's Aunt Monica, who was married to Harry Wills, founder of the Imperial Tobacco Company.[328]

That autumn, Bragg and Alice attended a conference in Como, Italy, commemorating the centenary of the death of the physicist Alessandro Volta. The conference seems to have been planned as a showcase of Fascist Italy, as the hospitality was unusually lavish. Alice tells us: "I think the conference part was rather languidly conducted. It was wonderfully warm, there was swimming and boating on the lake of Como, steamer trips for us all, and wonderful fireworks displays at night. Great houses around were open, and Italian hostesses entertained us."[329] The government paid for travel to Italy for both Bragg and Alice as well as unlimited accommodation and travel within the country. The Braggs took full advantage of the opportunity, staying at Lenno on Lake Como before the conference and Rome afterwards. They intended to go on to Naples, Florence, and Venice, but they had to return to England when Alice contracted paratyphoid—as did Rutherford, both having made the mistake of drinking orangeade rather than champagne.

In Rome, the Braggs stayed at the Hotel des Ambassadeurs: "It was a very grand hotel and we tried to preserve the labels on our bags for as many years afterwards as possible." It is not usual that the social program of a scientific conference includes a meeting with the head of government or state, but invitees to the Volta centenary were offered the choice of meeting Mussolini or the Pope. The wives selected Mussolini, as they did not have the attire appropriate to a Papal audience. The *Duce* addressed every guest in his own language. Bragg was "very impressed."[330]

It does not seem to have occurred to Bragg, either then or later, that an unscrupulous dictator might be using him as a propaganda stooge.[v] He was not a political man and his absorption in research questions often blinded him to outside events.[w] According to his daughter Patience, "My father was really a liberal, with a small 'l' not a big 'L'. He thought politics were rather 'sordid'. I think he really didn't like politics of any kind whether national or university."[331] As a Nobel Laureate he was often asked to sign petitions on political or humanitarian issues; usually, but not always, he refused. In 1970, declining Kendrew's request that he sign a letter written by French biologist Jacques Monod protesting the treatment of geneticist Zhores Medvedev by the Soviet government, Bragg wrote: "However strong their individual feelings, Nobel Laureates have no right as a body to deal with political questions."[332]

Bragg had been a member of the Manchester Literary and Philosophical Society since he had moved to the city in 1919. In 1927–8, he served as the

[v] To be fair to Bragg, he was not alone in this respect. Nevill Mott, who met Mussolini at a conference in 1932, wrote of him: "I think we were all impressed. I remember that Bohr was. A veritable renaissance patron of the sciences, we thought him, and the man who made the trains run to time. Or so we believed." [Mott, N. (1986). *A Life In Science*, p. 42. Taylor and Francis, London.] As the last sentence quoted indicates, Mott, unlike Bragg, appears to have later changed his mind about *il Duce*!

[w] For example, during the winter of 1937–8, Bragg had a ski trip in Austria with Herman Mark, a physical chemist whose mother was Jewish. Mark enquired about the possibility of getting a position in England, but Bragg was "too dumb" to realize the urgency of the situation (RI MS WLB87, p. 89).

Society's President, a position formerly occupied by the chemist John Dalton. As part of his duties, Bragg delivered a presidential address entitled "Some Views on the Teaching of Science."[333] In this he recalled a previous talk on science teaching, one that he had given to the student Science Federation of Manchester University, in which he had "asked the students to forget for a while that we were those hereditary enemies, examiner and examinee, and to give up the idea that it was my duty to cram as much knowledge into them as could be managed in three years, and theirs merely to retain that knowledge for a sufficient length of time to enable them to be successful in a test at the end." At the end of that earlier talk, a student had punctured Bragg's balloon by pointing out that it was well known among his fellow students that anyone memorizing a certain physics textbook was sure to get a first-class degree.

In his talk to the "Lit. and Phil.," Bragg still had no means of dissuading students from rote memorization: "The student hates such tests [on general principles], for he feels uncertain how well he is going to be able to satisfy them, whereas in answering standard questions he feels on safe ground." However, by training students to think rather than memorize, "we may not produce so learned a man as the bookworm, but we will produce a more able man . . . I cannot help feeling that we have not freed ourselves from the mediaeval tradition that we exist to produce the learned man." Bragg's membership in the Lit. and Phil. exemplified his belief that the arts and science were of equal importance, and he promoted this view whenever possible.

In early 1928, Bragg went to the Massachusetts Institute of Technology (MIT) as Visiting Professor. His research group was left in the capable hands of James and Will Taylor. Taylor had obtained a B.Sc. in Honors Physics at Manchester in 1926. Bragg considered him "the best student I ever had at Manchester; I remember it was difficult to think of any reason for not giving him full marks on practically all his papers in the Finals." He became an assistant lecturer in the Physics Department in 1928.[334]

Bragg was invited to MIT by Samuel Stratton, the President, because Charles Norton, Professor of Physics, and the other physics faculty were heavily engaged in industrial contract research to the detriment of their academic duties: Norton "spent most of his time investigating bricks; his office was full of them." When Bragg arrived in the Physics Department, Norton took him to a well-equipped office with Bragg's name on the door. "We pressed a bell, a beautiful blonde secretary appeared, and [Norton] waving his arm towards her said 'She's yours.' " Bragg taught a well-attended crystallography course. For this he required models, and a research student, Bert Warren, was given the job of making them.[335]

Warren soon became a friend: "One day Professor Bragg dropped the remark that being entertained and invited out to dinner nearly every night in the week was rather strenuous and tiring. So with the naiveness of youth, I promptly asked if he would like a change such as going to the hockey game that night between the Canadian Mapleleafs [sic] and the Boston Bruins. He eagerly accepted and we went, and it was an evening which I shall never forget. First he had to know which was the home team, and from then on he cheered as loudly and enthusiastically for the Boston Bruins as any Bostonian in the audience. Another time we were taking a Sunday hike in the mountains of New Hampshire. After a long hike in the snow we built a camp fire to cook the hot dogs which we had planned for lunch. On opening the packsack, I found that I had brought the rolls but had forgotten the dogs (frankfurters). Professor Bragg . . . had never supposed that any person could have a face as long and sad as mine when the awful discovery was made."[336]

Alice joined him in April, suffering from the after-effects of a rough crossing and the paratyphoid she had contracted in Italy. Bragg surprised her by driving her from the docks to Cambridge, having "learnt to drive in the dextro—as against the laevo—manner."[337] On Alice's first night in Boston, she and her husband had dinner at the President's house with Stratton and the Governor of Massachusetts, Alvan Fuller. According to Alice, dinner was served on a service that had belonged to Napoleon. As Prohibition was still in effect, the Braggs were rather surprised that Stratton served wine and other alcoholic drinks in the presence of the Governor! "I think there was a polite convention that such drinks were remnants left over from the pre-Prohibition days, but of course this was impossible after so long a lapse of time."

During their stay in Boston, the Braggs stayed in a small apartment in a hotel just off the Common. They explored the countryside by car, having picnics and visiting "picturesque New England villages" that they were disappointed to find now occupied by Poles. They visited friends in Montreal and then went skiing in the Laurentians mountains of Quebec, Alice's first time on skis. Apparently not missing Stephen and David too badly, Bragg wrote to Pauling: "We are thoroughly enjoying the casting off of all responsibilities associated with Manchester and children and in fact feel we are having a second honeymoon." Alice loved MIT: "the look of Boston under snow, with the old purple glass in the windows, the Sargent water colours in the picture gallery, the gorgeous lobsters and clams we ate." The Braggs were back in England by June.

While at MIT, Bragg was unable to satisfy the request that he find research projects for all 40 of Norton's assistants. However, he did come up with an excellent project for Warren. Bragg had brought with him from Manchester a set of quantitative X-ray measurements on diopside, $CaMg(SiO_3)_2$. Sections of the crystal cut perpendicular to the principal axes had been supplied by the ever-obliging Arthur Hutchinson. West had mounted these in the spectrometer such that the perpendicular axis was vertical and rotated the section while taking intensity measurements. This gave a set of readings for all planes in a crystallographic zone—essentially the spectrometric equivalent of a rotation photograph. For example, the crystal section cut perpendicular to the *b* axis was used to measure all planes with Millerian indices $(h0l)$. Bragg now set Warren to work on the analysis of the diopside measurements and it was he who "had the brain-wave which provided the solution."[338]

Initial studies showed that diopside was going to be a very difficult structure to solve. There were four molecules of $CaMg(SiO_3)_2$ in the unit cell, the symmetry elements of which multiplied each atom in a general position by eight. As there are only four Ca and four Mg, these had to lie on centers of symmetry or on two-fold rotation axes. However, there were no restrictions on the Si or O atoms.

The parameters parallel to the (010) plane (formed by the *a*- and *c*-axes) seemed easier to start with. Because of symmetry, only one quadrant of the (010) projection, containing one molecule of $CaMg(SiO_3)_2$, needed to be considered. In this quadrant, Ca and Mg atoms lie either at points which are projections of the two centers of symmetry or at the point which is the projection of the two-fold axis. Warren and Bragg started by finding the two parameters that define the position of the two Si atoms in the (010) plane. They showed that certain locations of the Si atoms in the quadrant could not explain the observed F values no matter where the O atoms were placed. This allowed them to narrow down the locations of the Si atoms to four small areas of the quadrant. Three of these could then be excluded because they brought neighboring atoms too close together.

The relatively low intensity reflection from the (400) plane, which contains the Ca and Mg atoms (whether these lie on two-fold axes or the center of symmetry), indicated that all the O atoms must be approximately halfway between these planes. The O atoms were therefore in general positions—three types each defined by three parameters. The *a*-axis coordinate of each type of O being known approximately, trial-and-error methods could be used to determine their *c*-axis coordinates.

Next, Bragg and Warren found that, if the Ca and Mg atoms were on centers of symmetry, no positions of the Si atoms would give certain observed values of $F(hk0)$; therefore, the metal atoms must be on the rotation axes. Each of the four rotation axes of the unit cell thus had a Ca and Mg atom on it and three paired O atoms around it; the arrangement of atoms along two of the axes was reversed compared to the other two. There was just enough room along the *b*-axis to fit these atoms, so they had to be ordered either Ca—Mg—2O—2O—2O—(and its reverse) or Mg—2O—2O—Ca—2O—(and its reverse), the former being exceedingly improbable. Having fixed the relative positions of atoms along the rotation axes, only two parameters remain: One for Ca (say) along the rotation axis and the other the undetermined coordinate for Si (parallel to *b*). It was easy to determine these by trial-and-error methods.

Warren's and Bragg's structure of diopside was published in September 1928.[339] Each Si atom was surrounded by four Os almost at the corners of a regular tetrahedron. Warren's "brain-wave" was that two of the O atoms around each Si atom were shared by neighboring Si atoms, creating chains along the *c*-axis (Figure 4.6(a)). As Bragg put it, "The new idea, the 'repugnant-to-common-sense idea', was an infinite negative ion."[340] The Ca and Mg atoms lay between these chains, with each Ca surrounded by eight O atoms and each

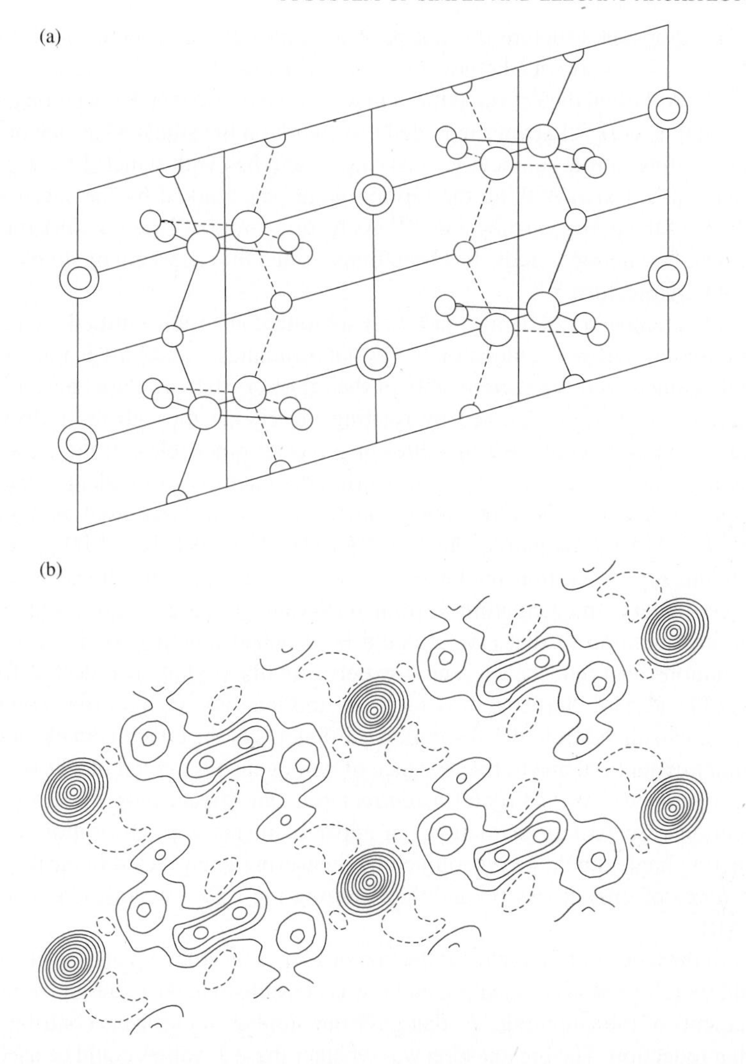

Fig. 4.6 Structure of diopside. (a) Warren and Bragg's 1928 structure of diopside. Projection on the (010) plane. Large open circles: Ca; small filled circles: Mg; large filled circles: Si; small open circles: O. (b) Bragg's 1929 two-dimensional Fourier map of the (010) projection of diopside. Adapted, with permission, from Figure 2 of Bragg, W. L. (1929). The determination of parameters in crystal structures by means of Fourier series. *Proceedings of the Royal Society of London A* **123**, 537–59. Published by the Royal Society

Mg by four. This structure immediately explained a physical feature of the diopside crystal—its tendency to fracture parallel to the *c*-axis. The cleavage planes were parallel to the Si—O chains and therefore cleavage did not involve the breakage of any of the strong Si—O bonds.

The diopside structure did not have an unusually large number of parameters—only 14. It was the novelty of the structure that made its analysis so difficult. Although the key breakthrough was made by Warren, Bragg took great pride in this work: "I always regarded this [work on the silicates] as one of the most exciting and aesthetically satisfying researches with which I have been associated... I always think the turning point was marked by the successful analysis of the pyroxene diopside."[341] As *Nature* wrote of Bragg's work on the silicates: "A chemical riddle has been transformed into a system of simple and elegant architecture."[342]

In September 1928, Bragg and West submitted an article entitled "A Technique for the X-Ray Examination of Crystal Structures with Many Parameters" which summarized the current state of the art of crystal-structure analysis.[343] Integrated intensities obtained by rotating the crystal through the reflecting position could be converted to values of F, as all other relevant factors were known. To avoid the difficulty of measuring the intensity of incident radiation for each reflection, the (400) reflection from rock salt was used as a standard. For Mo K_α radiation, this is 0.98×10^{-4}; for Rh K_α, 1.09×10^{-4}. Optimum crystal section thickness was $1/\mu$, where μ is the linear absorption coefficient. In calculating F from ρ (the integrated intensity), a number of terms common to that crystal could be grouped together to simplify the calculations. μ could be measured experimentally (radiation reflected from the (111) face of fluorspar was used, as this has no second-order component) or calculated from the absorption coefficients of the constituent elements. Primary extinction, due to the presence of homogeneous crystal blocks of sufficient size, was very difficult to correct for, but affects only the strongest reflections. Secondary extinction, an apparent increase in absorption at the reflecting angle, could be determined by measuring uncorrected intensity (ρ') at a series of thicknesses (t) and then plotting log (ρ'/t) against t, as shown by BJB.

All the factors affecting the reflection of X-rays by atomic planes in crystals could therefore be accounted for, and it was possible to make quantitative measurements of the amplitude, F, that gave the number of electrons contributing to that reflection. The big question was whether these F values could be used to solve the structure of the crystal directly, instead of by trial and error. In theory, "the spatial distribution of scattering matter is given by a trebly infinite [Fourier] series with all values $F(hkl)$." In practice, however, lack of knowledge about the phases of the scattered radiation (or the signs, in the case of a centrosymmetric crystal) created an apparently insuperable obstacle: "It would appear that the crystal structure must first be analysed before the Fourier series can be used... There would appear to be no way of avoiding trial and error methods in the general case [as opposed to special cases, e.g., simple crystals of high symmetry]; it is forced on us by the impotence of the X-ray measurements to determine phase relationship between the scattered waves." Bragg and West explicitly described the Holy Grail of X-ray analysis: "Crystal analysis would become a matter of routine if a general formula could be found, which when the experimental measurements where [sic] substituted, yielded by automatic computation the required answer of the crystal structure."

When he was in Montreal in the spring of 1928, Bragg had discussed Fourier methods with a former student of WHBs, Arthur Lindo Patterson. In February of the following year, Bragg wrote to Patterson: "My father aroused my interest in it again by some remarks he made about the Fourier series outlining some of his organic crystals, and by a discussion we had about my paper with West. This made me sit down and thrash out the application of the two-dimensional Fourier series, a thing which I had rather funked tackling at the time I had my talk with you because it seemed so lengthy. However the calculations were not nearly so bad as they looked at first sight and we have now applied the two-dimensional Fourier series to several examples."[344]

A more detailed description of the origins of two-dimensional Fourier analysis was given in a paper submitted in April 1929: "A few months ago we [WHB and Bragg] discussed the possibility of making a more satisfactory use of the Fourier method. We had approached this problem along slightly different lines. He had attempted to apply two dimensional and three dimensional Fourier series to an organic substance, diphenyl, and had found that the first few terms of the Fourier series did in fact outline the general massing of scattering matter in the unit cell if certain assumptions about phase were made. I had been interested in the analogy between X-ray analysis and optical image formation, and the general relation of Fourier analysis to analysis by trial and error, as outlined briefly in a recent paper in the Zeitschrift [*für Krystallographie*] by Mr. West and myself. We had, in our extensive measurements on inorganic crystals, sufficient information to make very faithful projections of the crystal by using Fourier series, and my father's results led me to take up the problem of computing the series for all planes around a crystal zone."[345]

The work to which Bragg referred was completed in February 1929.[346] As described above, Bragg had performed one-dimensional Fourier syntheses along several axes of beryl; his group and his father's had subsequently used a similar approach to study topaz and alum, respectively. These one-dimensional syntheses only gave electron density along an axis; to solve a crystal structure, however, one really needed a three-dimensional synthesis. As the calculations had to be performed by hand or with primitive mechanical calculators, the summation of hundreds of terms for every point in the three-dimensional space of the unit cell was prohibitive. Bragg decided to compromise by doing two-dimensional Fourier syntheses of projections onto each face of the unit cell of diopside. To obtain the intensity measurements required, he used a neat trick. A crystallographic zone is defined as a set of planes that are all parallel to a particular axis; if Bragg chose the zone axis such that it was perpendicular to the projection plane of interest, then the Fourier syntheses from the reflections of that zone would form a grid with the atoms located at the intersections.

If, for example, Bragg measured the intensities of all reflections of the type $(0kl)$, representing planes parallel to the a-axis, the resulting Fourier would be a map of electron density of the projection on the (100) face. Similarly, reflections from the ($h0l$) and ($hk0$) zones would give electron-density maps of the (010) and (001) projections, respectively. To get the intensities, it was now much easier to use photographic methods rather than the X-ray spectrometer, even though the former were less quantitative.

To construct the two-dimensional Fourier map of the projection on the (100) face of diopside, Bragg used the following formula:

$$\rho(y, z) = (1/bc \sin \alpha) \sum_{-\infty}^{+\infty} \sum_{-\infty}^{+\infty} F(0kl) \cos 2\pi(ky/b + lz/c)$$

where $\rho(y, z)$ is the electron density at the point with coordinates (y, z), b and c are two of the unit-cell dimensions, α is the angle between the b- and c-axes and k and l are the Miller indices of the planes involved. The resolution of the electron-density map—and the amount of calculation involved—depended on how finely the b- and c-axes of the unit cell were sliced. (Because diopside is a monoclinic crystal, the b- and c-axes of the unit cell are unequal in length.) Bragg divided b into 24 and c into 12, giving 288 points in the projection. For each point, a Fourier series of about 40 terms had to be summed. In other words, for each point in a projection, Bragg had to calculate $F(0kl) \cos 2\pi(ky/b+lz/c)$ for each of about 40 ($0kl$) intensity measurements. This then had to be done a total of 288 times to create the entire projection. Finally, the whole operation had to be repeated twice to generate the other two projections. So calculations of the type $F(0kl) \cos 2\pi(ky/b + lz/c)$ (or its counterpart for the other projections) had to be done a total of 34,560 ($40 \times 288 \times 3$) times.

What these calculations produced was a series of numbers, representing electrons per unit area, in the plane of projection. To visualize the structure, Bragg drew contour lines at intervals of 400 for each projection—the first electron-density maps. After all this work, he must have been highly relieved when these maps corresponded closely to the projections West and he had determined by the trial-and-error method (Figure 4.6). In the (010) projection, for example, the superimposed Ca and Mg atoms could be clearly seen as circular areas with tightly packed contour lines; the atomic centers of the Si and O atoms could also be resolved, although less clearly. To determine values for parameters from the Fourier maps, Bragg considered an atom being located either at a maximum electron-density value or at the "center of gravity" of a peak. Ignoring atoms that overlapped others, the average difference in the values of the 14 parameters of diopside as measured by the Fourier method and the "classical" method was only 0.5%. Summing up the number of electrons present in the prominent peaks gave a total of 28.3 electrons, in good agreement with Ca (atomic number 18) plus Mg (atomic number 10). For Si and O, the electron values were generally 11.7 and 8.3, respectively, suggesting "that both the oxygen atoms and the silicon atoms are only partially ionised." However, Bragg noted that there were several sources of error in these electron determinations: The difficulty in estimating extinction in calculating F, the indistinct contour boundaries for O, impurities in the crystal, and the fact that the Fourier series terminated prematurely (because only orders corresponding to angles of incidence less than 90° can be measured).

The formula given above could be used for a unit cell of any shape so long as it had a center of symmetry. However, the amplitude $[F(hkl)]$ values could be positive (if the scattering matter in the unit cell diffracts in the same phase as an electron at the origin) or negative (if the scattering matter in the unit cell diffracts in the opposite phase to that of an electron at the origin). This information was not provided by X-ray analysis. As Bragg wrote in his paper describing the two-dimensional Fourier method, "As was emphasised by Duane, any given set of X-ray results may be explained even in the case of a centrosymmetrical crystal by as many different arrangements of scattering matter as there are permutations of signs in the Fourier coefficients."

This ambiguity of sign was not a problem in the two-dimensional Fourier of diopside, as it had not been in the one-dimensional analysis of beryl, because its structure had already been determined by trial and error. If Bragg had performed the Fourier analysis of diopside, which involved about 40 reflections, without any knowledge or assumptions about its structure, the number of possible permutations of sign—and therefore the number of possible structures—would have been 40!, or 8×10^{47}. Most of these "arrangements of scattering matter" could have been immediately rejected—those that had negative density values, or "atoms" with the wrong numbers of electrons—but it was completely impractical to do a Fourier synthesis 8×10^{47} times, even if the correct structure could then be identified.

This raised again the awkward question—of what use was Fourier analysis for solving crystal structure if one had to know the structure in order to do the Fourier analysis? Bragg had a couple of answers. For one thing, the two-dimensional Fourier analysis of diopside essentially proved the West and Bragg structure by showing that the signs determined by trial and error gave consistent projections on all three planes, the correct number of atoms per unit cell and the correct number of electrons per atom. Another advantage of the Fourier analysis is that it compensates for uncertainties in the measurement of extinction: "The density is determined by a large number of [high-index] F values not affected by extinction, and the few reflexions at low angles which are so affected are relatively unimportant." Fourier synthesis was still a complement to, rather than a replacement for, the trial-and-error method.

However, Bragg's Fourier analysis of diopside provided a glimmer of hope that the "phase problem" could perhaps be circumvented. The scattering contribution of the Ca and Mg atoms superimposed in the (010) projection is so great that it swamps those of all other atoms, and therefore all terms in the $F(h0l)$ Fourier series are positive at that point in the unit cell. If this had been realized during the trial-and-error analysis, "it would then have been possible to use the Fourier series to find the projection on (001) and so the x and z coordinates of all the atoms."

The Fourier analysis of diopside was authored by Bragg alone, but it contained an unusually long and detailed acknowledgment: "It is with great pleasure that I acknowledge my indebtedness to my father, Sir William Bragg, for suggestions which materially contributed to the work described in this paper. At the time when I was following up the connection between our usual methods of analysis and the analysis by Fourier series, a connection briefly treated in the paper by Mr. West and myself, my father showed me some results which he had obtained by using relative values of the first few terms of two- and three-dimensional Fourier series to indicate the general distribution of scattering matter in certain organic compounds. It was largely as a result of his suggestions that I was encouraged to make all the computations for these two-dimensional series, using the extensive absolute measurements which we had made on certain crystals." Despite this lengthy explanation, Bragg always felt guilty that he had not insisted on giving WHB his share of the credit. He wrote in 1961: "I wished to publish a joint paper with my father, but he insisted I should publish it alone. The first two-dimensional Fouriers appeared therefore as the three principal plane projections of diopside, though credit for its start is due entirely to my father."[347]

Bragg's two-dimensional Fourier analysis of diopside is a watershed in the history of structural crystallography, as the same approach—electron-density maps of centrosymmetrical projections—would play a crucial role in solving the structures of proteins, vastly complicated molecules with thousands of parameters for which the trial-and-error method would have been unthinkable. As Guy Dodson has written, the electron-density map is "at the heart of X-ray crystallography."[348] In Bragg's long career, the development of the two-dimensional Fourier technique ranks second only to his discovery of the basis of X-ray diffraction by crystals. These two great breakthroughs have much in common. In both cases, Bragg did not have the original idea: In 1912, the idea was Laue's, in 1929 it was WHBs. Rather, Bragg's critical contribution was in finding a way of making the original idea into a powerful technique for analyzing crystal structure. In 1912, this paid immediate dividends; in 1929, it would take much longer.

It is something of a mystery that a busy department head with a young family could find the time to perform the mind-numbing number-crunching necessary for a two-dimensional Fourier synthesis. Two years later, Arnold Beevers took 90 minutes to sum the Fourier series for *one point* in the unit cell of copper sulfate. Assuming a similar rate, it would have taken Bragg almost 1300 hours of computation to complete the three face-projections of diopside. According to Henry Lipson, who later worked with Bragg: "he would not tell us how he summed the [Fourier] series for diopside; he merely said that it was useful to have a billiard table—as he had in his house—for laying out his papers!"[349] However, Bragg only moved to a house with a billiard room in 1933. He later wrote of the Fourier analysis of diopside: "I worked out all the values of $F \cos \theta$ with a slide rule—it took about a week."[350]

Linus Pauling's laboratory at Caltech had rapidly emerged as a major center for crystallographic analysis. In February 1929, Bragg essentially had to retract a structure of cyanite published by Taylor and W. W. Jackson in his group after Pauling pointed out that it brought certain silicon and aluminum atoms too close together.[351] In April, Pauling dropped a theoretical bombshell when he published a paper entitled "The Principles Determining the Structure of Complex Ionic Crystals."[352] This contained a series of rules of how ions pack in crystals that could be used to generate plausible structures for testing by X-ray analysis—not the direct method that Bragg wanted, but rather a major short-cut for the trial-and-error method.

Pauling claimed that these rules applied to crystals with small, usually tri- or tetravalent cations and large, non-deformable, uni- or divalent anions (O or F). For the sake of simplicity, it will be assumed here that the anion is oxygen. Pauling's first rule provided a means of predicting how many oxygen atoms would form a coordination polyhedron around a metal ion from the ratio of the metal ion radius to that of oxygen. The four possible polyhedra are tetrahedron (coordination number 4), octahedron (coordination number 6), and cube (coordination number 8). The second rule stated that the electrical charge of an oxygen ion (2−) would be compensated by the fractional charges of the cations with which it forms bonds. The third rule was polyhedra were less likely to share edges than corners, and even less likely to share faces. Rule 4 applied to crystals with different cations of large valence and small size, and stated that their polyhedra would not share even corners. Rule 5, the "principle of parsimony," stated that all cations of a particular element should form the same coordination polyhedra, even if these cations were crystallographically non-equivalent.

Pauling's rule 1 provided a theoretical explanation for a phenomenon that Bragg and others had previously noted. Rules 4 and 5 referred to special cases. It was rules 2 and 3 that were of greatest importance. Rule 2 took an obvious idea—that in a stable crystalline structure the numbers of positive and negative charges must be equal—and made a corollary that was not obvious: That there must be equal numbers of positive and negative charges locally as well as globally. In beryl, for example, there are two types of O atoms, one shared by two Si atoms and one shared by Si, Al, and Be. Si is tetravalent and lies between four O atoms, so Si—O is given the valence value of one. Be is divalent and lies between four Os, so the Be-O bond has valency $\frac{1}{2}$. Al is trivalent and lies between six Os, so the Al—O bond has valency $\frac{1}{2}$. The first type of O atom has its valence balanced by two Si (1+1), the second by Si, Be, and Al ($1 + \frac{1}{2} + \frac{1}{2}$). Pauling's third rule, that contact between polyhedral elements should be minimized, was really a way of stating that the stability of the structure would be increased if the cations that lay at the centers of these polyhedra were separated from one another by as great a distance as was consistent with the other rules. For example, "The sharing of an edge between two regular tetrahedra brings the cations at their centers to a distance from each other only 0.58 times that obtaining in case the tetrahedra share a corner only; and the sharing of a face decreases this distance to 0.33 times its original value."

In his biographical memoir of Bragg, David Phillips wrote: "Linus Pauling, who was a visitor in 1929–30, had proposed a method of describing the structures in terms of SiO_4 and other polyhedral units and he went on to propound a set of principles governing the assembly of ionic compounds which depended heavily on the evidence accumulated in Manchester, as Bragg rather sadly remarked, and subsumed the less-well-developed rules that had been evolved there."[353] Phillips' account of this episode makes it sound as if Pauling visited Manchester, learned Bragg's approach to solving crystal structures and then published these under his own name. Indeed, this was the general impression of several people who later worked with Bragg.[354] For example, Aaron Klug heard the following version of the story from James: "Pauling came on a visit and Bragg showed him their new structures, and he showed him very freely, and so the rules which later followed, Pauling's three rules, Bragg already had two of them and was planning to write up his generalization of silicate structures and he was furious when Pauling, having the advantage of the structures he had solved and which had not been communicated to Bragg, was able to see much more."[355]

Pauling's version of the story was quite different. He told Horace Judson in 1975: "Bragg and members of his school . . . began studying some of the silicate minerals at about the time that I was studying the silicate minerals, and of course I formulated a set of principles about crystals of that type and Bragg made use of them, wrote a long paper in which he applied these principles."[356]

In view of these diametrically opposite accounts, and because of the importance of the relationship between Bragg and Pauling in the development of macromolecular crystallography, it is worthwhile to examine the episode of Pauling's rules in more detail. In January 1928, a letter from Pauling to Bragg mentioned that, as a result of his studies on brookite, one of the polymorphic forms of titanium dioxide, "we think that we have found some interesting general principles regarding the structure of coordination compounds."[357] Pauling sent Bragg a copy of the brookite manuscript in May, and a preliminary account of the "general principles," written for a Sommerfeld *Festschrift*, in August. Therefore, Pauling's visit to Manchester came over a year after his paper on the principles governing ionic packing in crystals was published, and a year and a half after he had drawn these principles to Bragg's attention.

The letter Bragg wrote to Pauling in response to the "general principles" freely acknowledged the novelty of the American's approach: "I like your way of looking at the coordination compounds very much indeed. One's point of view is affected by the way in which one arrives at it, and I have got accustomed to thinking of the packing together of the atoms . . . It seems to me that the two conceptions, that of atomic packing and that of linking up coordinated polyhedra, back each other up, each having its advantages."[358] As early as October 1929, Bragg referred to "Pauling's rule for electrostatic valency"[359]—not a likely form of words if he felt that he had discovered it first.

Bragg's own approach was, in fact, quite different from Pauling's. As he described it in a 1929 paper,[360] the first step in solving a crystal structure was determining its space-group, after which "The atoms of the crystal are grouped in accord with the symmetry elements." Next, the fact that crystals consist of "atoms of definite size packed together like small spheres" limits possible structures greatly and allows a "tentative model" to be generated in which the symmetry requirements are met and no atomic overlap occurs. Finally, quantitative measurements of diffraction allow the investigator to determine "the fraction of the crystalline unit which conspires to scatter radiation in a given direction." If this agrees with the tentative structure, the structure is very probably correct. Bragg had shown that silicon atoms were surrounded by four oxygens in a tetrahedral arrangement and aluminum atoms by six oxygens in an octahedral arrangement, but there is nothing in his writings on mineral structure akin to Pauling's geometric view of linked polyhedra. According to J. G. Burke, the followers of Pythagoras viewed solids arising from points, lines, and planes, whereas Plato thought matter was constructed from isosceles and equilateral triangles.[361] If so, Bragg was a Pythagorean, Pauling a Platonist.

Nor did Bragg propose anything similar to Pauling's idea of local compensation of charge; he was always weak on chemistry and valency was something he struggled to incorporate into crystal structure, from the Pope–Barlow valence volume theory to the mis-estimates of atomic radius to the nature of the silicon "ion." There can be no doubt that Pauling's rules were arrived at quite independently of any theories of crystal packing emanating from Manchester, although they were of course based in part upon specific structures determined by Bragg's group.

Publicly, Bragg seems to have felt it necessary to minimize the novelty of Pauling's approach. In his 1930 paper entitled "The Structure of Silicates," he wrote: "In contrasting his method [of representing structures as polyhedra joined by corners, edges or faces rather than close-packed groups] with that adopted by the author, Pauling has laid an emphasis on the difference between them which is perhaps rather excessive; it is only a matter of convenience of description to replace the idea of regular groups of large anions around small cations with that of tetrahedra and octahedra, and of linking by sharing oxygen atoms with the sharing of corners and edges." Pauling's other innovation, the idea of local compensation of charge, was also treated rather off-handedly by Bragg: "Pauling has made the important step of pointing out that is true in detail, for anions in different situations in the crystal, as well as in sum for the crystal as a whole." Bragg also claimed a share of the credit by pointing out that "Pauling has made frequent reference to the structures analysed by the author in illustrating their application."[362]

It appears that Bragg's dismissive reaction to Pauling's rules of ionic packing in crystals was conditioned not by any feeling that the American had stolen his ideas, which would have been patently absurd, but rather by the uncomfortable realization that his pre-eminence in the X-ray analysis of inorganic crystals was being threatened by a very powerful competitor. Pauling was younger, unconstrained by British scientific mores and understood chemistry—he could already be described as the leading chemist of his day—and, to add insult to injury, Pauling, unlike Bragg, understood the quantum mechanics that was revolutionizing Bragg's own field of physics. The sudden emergence of a scientific Titan within the field that Bragg thought of, not unreasonably, as his own, as shown by the black eye he received over the structure of cyanite and the enthusiastic reception of Pauling's rules, may have contributed to Bragg's abandoning of the study of silicates soon after. However, it would not be too long before he found himself in competition with Pauling again.

Bragg had now been in Manchester for 10 years. He had built up a strong research group centered on James and Taylor and including a number of outstanding visiting research workers. Frederik Zachariasen, who studied the silicates, "had the reputation in the laboratory of turning out a crystal structure a day when most people took months to complete an analysis."[363] Another visitor was Felix Machatschki, who worked as a research fellow with Bragg in 1928–9. His analysis of the structure of danburite, $CaB_2Si_2O_8$, was still incomplete when he left. "He was given a farewell party in the laboratory, with a central feature of a large cake. Machatschki, asked to cut the first slice, had his hand guided to the right spot. When the section of cake was removed, it revealed the structure of Danburite depicted by cherries, raisins and angelica baked into the cake—or at any rate what the lady who baked the cake thought Danburite ought to look like."[364] In 1929, Bert Warren arrived from MIT, where he had moved on from diopside to other types of silicates (pyremenes and amphiboles). That same year, the theoretical physicist Ivar Waller came from Uppsala and collaborated with James and Hartree in calculating the zero-point energy of rock salt.

In 1928, Bragg, Patrick Blackett, and George Thomson went on a sailing holiday. Starting from the Clyde, they sailed through the Crinan Canal and on to Mull: "We found some very remote and romantic anchorages by the shores of the many small uninhabited islands in those parts."[365] In the spring of 1929, Bragg and Alice went to Holland, where he gave talks at a number of universities, including those at Utrecht, Delft, Amsterdam, and Groningen, and visited the Philips company in Eindhoven. Their summer holiday was again spent in Pensarn. These latter events occurred under a shadow, as Gwen Bragg was seriously ill. "Poor Dad could not bear to talk about Mother's illness—his face became blank when it was mentioned. I remember the surgeon taking me aside once because he thought that he had quite failed to convey to my father the hopelessness of the case, and he asked me to help. Of course this was not so, it was just that my father so completely concealed his feelings."[366] Gwen died in September 1929.

Bragg and his mother had never been close. In a biographical sketch of his father, Bragg described Gwen as: "A woman of wit and great understanding, with really no interest in intellectual attainments (her genius was expressed in her painting), she made friends in the R. I., brought friends there, and people near her made friends with each other."[367] He was more candid in his autobiography: "She had grown up in the rather haphazard way peculiar to the large Todd family . . . she had resisted with complete success any attempt to educate her . . . Nothing could persuade her that Cambridge was not south of London."[368] With the exception of painting, she had little in common with her intellectual and introspective son.

It is clear that Bragg found Gwen exasperating: "She could sense halfway through a sentence from the listener's reaction when she was on the wrong tack and change what she was going to say in a flash . . . it always had to be quite clear that any family decision was not hers, but that she was giving way to the wishes of others and doing what she thought they wanted, although actually Mother very effectively guided decisions in the direction she thought best."[369] In his sister's opinion, Bragg suspected that Gwen had always preferred Bob. Bragg and Gwendy both thought that Bob handled their mother's peccadilloes much better: In Gwendy's words, "Willy was inept with her, too earnest, would take her exaggerations too seriously when he ought to have laughed and challenged them. They gave me a lot of trouble!"[370]

The death of Gwen came as a devastating blow to WHB, who, in addition to their personal bond, had relied on her professionally for the social activities of the Royal Institution. Fortunately, others were willing to fill the gaps in his life. Gwendy accompanied WHB on visits to the United States in 1930 and South America in 1932. On their return from the latter trip, she married the architect Alban Caroe. Lorna Todd, Bragg's favorite aunt, came over from Australia to keep house at the Royal Institution. After 18 months, the Caroes moved into the flat "over the shop."[371]

Another personal problem was clouding Bragg's life at this time. All the exotic travel and hobnobbing with famous people did not console Alice for the fact that, at the end of each trip, she had to return to wet and grimy Manchester. The theoretical physicist Rudolf Peierls, who joined Bragg's department in 1933, had the following first impressions of Manchester: "The buildings had been erected mostly during the Victorian period and were in poor taste, and there were many slum areas. The new part, where we lived, consisted of cheap houses put up by speculative builders. Most of the older houses were black with soot, so that even the few attractive buildings in the centre were not easily distinguished."[372] There were frequent fogs of Dickensian density; these permeated indoors, making it difficult to see the stage from the rear of a theater. When Alexander Todd was appointed to the Sir Samuel Hall Chair of Chemistry at Manchester University in 1938, he wrote to Pauling: "The only snag is one which you will appreciate from personal knowledge—the somewhat depressing industrial area in which Manchester lies." He also referred to "the notorious climate of Manchester." However, as Todd pointed out, he could hardly complain—he had been brought up in Glasgow![373]

Bragg was under great pressure to find a position in surroundings more acceptable to his wife. In June 1929, he was desperate enough to write to Rutherford about "the possible post at Cambridge which you mentioned to me some time ago."[374] This seems to have referred to a proposed professorship of crystallography. Bragg suggested "an additional chair of experimental physics with especial charge of the physics of the solid state." This suggestion went nowhere, but it is a measure of Bragg's desperation that he would consider giving up the second job in British physics to become Rutherford's underling at Cambridge.

A more viable possibility was the Professorship of Physics at Imperial College London, which Bragg was offered in the spring of 1930. He agonized over the offer during a family holiday at a farm on Derwentwater. On their return to Manchester he had "some kind of nervous breakdown . . . It was precipitated by worry about an invitation I had received to go to Imperial College, though I expect the root causes were much deeper." Torn between two loyalties—to his wife and to Manchester University—Bragg went to discuss the situation with WHB, "but it was one of those times when it was quite impossible to get him to talk." Bragg's emotions did not improve when he made the decision to decline the position at Imperial College, which went instead to his friend George Thomson: "I was in a bad way and caused Alice much distress." The family had a summer holiday at WHBs cottage, Watlands, during which Alice and WHB came up with a plan for Bragg to take a term off.[375]

According to Phillips, Bragg's breakdown may have been precipitated in part by the strain of overwork. He was at this time involved in planning a new building, writing a new edition of *X-Rays and Crystal Structure* (published in 1933 as *The Crystalline State*) and preparing new lectures for industrial physicists. The onset of the Great Depression, which severely affected the industrial north-west of England, may also have been a factor.[376]

It was around this time that Pauling spent a month in Manchester. Bragg arranged a house and maid, and gave Pauling an assistant to help him in making measurements on the silicate epidote ($Ca_2[Al, Fe]_3Si_3O_{12}OH$). "Despite Bragg's helpfulness in these ways, the stay in Manchester was a disappointment to me. I had determined the structures of some silicate minerals, as had also Bragg and his co-workers, and I had published a paper about structural principles for silicates and other minerals in which the bonds have a considerable amount of ionic character, amplifying and extending the principle of close packing of oxygen atoms that had been formulated by Bragg. I had anticipated that there would be discussions between Bragg and me about these matters . . . however, Bragg did not ever ask me to discuss scientific matters with him and I, his junior by eleven years, did not have courage enough to ask for such a discussion with him."

Pauling was also disappointed that there were no seminars during the month he spent in Manchester, and contrasted the atmosphere unfavorably with the "livelier" one at Cambridge, where he visited Bernal. "At that time, although still quite young, he [Bragg] held an important position in science, involving administrative and teaching duties, as well as the direction of research. As a result, when quantum mechanics was discovered he was not in a position to devote enough time to this rather complicated and somewhat abstruse subject to master it. I suggest that he may have felt handicapped by this lack, and that it may have kept him from entering into lively scientific arguments and discussions."[377]

Pauling may not have realized, at the time of his visit or when he wrote the above account, that Bragg was going through a period of great personal turmoil. However, there is probably some truth to Pauling's feeling that quantum mechanics created a barrier between the two men. Bragg seems to have been going through a bit of a scientific identity crisis at this point. In January 1930, he wrote to Harold (?) Robinson: "I always wish I were in a field which had more right to be called Physics."[378] In the June 1929 letter to Rutherford mentioned above, Bragg wrote: "I do not want to label myself a crystallographer as against a physicist."[379] Embarrassed at talking the old-fashioned language of classical physics, he may well have avoided any substantive discussions with Pauling.

If this first face-to-face meeting did not lead to any great friendship between Pauling and Bragg, another event at this time may have worsened things. In August 1929, Bragg had written to Pauling: "I am rather keen now to have a shot at the micas and felspars and some of the zeolites."[380] If this was meant as a hint that Pauling should do the gentlemanly thing and leave these crystals to Bragg, it did not work; Pauling wrote back: "I have begun work on the zeolites too, for no particular reason except that I had some good crystals at hand."[381] He also had his fingers into mica. In April 1930, Bragg complained to Zachariasen that Pauling had beaten him to the punch: "His mica structure is essentially the one we talked over last summer and we were really very slow to have let him get it out first. West was interested in another problem and was just starting to clear it up when Pauling's paper appeared."[382]

By March 1930, when Bragg submitted a review article entitled "The Structure of Silicates," 35 silicate minerals had been analyzed, about half of them by his own group. "the formula of a silicate is best expressed in the following way:—The unit of structure contains a definite number N of oxygen atoms. It contains X positions that may be occupied by silicon or aluminium, with a general restriction on the extent to which aluminium can replace silicon. Y positions can be occupied by Mg^{++}, Fe^{++}, Al^{+++}, Fe^{+++}. Z positions may be occupied by Na^{+}, Ca^{++}, and sometimes larger ions such as K^{+}. In addition, the sum of positive valencies must balance the sum of negative valencies. With these limitations, variations in compositions are possible. Fluorine and the group OH behave like oxygen, and their total number is a constant." As oxygen is the largest and most abundant ion, the lattice dimensions are determined by "the universal characteristic distance between oxygen atoms." This is always around 2.7 Å, but can be reduced to 2.5 Å in Os of a tetrahedral group around Si, and can increase to 2.8 or 2.9 Å if the O atoms are not bound to the same ion. The metal and silicon atoms occupy the interstices between oxygen tetrahedra and octahedra. Some of the oxygen ions may be shared by two polyhedra, leading to Pauling's geometric conception of the silicate structure. This sharing of oxygens explains the fact that both in the orthosilicates, with an O/Si ratio of four, and in the metasilicates, with an O/Si ratio of three, each Si is surrounded by four oxygens. "The silicates are distinguished . . . by the way in which (SiO_4) groups can be linked together to form silicon–oxygen complexes with indefinite extension in space. It is this feature which gives rise to the variety of silicate structures, and which has caused the role which silicon plays in the inorganic world to be compared to that which carbon plays in organic chemistry."[383]

Bragg was now losing interest in the silicates. In April 1930, he wrote to Zachariasen: "I am rather wondering what line of compounds to go on to now. The inorganic field is really getting pretty well worked out."[384] The following month, he wrote to Warren: "We must look for fresh worlds to conquer."[385] In October, Warren and Bragg submitted a manuscript on the structure of chrysotile, a form of asbestos with the empirical formula $H_4Mg_3Si_2O_9$.[386] They showed that the fibrous nature and flexibility of the material could be explained by the presence of Si—O chains running in the fiber-axis direction, with areas containing only "weak secondary forces" between the chains. It was the last mineral structure on which Bragg worked directly. His group continued the analysis of silicate and other minerals—Will Taylor went on to tackle the complex feldspars (granites)—but Bragg had decided that his personal research interests would change.[x]

As he had made the analogy between silicon in minerals and carbon in organic molecules, it surely must have occurred to Bragg to take up the ultimate challenge in X-ray analysis—the structure of biological macromolecules. This analogy was drawn to his attention by William Astbury, a former student of WHBs who was now lecturer in textile physics at Leeds University. Astbury was using X-ray methods to study stretched and unstretched forms of the fibrous protein keratin. In October 1930, he wrote to Bragg describing the similarity between "spreading of the layer lines" in chrysotile and a similar phenomenon in steam-extended keratin: "the disturbance in the keratin lattice is exactly analogous to that in the chrysotile lattice, in that the distortion takes place *in one direction only*."[387] (Emphasis in original.) The same month, Robert Robinson, Waynflete Professor of Chemistry at Oxford University, wrote to Bragg about a structure of the plant polysaccharide cellulose proposed by Kurt Meyer and Herman Mark.[388] Bragg replied that a problem for the X-ray analysis of biological substances was that the lengths of organic bonds had not been well characterized.[389] Several other factors might have dissuaded Bragg from attempting the X-ray analysis of proteins *c*.1930. His organic chemistry was weaker than his inorganic chemistry and there was no solution in sight for the phase problem in Fourier analysis. In any case, organic

[x] At the British Association meeting in 1934, Bragg gave an overview of mineral structure in which he concluded: "The fortunate existence of a raft of rock on which life is possible is seen to be a result of the geometric properties of tetrahedra and octahedra." It was again clear that he regarded this as a mature area of research: "with the recent analysis of the felspars [granite] it may be claimed that the main survey [of the mineral world] has been completed." [Bragg, W. L. (1934). Exploration of the mineral world by X-rays. *Nature* **134**, 401–4.]

molecules were his father's domain. Also, no crystalline proteins had yet been shown to diffract X-rays.

Bragg, as described above, wished to be thought of as a solid-state physicist rather than a crystallographer. The research direction he chose in the early 1930s was consistent with that self-image—he decided to work on the structure of metals.

5
Plus–plus chemistry: Manchester, 1931–7

The solution that Alice and WHB had come up with for Bragg's 1930 nervous collapse was for him to get away from Manchester for an extended period. It was therefore arranged that he would spend the first 3 months of 1931 at the Institute for Theoretical Physics in Munich. Alice, now pregnant again, accompanied her husband, but the boys were left behind in England.

It was fitting that Bragg would spend his leave of absence in the institute where the science of X-ray analysis had begun. Indeed, he visited Café Lutz, where Laue, Friedrich, and Knipping had hatched their plan, and was pleased to find that physicists still used the tables as blackboards.[390] However, Bragg did not use the opportunity of being at Arnold Sommerfeld's institute to learn quantum physics, as Pauling had done 5 years earlier. He did accompany Sommerfeld on his regular weekend ski-trips; as noted above, it was on one such occasion that Laue proposed the use of crystals as X-ray diffraction gratings. The physicists took a mountain railway and then were led on foot to a cow byre by Sommerfeld's assistant, Karl Selmayr, bearing a blazing torch. Sommerfeld was short and fat, and by the time he was dressed for skiing he was "practically spherical." Apart from lighting the way to their makeshift accommodation, Selmayr's job was to ski behind Sommerfeld and help him up when he fell over. In the evenings the students would be exhausted from the day's skiing, but Sommerfeld would light a cigar and initiate a discussion on some abstract problem of theoretical physics.

While in Munich, Bragg and Alice stayed in a boardinghouse and went to the opera in the "perfect little Mozart theatre"—presumably at Alice's instigation, as Bragg was completely unmusical. They also took walks in the *Englischer Garten*, went on expeditions to Garmisch and the Zugspitze and enjoyed the traditional pre-Easter festivities of *Fasching*. It appears that Bragg did not learn much theoretical physics, but Alice's plan was successful: The Munich period "did a great deal to put me right again."[391]

Although depression would not incapacitate Bragg again, black moods would prove a lifelong affliction. His daughter Patience remembered that: "he sometimes got into blind, black rages. He would get really angry about things and then afterwards he would actually hit his head and say 'Why on earth did I say that?' He could really get furious and not even be able to speak coherently, he would be so angry."[392] One such episode was triggered by a seemingly trivial incident in which the local vicar referred to Australia as "the land of the convicts."[y] His sister Gwendy wrote that Bragg's "brain-storms . . . happened in minor form quite often."[393]

Back in Manchester, Bragg concentrated on writing his part of *The Crystalline State*. At the end of April, he delivered the Kelvin Lecture of the Institution of Electrical Engineers on "The Architecture of the Solid State."[394] He was awarded the Royal Society's Hughes Medal, which recognizes original discoveries in the physical sciences, particularly electricity and magnetism. He was also elected to a three-year term on the council of the Royal Society.[395] Bragg and Alice's third child and first daughter, Margaret Alice, was born on June 23.

In 1932, the Bragg family spent their summer holiday at WHBs cottage, Watlands, in Chiddingfold, Surrey. Stephen Bragg remembers Watlands as a house set on a large plot of land with a wood and a large pond. He and David dug a clay pit to make pots and assembled a wigwam that WHB had brought back from the United States. Also there were Gwendy and her fiancé, Alban Caroe, who designed a hut for the boys to build in the woods.[396]

Later that year, Bragg and Alice went to the USSR to attend a conference at the Röntgen Institute in Leningrad. Together with a group of Cambridge physicists, they sailed through the Kiel canal in a motorboat. Conditions on board were rather primitive, but the crew obligingly let the passengers set up a deck-tennis court on the bridge, provided they paused when a sighting had to be taken. Conditions in Leningrad were just as bad as on the boat; according to Alice, "There were broken windows patched if at all with rags, peeling paint, scarcely a car to be seen, and trams so packed with people that they sometimes dropped off." Bragg and Alice stayed at a "large but derelict hotel reserved for foreigners," where they caught fleas and found the food "very queer." To make things worse, the conference had been cancelled. While Peter Kapitza, a Soviet physicist working in the Cavendish Laboratory, managed to arrange an ad hoc conference, the Braggs visited Tsarke Seloe, the former Imperial palace about 20 miles away, which was "fantastic in its sumptuous and barbaric taste" and where they saw the legendary amber room and a chandelier made entirely of diamonds. Alice was the only foreigner to attend a concert celebrating the centenary of the Alexandrinsky Theater.

After the conference, Bragg and Alice travelled to Moscow, and then on to a dacha owned by Allan Monkhouse of Metropolitan Vickers. Moscow was "very shabby," but the dacha was "very civilised." As Bragg and Alice had no tickets home, Monkhouse helped them get a train to Berlin via Warsaw. Shortly thereafter, he was arrested on charges of espionage and sabotage and sentenced to deportation after a sensational show trial.[397]

[y] South Australia was never a penal colony.

In early 1933, the Braggs moved to Windy Howe, Alderley Edge, about 15 miles south of Manchester, which they rented using money Alice received from the estate of Aunt Monica. It was a large house with five family bedrooms plus others for the servants. There was a billiard room that was usually used by Bragg as an unofficial office, his papers covering the surface of the table. A large basement was used as a playroom by the children. Windy Howe had a magnificent location, above the fogs of Manchester and overlooking the Cheshire plain to the Welsh mountains beyond: "to pull up the blind in the morning and see the plain lit by the early sun was something I have never forgotten . . . Altogether it was as if we had been transplanted to a new world." Bragg was also delighted by the garden. In stark contrast to the "sad dirty one" at 45 Pine Road, hundreds of violets bloomed around the rose beds and "snowdrifts of arabis" made "a kind of fairyland." Star of Bethlehem grew in profusion in the surrounding fields, and water-violet, a rare species of primrose, grew down the lane. There was abundant bird-life, and Stephen became a keen butterfly-collector. The Braggs played tennis on the court at the bottom of the garden, picnicked on the lawn, and went on country walks.[398]

Even Alice, who had been brought up in a very comfortable upper-middle-class environment, thought Windy Howe was "definitely rather pretentious." The household included no fewer than six servants: Three maids, a cook, a nanny, and a full-time gardener. (In his history of Cambridge University, Christopher Brooke noted that in 1931–2 his parents paid a total of £141 per year for a nurse, cook, and house-parlor maid.[399]) Nonetheless, Alice felt that they were rather looked down on by the neighbors, who were mainly in industry.

Stephen and David had been attending Alice's old school, Ladybarn House. They now moved to Harden House preparatory school, Bragg being of the opinion that it was "unnatural" to send such young children to boarding schools. Harden House was "not ideal" but mathematics was well-taught by a former student of Bragg's. With the boys in school and a house full of servants, Alice had a lot of time on her hands. She studied law and passed the Part 1 examination, but because of the family's departure from Manchester and the subsequent outbreak of war did not pursue the degree further. She also did a correspondence course in journalism, and had several articles published.[400]

Bragg was an excellent father to his young family. He built Stephen a model railway with rails made of wooden beading and points operated by string. "To illustrate the Christopher Robin stories of A.A. Milne, WLB constructed a shadow theatre, consisting of a model stage covered with a thin sheet of translucent paper which was lit from behind. The audience then saw the silhouettes of cardboard cut-out figures of the principal actors performing a shadow play: sixty years later I can still remember the realism of the flood water rising around the marooned Pooh Bear while he sat astride a bough with his few remaining jars of honey."[401]

Soon after the move to Alderley Edge, Bragg, again accompanied by Alice, went to Madrid to give a lecture to the Spanish Academy of Sciences. After the lecture, they visited the Asturias region, Toledo, and the Guadarrama hills. Around this time, Bragg noticed that his left hand was becoming paralyzed.

A surgeon diagnosed a pinched nerve at the back of the elbow that Bragg had damaged as a child. An operation stopped the paralysis, but Bragg never regained full use of his left hand.[402]

In March 1933, Bragg signalled his change of research direction by giving a Friday Evening Discourse on "Structure of Alloys" at the Royal Institution.[403] It had been known since prehistoric times that combining two or more metals could produce an alloy with physical properties quite different from those of either of the parent elements. X-ray diffraction now allowed the structures of metals and alloys to be determined. Of particular use was X-ray powder diffraction, a technique that had been developed by Peter Debye and Paul Scherrer in 1916. Grinding a crystalline substance to a powder results in crystallites with random orientation. When such a crystalline powder is irradiated with X-rays, the planes that satisfy the condition for diffraction will be randomly oriented in the other two dimensions, and therefore the diffraction pattern will consist of rings rather than spots; as Bragg later put it, this effect is similar to the halo seen around the sun during an ice fog. The distance between these rings is related to the *d*-spacings of the crystal planes present in the powder by Bragg's law. This technique had revolutionized the understanding of alloys.

As Bragg later put it to a non-scientific audience: "Broadly speaking, the three main divisions of Chemistry may be called plus–minus Chemistry, minus–minus Chemistry, and plus–plus Chemistry." "Plus–minus chemistry" is that of compounds consisting of positive and negative ions associated by electrostatic bonds, such as many of the minerals Bragg had previously studied. "Minus–minus chemistry" is organic chemistry, which involves compounds formed by covalent (shared-electron) bonds between electronegative elements such as carbon, hydrogen, nitrogen, and oxygen. "Plus–plus chemistry" is that of metals, which consist of lattices of positively charged ions surrounded by a "sea" of electrons: "Each electropositive atom, of whatever kind, brings its passport in the form of loosely held electrons to add to the common stock, and all are welcome to the association."[404]

Metal physics had long been a sideline at Manchester. Albert Bradley, who Lipson described as "perhaps the most single-minded person that I have ever met"[405], was Bragg's first research student at Manchester. In 1926, after completing his doctorate, Bradley was sent by Bragg to Arne Westgren's laboratory in Stockholm for a year to learn the use of the powder diffraction method to study metals and alloys. Westgren, it will be recalled, had acted as the Braggs' tour-guide when they attended the Nobel ceremonies in 1922. On his return to Manchester, Bradley used the powder method to solve the structure of γ-brass. It was a commercially useless alloy, but the 52-atom unit cell of γ-brass proved to be shared by a number of "γ" alloys.

An important theoretical advance in metal physics had come from the Oxford chemist William Hume-Rothery, who proposed that distinct alloy phases correspond to particular ratios of valence electrons. For example, the β phase, which is body-centered cubic, corresponds to three valence electrons per two atoms. The γ phase has an electron/atom ratio of $21:13$, and the hexagonal ϵ phase a ratio of $7:4$. The different atoms of an alloy may occupy "different" positions, such as the cube corners and body centers in CuZn, or random positions, as in Cu_5Si, in which the number of atoms in the unit cell (20) is not a multiple of the empirical formula. Evan Williams, who had done a Ph.D. with Bragg and was now an assistant lecturer at Manchester, had shown that the latter effect must be due to an increase in energy of structures in which atoms of the same type are at the corner and center of the same cube.

In the period 1928–32, Bradley was paid by the Metropolitan-Vickers company. Charles Sykes, who as liaison between Metropolitan-Vickers and Bragg's group visited Windy Howe on several occasions, wrote: "when we got down to the serious discussion he [Bragg] would adjourn to the billiard room which contained a full-size billiard table. On all my visits to this room, I never saw any balls on the billiard table; it was covered with reprints of papers and, as the argument developed, Bragg would get up from his chair, wander round the table, pick out the appropriate reprint, and we would then examine it in terms of the ideas we were discussing at the table."[406]

Bragg described in a 1945 letter to Patrick Blackett how he got involved in the work on alloys: "Sykes at Metropolitan-Vickers had been experimenting with an iron-aluminium alloy as a possible cheap material for electric fires and heaters. He found queer anomalies in the electrical resistance before and after heat treatment. Bradley then examined the alloy with X-rays, and discovered that it could be changed from an ordered to a disordered structure by thermal treatment. I could sense the general way in which one ought to be able to apply statistical methods to deduce theoretically the variation of order with temperature, but my mathematics were too weak for it to be possible for me to deduce the necessary formula. It was at this point that I asked Williams to help. I retain a most vivid impression of the speed with which he worked. I first went over the problem with him one afternoon, and he appeared next morning with masses of formulae of the most ingenious type. They were not right at this first stage, we had to modify them a great deal, but the energy with which he threw himself into the job was characteristic of Williams . . . Williams' elegant treatment of collision problems was beyond me mathematically, but I could see how attractive his way of tackling them was."[407]

Between Bradley, who "was a wizard at sorting out complex ternary equilibrium diagrams, and the structures of complex alloy phases"[408] and "that volatile genius, E.J. Williams,"[409] Bragg had the firepower for a strong research program in metal physics. For additional help, he turned to Hans Bethe and Rudolf Peierls, two brilliant theoreticians then in Manchester.

The problem Bragg and Williams were addressing concerned the changes that occur in alloys during cooling. It was common empirical knowledge that in many cases the rate of cooling had an important bearing on the physical properties of an alloy. Rapid cooling, or quenching, often resulted in an alloy with quite different mechanical properties from one subjected to slow cooling, or annealing. It was suspected that these differences were due to variations in the positions that the two types of atoms occupied in the crystal lattice. In a body-centered cubic lattice, for example, there are two types of atomic location—cube corners and body centers. These positions are crystallographically equivalent—depending on where the origin of the unit cell is set, the same atom can be located at either a cube corner or body center. However, several alloys had been shown to consist of a lattice in which atoms of one metal occupy cube corners and atoms of the other metal occupy body centers. Such structures, in which different atoms occupy different relative positions, were given the name superlattices.

Bragg and Williams defined a completely ordered lattice as one in which every atom is in its "proper" position—that is, all atoms of type *A* occupy one set of positions and all atoms of type *B* occupy the other set of positions. In a completely disordered lattice, atoms of both types are randomly distributed between the two sets of positions. The rate at which atoms change positions varies directly with temperature. By considering the change in potential energy which occurs when one atom is moved from an "ordered" to a "disordered" position, Bragg and Williams showed that a state of stable equilibrium, in which a preexisting degree of order is maintained even as atoms change positions, can only occur below a critical temperature. On cooling the alloy, there is a sudden onset of order below the critical temperature, followed by a more gradual increase in ordering until complete order is achieved. Because of a hysteresis effect, the critical temperature is higher when the alloy is being heated than when it is being cooled. As Bragg later described this process, a larger amount of energy is required to move an atom from a "right" to a "wrong" position in an ordered lattice than in a partially disordered one because of "public opinion." As the lattice becomes more disordered, "demoralization sets in."[410]

The difference between annealed and quenched alloys was shown to depend upon the relaxation time, which is a measure of how quickly an alloy reaches equilibrium as it cools. This allowed Bragg and Williams to make quantitative predictions about annealing and quenching. Knowing the relaxation time for a particular alloy, the time it will take to anneal completely at a particular temperature could be calculated. Bragg and Williams were able to define, in terms of relaxation time and temperature, conditions under which it is impossible to quench an alloy quickly enough to produce a particular equilibrium state. Conversely, there were conditions under which an alloy could not be annealed slowly enough to produce a particular equilibrium. The theoretical limits of quenching and annealing were thereby established.[411]

Bragg later compared the formation of a superlattice to a dinner party: "The pattern of phase sites is represented by the pattern of chairs set around the table. It remains constant throughout the phenomenon; atoms, or diners, may be placed at these positions but at no others. The equal numbers of atoms of two kinds, A and B, we will represent by equal numbers of ladies and gentlemen at our party. At the beginning of the dinner, when appetites are keen (high temperature) no regard is paid to the order of seating of the guests. Ladies and gentlemen occupy the seats quite at random. As hunger becomes less keen (cooling of alloy) manners are remembered and a reshuffling takes place into the ordered scheme lady-gentleman-lady-gentleman round the table."[412] Vivid analogies of this type were to serve Bragg well when, as Director of the Royal Institution, he had to give many lectures on science to lay audiences.

Bragg was very proud of his work on the order–disorder transition in alloys, writing of it in 1961: "It is no exaggeration to say that the principles of metal chemistry for the first time began to emerge."[413] He published a follow-up paper[414] with Williams in 1935 and occasionally other papers in the area of metal physics. For the most part, however, he was content to delegate metals research to others, as he had earlier with minerals. Increasingly, other commitments were distracting Bragg from personally directing research. Now in his mid-forties, he no doubt realized that his most significant contributions had already been made. Like Svante Arrhenius before him, he was fated to spend less time making famous theories and more time accepting honorary degrees.

There were many opportunities for a Nobel Prize-winning physicist to travel the world. One attractive offer Bragg received was to become visiting professor in the Baker Laboratory of Chemistry at Cornell University in Ithaca, New York. After having to reschedule twice, he eventually spent the first 6 months of 1934 at Cornell. His duties were light—he gave a series of lectures on crystal structure and spent the rest of his time writing a book, *The Atomic Structure of Minerals*. Bragg also gave an "introductory public lecture" on "The Physical Sciences." This was rather more philosophical in tone than Bragg generally cared to be, discussing the indeterminism of many physical processes and the Heisenberg Uncertainty Principle. The ideas were not original to Bragg, but he did present them in elegant terms: "Physical processes may destroy the miraculous, but can not create it"; "Nothing can exceed our instinctive horror of the finite, our revulsion at the idea of being entrapped in a mechanical web."[415]

For the first few months, Bragg stayed at the exclusive "Telluride" fraternity house, where the other guests were an expert on witchcraft and a "Professor of Pomology." The mirror for the 200-inch reflecting telescope on Mount Palomar was then being made at the nearby Corning glass works. Bragg, who after all had a professional interest in ordering of solids during cooling—not to mention reflection phenomena—went to see this operation. He was amused to find that the quality of the molten glass was determined by an old man who decided whether it would be poured or put back into the furnace.

In early spring, Alice arrived by boat in Boston, and she and Bragg went to Charlottesville, Virginia, to stay with a Cornell professor. They then spent 3 days—the maximum period a woman was allowed—in Telluride, before moving to an apartment on the edge of campus. There they "had tremendous fun looking after ourselves and shopping"—jobs that servants did for them in England. However, the "tremendous fun" of housework did not extend to "heavy cleaning," which was done by a student. They had the use of a car, with which they made excursions to the Finger Lakes region, where glacial action had created a series of steep-sided lakes, each fed by its own waterfall. There Bragg and Alice would eat a picnic supper and explore the canyons, filled with unfamiliar flowers and wildlife. For "May Week," a long weekend in which girls were invited to stay and dances were held, the men of Telluride invited Alice to act as chaperone. Bragg danced with a girl who was majoring in Philosophy and sanitary engineering, "which always seemed to me the ideal preparation for married life."[416]

During the period of his visiting professorship at Cornell, Bragg attended a meeting of the American Physical Society in Washington, DC, which stimulated his interest in Fourier analysis. Arthur Lindo Patterson, who had worked with WHB for two years before taking up a position at McGill University in Montreal in 1927, was obsessed with finding a way of liberating Fourier analysis from the yoke of the phase problem. Bragg had discussed this with Patterson when he visited Montreal the spring of 1928; subsequently Patterson had written to WHB: "I am still working spasmodically on the properties of symmetrical Fourier series. I might really say that they have filled the place in my heart previously occupied by cross-word puzzles, and at the moment, they are of about the same value."[417]

At the American Physical Society meeting of 1934, Patterson presented something of more value than crossword puzzles. He had come up with a mathematical function, A, that described the distance and direction between two scattering masses in a crystal. For a two-dimensional Fourier synthesis of a projection on the (100) face, the applicable form of the Patterson function was:

$$A(yz) = \sum_{k\,l=-\infty}^{\infty}\sum |F(0kl)|^2 \cos 2\pi\ (ky/b + lz/c)$$

where y and z are the coordinates of A, k and l are Miller indices, and b and c are the lengths of the unit-cell axes in the plane of projection.

The great advantage of the Patterson function was that it was related to F^2, the square of the structure amplitude, rather than F. As it is always a positive number, F^2 can be determined from intensity measurements, whereas the sign of F cannot be. In that sense, Patterson had solved the phase problem that provided the great obstacle to the use of Fourier methods in X-ray analysis. However, there was a great disadvantage to the Patterson method, too. A two-dimensional Fourier synthesis produces an electron-density distribution on a projection plane—literally a map of the atoms of the crystal. A two-dimensional Patterson map is a very different thing—a graph in which the position of a peak represents the vector between two atoms in a crystal and the weighting of that peak represents the total number of electrons in the two atoms (Figure 5.1).

A Patterson map contains all the information present in a Fourier map, but that information is scrambled. The position of a peak in a Fourier map represents the position of an atom relative to a corner of the unit cell; the position of a peak in a Patterson map represents only the distance and direction between two atoms that might be anywhere in the unit cell. The volume of a peak in a Fourier map represents the number of electrons in the atom at that position; the volume of a peak in a Patterson map represents the product of the number of electrons in the atoms contributing to that vector. Any set of intensity measurements could therefore be used to construct a Patterson map—but whether that map would

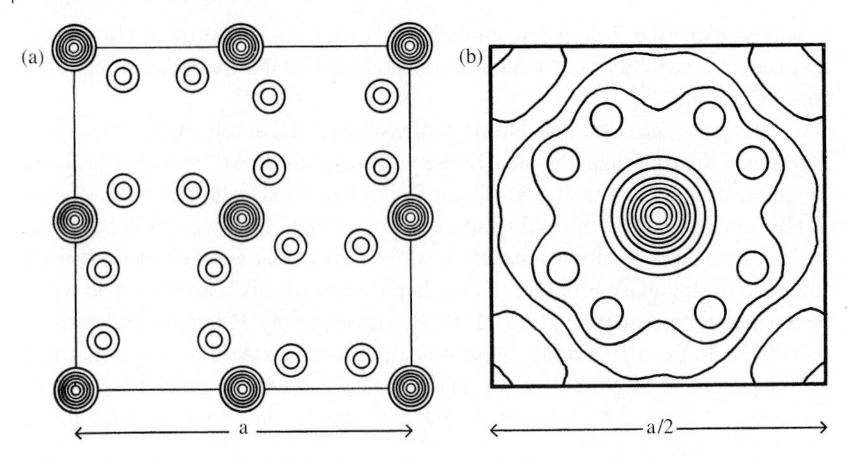

Fig. 5.1 Structure and Patterson map of potassium dihydrogen phosphate. (a) Electron-density map of KH_2PO_4, projection on (001) plane. Large circles are potassium or phosphorus atoms, small circles are oxygen atoms. (b) Patterson map of KH_2PO_4, projection on (001) plane. Reproduced, with permission, from Figure 2 of Patterson, A. L. (1935). A Fourier series method for the determination of the components of interatomic distances in crystals. *Physical Review* **46**, 372–6. Copyright 1934 by the American Physical Society

help solve the structure of the crystal was quite a different matter. Patterson's crossword puzzle was a cryptic one indeed.

In the publication that followed his talk in Washington, Patterson illustrated the new method with a map of the (010) projection of hexachlorobenzene (C_6Cl_6), using intensity measurements of ($h0l$) reflections made by another WHB protégé, Kathleen Lonsdale (neé Yardley). Making the reasonable assumption that the main peaks are due to Cl—Cl distances and lesser ones to Cl—C and C—C ones (the atomic numbers of Cl and C are 17 and 6, respectively), and knowing the space-group, Patterson could infer a structure with an outer hexagon of Cl atoms and an inner one of C atoms. This allowed him to assign the peaks in the map to specific interatomic vectors, and thereby to determine the bond lengths and angles in the plane of projection. These values were in good agreement with those of Lonsdale. Patterson concluded: "The final values for the interatomic distances must be obtained from a Fourier analysis or a parameter determination of the usual [trial-and-error] type but the approximate information provided by the F^2-series will eliminate a great many of the possibilities which normally have to be tested by trial and error."[418]

An important refinement to the Patterson method was introduced in 1936 by David Harker from Pauling's department at Caltech. Harker utilized the symmetry of the crystal to simplify the calculations involved.[419]

The Patterson method therefore seemed to simplify the trial-and-error method for simple crystals like hexachlorobenzene, with only two types of atom (or potassium dihydrogen phosphate, a crystal with three parameters also analyzed by Patterson), but it did not seem likely to unlock the full potential of Fourier methods.

A more modest but more practical advance in Fourier analysis of crystals came from Arnold Beevers and Henry Lipson, two research students in the Department of Physics at Liverpool University. In 1931, they visited Manchester for advice on X-ray methods. Bragg was "very helpful" and Will Taylor assisted them with the analysis of two crystals of high symmetry. Beevers and Lipson then decided to try to solve the structure of copper sulfate pentahydrate—the crystal with which Laue, Friedrich, and Knipping had first shown diffraction of X-rays. Even 20 years later, copper sulfate was still a difficult structure, because it belongs to the crystal system of lowest symmetry (triclinic) and its unit cell has a crystallographically unusual number (five) of water molecules. As they were able to locate the Cu and S atoms but not the O's by trial and error, Beevers suggested trying a two-dimensional Fourier analysis. For this, Bragg allowed them to spend a month in Manchester making quantitative measurements. If Bragg had a secret short cut for summing Fourier series, he did not confide it to Beevers, who, after summing the data for one point in the unit cell, estimated that the full synthesis would take 9 months.

Lipson then realized that factoring the Fourier summation into separate sine and cosine terms would greatly simplify the calculations—essentially, this considered each point on the electron-density map to be the sum of two one-dimensional electron distributions: One parallel to the x-axis, the other parallel to the y-axis. He and Beevers used trigonometric tables to calculate these terms for all relevant values of h and k and 60 different values of x/a and y/b (where x is a fraction of the a-axis length and y is a fraction of the b-axis length), and recorded these on strips of paper. It was then a relatively easy matter to multiply each F value by the corresponding sine and cosine terms and sum the series for that point in the unit cell.

Using Beever–Lipson strips, the Fourier synthesis for the (001) projection of copper sulfate took only 4 weeks. Because the numbers on the strips were independent of actual intensity measurements, they could be used for any crystal. Bragg arranged for Manchester University to lend Beevers and Lipson £200 to get their strips printed and mass-produced—a sum which they were soon able to recoup, as the strips became a standard tool of X-ray analysis until digital computers rendered them obsolete.[420]

Also in 1934, Bragg began to become actively involved in the public explanation of science, giving a series of six radio lectures on "Light."[421] At the year's end, he presented the Christmas Lectures "for a juvenile auditory" at the Royal Institution. At Alice's suggestion, the topic of the latter was "Electricity." These lectures for an audience of school children had been established in 1826 and had acquired a tradition of vivid demonstrations of scientific phenomena. With the assistance of William Kay, the laboratory steward he had inherited from Rutherford, Bragg came up with a series of lectures that more than lived up to this tradition. The theme of one lecture was that a dynamo is an electric motor "turned inside out." Bragg used a model to illustrate what happens inside the cylinder of a motorcar and compared the commutator, which reverses the direction of the current, to the carrot held in front of the donkey's nose. He magnetized a strip of silicon iron, which is particularly susceptible to weak currents, by holding it up in the magnetic field of the Earth. In a replica of the control room of a power station, Bragg used the motor to lift weights with a crane and as a dynamo to light a lamp when the mains electricity was turned off.[422] His son Stephen was pressed into service, being placed on a platform hung by a wire from the roof and pulled around by an electrified ebonite rod.[423]

In another lecture, Bragg combined his passions for art and science to demonstrate the difference between direct and alternating current. He held the negative lead from a battery against the corner of a piece of starch paper impregnated with potassium iodide and used the positive lead, which turned the iodide violet-black, to draw an elephant and an "ornamental fish"; when he used mains electricity, a dotted line was produced. A line drawn with alternating current for one second consisted of 46 dots, in reasonable agreement with mains frequency of 50 Hz. Bragg also demonstrated the principle of the transformer, and described how the Battersea Power Station sent out electricity at 132,000 V, which was gradually stepped down to 230 V for domestic use.[424] The success of these lectures must have made Bragg realize that he had a talent for demonstrating scientific principles to children—his lectures to lay, particularly young, audiences were always more highly acclaimed than those to audiences of university students.

Bragg seems to have enjoyed teaching but found some aspects of it frustrating. At a prize-giving ceremony at Leys School in Cambridge, he acknowledged that students subsequently forgot 99% of what they learned at school, but "it is an acquisition of good style and the ability so to arrange your thoughts as to secure the greater effectiveness."[425] His main concern about the educational system was the effect that examinations had on selection of students and on their attitude to learning. Bragg had doubts about the Higher School Certificate examinations that were used as the main criterion of university admission. On the one hand, they were national and therefore even-handed, but on the other no examination could determine a child's potential so well as personal knowledge could.[426] In his university teaching, he was discouraged by the inability of physics students to answer questions requiring originality or breadth of knowledge. He therefore contented himself with stressing fundamental ideas and reasoning: "The student who tries to make up for his lack of understanding by memorising sets himself a stupendous task and is easily surpassed by the man who has a grasp of essentials."[427]

Despite his success as a populizer of science, his administrative load and busy travel schedule, Bragg was not quite ready yet to give up trying to make new contributions to scientific knowledge. The question was which direction to go now. Metal physics never seemed to fully engage his interest, perhaps because it was too mathematical or because it was not as aesthetically satisfying as the beautiful patterns of the silicates. In the Bruce-Preller Lecture of the

Royal Society of Edinburgh given in February 1935, Bragg mentioned a novel area for X-ray analysis—biochemistry: "The X-ray investigation of structures produced by living matter has a very recent history, but it is perhaps the most interesting field of all." He gave as an example Astbury's "startlingly novel" and "highly significant" mechanism of stretching of the fibrous protein keratin, which exemplifies "the vistas which are being opened up."[428]

Proteins were known to be polypeptides, chains of amino acids linked head to tail. Studies dating back to the early 1920s had shown that fiber-forming proteins such as keratin can diffract X-rays. However, these exhibit periodicity only in one direction—parallel to the polypeptide chain. This suggested that the amino acids are organized in space such that similar atomic planes are situated the same distance apart. The *d* spacing corresponding to the most prominent reflection from α (unstretched) keratin—5.15 Å—led Astbury to propose that the polypeptide chain is folded into hexagonal loops.[429]

Most known proteins were globular, not fibrous. Many globular proteins—notably the blood protein hemoglobin—can form crystals and should therefore give diffraction patterns. Diffraction of X-rays by a protein crystal had, however, only been shown in 1934, when John Desmond Bernal obtained some crystals of the enzyme pepsin which were in their mother liquor (the solution from which the crystals had formed). By keeping these pepsin crystals wet, Bernal and his student Dorothy Crowfoot were able to obtain good diffraction patterns.[430] Interpreting these patterns, however, was another matter. By this time, it was generally believed that proteins are macromolecules; using his new ultracentrifuge, for example, Theodor Svedberg in Uppsala had reported that the molecular weight of hemoglobin was about 65,000 times that of a hydrogen atom—far, far higher than any substance successfully analyzed by X-ray methods.[431] As there were about 20 amino acids known to occur in proteins, the number of possible combinations of amino acids in a molecule as large as hemoglobin was almost infinite. If the amino acids were arranged in some sort of repeating pattern along the polypeptide chain, however, analysis of the structure would be much easier. Maybe the unit cell of a protein crystal would only contain a few amino acids, in which case X-ray analysis was at least conceivable.

In October 1935, Bragg took his first tentative step towards the X-ray analysis of macromolecules by writing to Albert Chibnall of Imperial College London and Emil Abderhalden at the University of Halle asking for samples of "proteins of known composition" and "di- and tripeptides," respectively. One of his students was using powder diffraction to study heteropoly acids such as phosphomolybdic and phosphotungstic acids, which Bragg thought might make "interesting compounds" with proteins and peptides.[432] Chibnall sent a sample of the milk protein casein,[433] but Abderhalden declined to part with his preparations.[434]

What of Bragg's deal with his father that precluded him from analyzing organic substances? This had become moot. Now in his seventies, WHB was no longer active in research, although his students, including Astbury, Bernal, and Lonsdale, were carrying his legacy forward. He had obtained the highest honors, in and beyond the world of science. In 1930, WHB was awarded the Copley Medal, the highest honor of the Royal Society. The following year, he became a member of the Order of Merit, which is limited to 24 members. In 1935, WHB became President of the Royal Society.[435]

Bragg may have been in a bit of a dilemma about which direction to take his career in general and his research in particular, but it was a very happy time in his personal life. Alice was pregnant again in the summer of 1935, so Bragg was "packed off" for a walking holiday with David Ritchie, an old friend from Cambridge who was now Professor of Philosophy in Manchester. They started from Killarney, walked on McGillycuddy's Reeks for a few days, then on to Waterville, Sneem, and Kenmare back to Killarney. Finding a place to spend the night was "a chancy business." On one occasion, they were refused accommodation at the local hotel, so they asked a woman at a shop if she knew of any other lodgings. She suggested the postmistress, but told them that if that did not work out they could stay with her, as her husband was away and it was a big bed! Bragg recalled this as "A memorable holiday because the country was so beautiful and fresh, although it was August, and the sea coast was magnificent." Bragg and Alice's fourth child and second daughter was born on September 11, and named Patience Mary.[436]

There was a good deal of turnover in Bragg's department in the mid-1930s. Will Taylor left in 1934, working with Bernal in Cambridge and WHB at the Royal Institution before becoming head of the Physics Department at the Manchester College of Technology. The following year, Joseph West went to Rangoon and Peierls to Cambridge. In 1936, Williams left and Bernard Lovell arrived from Bristol.[437]

Henry Lipson also joined Bragg's group in 1936. Lipson's first contribution was to assist Bragg in the development of a method for simplifying the trial-and-error method of X-ray analysis. This was based in part on the fact that there are far fewer two-dimensional lattices, or plane groups, than three-dimensional lattices, or space groups. Therefore, although a crystal can belong to any one of 230 space groups, a projection of that crystal onto a face of its unit cell has to belong to one of only 17 plane groups. Bragg and Lipson realized that each of these plane groups would correspond to a particular distribution of structure factor, or theoretical amplitude of diffracted radiation. For a primitive rectangular lattice, such as could result from the projection of a monoclinic unit cell onto its (001) or (100) face, the structure factor for $(hk0)$ reflections at a point with coordinates (x, y) will be $4 \cos(2\pi hx/a) \cos(2\pi ky/b)$.

For all combinations of low values of h and k, Bragg and Lipson plotted the value for structure factor on a graph with axes x/a and y/b and joined points of like value with contour lines. One could then take a transparent sheet with a projection of the proposed structure drawn upon it, superimpose it upon a structure-factor graph for, say, the (110) reflection, then read off from the contour lines the F value corresponding to that structure. Thus, for each of the three projections, the structure factors for each reflection could be compared with the measured intensity value. It was very unlikely that the structure factors and intensities would correspond at all well—this would be equivalent to guessing the structure immediately—but the great value of the graphs was that they showed which way the atoms would have to move in order to improve the agreement between structure factor and intensity. The name of the game was to come up with a structure consistent with symmetry that gave large structure factors for strong reflections and low structure factors for weak reflections. Bragg and Lipson therefore argued that the use of contoured structure-factor graphs would considerably facilitate the refinement of structures.[438]

In Alice's memoir, she wrote: "Looking back, I know that W.L.B. was very happy in the Alderley days. The look of a countryside was of great importance as a background to his life, the shape of trees, the stretch of water meadows below us in the plain, the rooks streaming home at night, counted for much."[439] However, he was still consumed with anxiety about forcing Alice to live in Manchester. In early 1934, Bragg corresponded with Bernal about the Jacksonian Professorship in Natural Philosophy at Cambridge, but seems not to have pursued the opportunity.[440] Founded in 1783, the Jacksonian was chronologically senior to the Cavendish Professorship then occupied by Rutherford, but far lower in prestige—the Cavendish Professor directed the Cavendish Laboratory and was head of the University Physics Department. As in 1929, Bragg may not have been able to bear the idea of being subordinate to Rutherford.

In 1937, a senior position in an attractive part of England finally was offered to Bragg—Director of the National Physical Laboratory (NPL) in Teddington, south-west of London. Bragg was "overjoyed" that Alice would be able to live in the south of England. Alice found Teddington to be "an unprepossessing dormitory town," but loved the Director's residence, Bushy House: "the most beautiful great Georgian house, overlooking the Park." Her positive verdict was "a tremendous relief" for Bragg.[441] Gwendy Caroe later wrote: "I believe that Willy snatched at the chance of getting Alice away from Manchester life to a lovely home at the NPL."[442]

Manchester was probably as sorry to lose Bragg as Adelaide had been to lose his father 30 years earlier. The University Senate passed a motion that, in addition to praising his achievements in research, read: "By his kindness and modesty, by the generosity of his appreciation of the work of others, and by the ease with which he could be approached, he made himself the trusted friend of his staff, his colleagues in research, and his students."[443]

Was it only to please Alice that Bragg abandoned his delightful life and successful career in Manchester to become a civil servant? In her memoir, Alice presents the move to Teddington as a career decision of her husband's, although she does admit that: "it was sometimes the wife who felt it [Manchester] was wet, dirty or too far away."[444] There were other factors. After 18 years in Manchester, Bragg felt that it may be time to move on. He was promised that the deputy director of the NPL would handle most of the administrative tasks. WHB, previously loath to give his son career advice, advised him to take the NPL position.[445] In addition, Bragg seems never to have overcome his disparaging view of life in the north of England. In February 1938, he wrote to Thomas Martin, General Secretary of the Royal Institution, whose nephew was planning to study in Manchester: "What, of course, your nephew will miss at Manchester is the social life. This is quite unique at Cambridge, and nothing in any provincial university can pretend to correspond. One cannot, in fact, deny that many of the students at a place like Manchester are rather 'rough-diamonds', indeed they must be when one considers their circumstances. I have always been impressed by the way in which the students overcome these disabilities, they are a very fine lot, but one must not count on the social side at all."[446]

6

Supreme position in British physics: The National Physical Laboratory and Cambridge, 1937–9

Bragg took up the position of Director of the National Physical Laboratory in November 1937, succeeding Joseph Petavel. He was not the NPL's first choice for the job; with his very limited experience of industrial research and abhorrence of administration, he was hardly the ideal candidate. Reg James, who earlier that year had left Manchester to become Professor of Physics at the University of Cape Town in South Africa, seemed to be unsure about the wisdom of Bragg's decision. He wrote: "Hearty congratulations and best wishes for the future in a difficult and strenuous job . . . I can't help a little regret too that another *real* scientific man is going to an administrative post."[447]

The NPL had been founded in 1899 with the mandate of standardizing instruments for physical investigation and determining and maintaining physical constants and standards of measurement. A site in Kew Gardens was initially proposed, but the new institution was eventually located in Bushy Park, Teddington, southwest of London. One of the first tasks of the NPL was the calibration of all clinical thermometers used in the United Kingdom, as well as barometers, chronometers, sextants, and other instruments of measurement. A wind-tunnel and a ship tank, holding 5000 tonnes of water, were constructed. In his history of the NPL, Edward Pyatt wrote: "By the 1930s the four main functions of the Laboratory had crystallised to be (a) assistance to industry in its immediate problems by advice and experiment, (b) research of longer range to open up new technical possibilities for industry, (c) the testing of instruments, and (d) the maintenance of the standards on which all physical measurements depend. . . ." It was organized into eight departments: Physics, Electricity, Radio, Metrology, Aerodynamics, Ship Research, Engineering, and Metallurgy.

Bushy House, where the Director of the NPL, like WHB at the Royal Institution, lived "above the shop," had been built in the seventeenth century by Edward Proger, a courtier of Charles II. After Proger's death in 1713, the house had a succession of aristocratic and royal occupants, including several Earls of Halifax, Lord North, William IV and his wife Queen Adelaide (after whom Bragg's hometown was named), and the exiled Duc de Nemours, son of

Louis-Philippe. It was returned to the Crown after Nemours' death in 1896 and obtained by the Government 5 years later for the use of the infant NPL.[448]

The top two storeys of Bushy House, used as the Director's residence, were renovated and redecorated to the Braggs' specifications and looked "very lovely indeed." The ground floor and basement, as well as several outlying buildings, were used by the laboratory. There was a tennis court and a garden of about 7 acres, beyond which lay the 1000-acre Bushy Park, where deer roamed and a magnificent avenue of chestnut trees led up to the house. Of the four Bragg children, only Margaret and Patience lived at Bushy House. They disliked the lack of privacy, but Margaret enjoyed riding a pony in the park. The boys had been sent to boarding school just before the move to Teddington: Stephen to Rugby, David to the Downs School in Worcestershire and then to Rugby.[449] Alice's brothers had all gone to Rugby; Bragg liked the headmaster and considered the science teaching "the best in England."[450] Alice spent much of her days writing articles in Queen Adelaide's former dressing room, including a history of Bushy House for "Harper's Bazaar."[451]

Of the three main activities he was involved in at Manchester, Bragg's order of preference was research, teaching, and administration.[452] At the NPL, there was little fundamental research, although Bragg may well have been interested in studies on the physical and mechanical properties of metals and alloys being conducted by the Metallurgy Department. There was no opportunity for teaching. Administration, Bragg's least favorite activity, occupied most of his time. According to Stephen Bragg, "There were a lot of committees, I can remember him saying, to get a little tarmac path up to the flagpole."[453] Bragg found the staff to be set in its ways: "many things were still being done which had long ceased to be useful." He also disliked the many formal occasions he was expected to attend.

The modest initiatives Bragg undertook at the NPL were unsuccessful. An attempt to bring eminent scientists in for lectures foundered for lack of an appropriate auditorium. Fearing German attacks as war loomed nearer, Bragg ordered bomb shelters built, a move he later regretted as a waste of precious resources. Even an invitation to report on sound-ranging, a subject still dear to his heart, turned out badly. Bragg found that the simple but effective system that had ended the First World War now "seemed to be hung around with as many gadgets as the White Knight." However, his critique was insufficiently diplomatic and the military authorities reacted defensively. Bragg's own verdict on his directorship of the NPL was harsh: "I have never felt proud of that year at the N.P.L."

Under the circumstances, it must have been a relief to get away. In the winter of 1937–8, the Braggs had a skiing holiday in Zell-am-See, Austria; in the spring, they holidayed at a farmhouse near Beaulieu, in the New Forest; that summer, they went to Studland in Dorset.[454] Their enjoyment of the "palmy times before the war" was tempered by a sense of *noblesse oblige*. In the summer of 1938, with war looming, the Dowager Marchioness of Reading asked Alice to help her in organizing a network of women to be involved in air raid precaution work. Alice agreed, and was given responsibility for northwest England.[455]

Bragg may have been "overjoyed" to get the NPL job, but there was one other that he would have preferred, both on professional and personal grounds—the Cavendish Professorship of Experimental Physics at Cambridge University. On October 17, 1937, before Bragg even took up his new appointment, this position became vacant when Ernest Rutherford suddenly died. The choice of a successor to Rutherford fell upon a group of "electors" that consisted of Henry Dean, the Vice-Chancellor of Cambridge University; Sir Frank Smith, Secretary of the DSIR; Ralph Fowler, Plummer Professor of Mathematical Physics at Cambridge; Charles Darwin, now Master of Christ's College; Henry Thirkill, President of Clare College; William Pope, still Professor of Chemistry at Cambridge; William Wilson, Professor of Physics at Bedford College London; Owen Richardson, Wheatstone Professor of Physics at King's College London; and Geoffrey Taylor, a Royal Society research professor at Cambridge. The electors met on February 11, 1938, and unanimously decided to offer the Cavendish Professorship to Bragg.[456] It is safe to assume that this decision was strongly supported by Darwin, Bragg's old friend and shipmate, and Pope, his "kind counsellor." Smith also was a supporter, although, as Bragg's boss at the NPL, "it's cutting off my nose to spite my face."[457]

The salary for the Cavendish chair, £1400 a year,[458] was probably considerably less than Bragg had been making at Teddington—the Director of the NPL earned £1500 per annum as early as 1918.[459] WHB opposed the idea, not only because of his positive view of the NPL position but also because he disliked Cambridge.[460] Although Alice would be sorry to leave her spectacular home and pampered lifestyle at Teddington, she would be closer to her parents and "was immensely relieved that W.L.B. was going to get back to university work, and above all research."[461] There seems little reason to doubt that the major factor in Bragg's decision to accept the Cavendish Professorship was that it was, as Edward Andrade put it, "the supreme position in British physics."[462]

Nevill Mott, a later Cavendish Professor, wrote: "Bragg was offered the Cambridge job by the electors, I have to assume because they [felt] the Cavendish needed a new line. Or did they just feel that Manchester should succeed to Cambridge, as York should (in the view of some) to Canterbury? I do not know, but I know of the tendency at that time of the nuclear fraternity to feel and express the view that what isn't nuclear isn't (fundamental) physics."[463] According to the low-temperature physicist Brian Pippard, "W.L. Bragg's election to the Cavendish chair of experimental physics in Cambridge was taken by many as a threat to the great tradition of fundamental physics research established by J.J. Thomson and, especially, Rutherford . . . The choice of a crystallographer, however distinguished, was a blow to many hopes."[464] In their biographical memoir of Norman Feather, one of the disgruntled nuclear physicists, William Cochran and Samuel Devons wrote: "The appointment of W.L. Bragg to succeed Rutherford was not popular with the nuclear physicists in Cambridge—it had been thought that [James] Chadwick might be appointed.

Bragg was a crystallographer, and that subject's second lease of life and crop of Nobel prizes lay well out of sight some 20 years in the future."[465]

In his history of the Cavendish Laboratory, James Crowther wrote: "The selection of an appropriate successor [to Rutherford] was difficult."[466] Under Rutherford, the Cavendish Laboratory had become a leading center for nuclear physics, with achievements such as Chadwick's discovery of the neutron and Francis Aston's development of the mass spectrograph. The selection of a non-nuclear physicist—indeed, someone who was felt not to be a physicist at all—could be seen as jettisoning Rutherford's legacy.

On the other hand, physicists in other areas of research had resented Rutherford's focus on the nucleus. Douglas Hartree, Bragg's former collaborator and now Professor of Theoretical Physics at Manchester, hoped that Bragg would be "more encouraging and sympathetic to theoretical aspects of physics than Rutherford ever gave the impression of being"—an opinion he claimed was shared by Patrick Blackett, who had succeeded Bragg as Langworthy Professor of Physics at Manchester University, and others.[467] The radioastronomer Edward Appleton, who was acting director of the Cavendish Laboratory, wrote to Bragg "You have always been my personal choice"[468]—rather disingenuously, as Appleton had aspired to the Cavendish chair himself. The anonymous editorialist of *Nature* made a telling point: "The Cavendish Laboratory is now so large that no one man can control it all closely, and Bragg's tact and gift of leadership form the best possible assurance of the happy co-operation of its many groups of research workers, while his brilliant lectures and personal charm ensure his success as a teacher of undergraduates."[469] In a similar vein, Bernal wrote to Bragg: "I think it will mean a great deal to Cambridge to have a man with a really broad view of Physics in its relations to other sciences and to practical things."[470] It would be difficult to satisfy all expectations.

Before leaving the NPL, Bragg replaced Rutherford in yet another position. In May 1938, he was elected Professor of Natural Philosophy in the Royal Institution. This position involved giving regular Discourses as well as afternoon lectures for University of London students, an innovation of WHB. Bragg had been giving occasional Discourses since 1920 and had given the Christmas Lectures in 1934–5, so his appointment to the vacant position was a natural one. It would give him a connection to the Royal Institution after his father's death.

Bragg took up the Cavendish chair in October 1938. Alice's father had found a house in West Road that belonged to Caius College. It had been rented to the Wollaston family: Mr Wollaston had been shot by a "demented King's undergraduate some years earlier." The Wollaston children hated to leave, and wrote "Death to the Braggs" in candlesmoke on the cellar ceiling. It was an early nineteenth-century cream brick house with a lead-and-slate roof, high ceilings, extensive outbuildings and an acre and three-quarters of garden. It was impossible to keep warm and expensive to run; Bragg and the children loved it, but for Alice it was a comedown from regal Bushy House.[471] Stephen Bragg recalls that 3 West Road was "on a very gracious scale, with a lovely big drawing-room, french windows into a nice garden, huge hall and some rather difficult bedrooms which one led out of another... It was a good house for entertaining in and had a lovely garden in which my father put a big herbaceous border. There was a greenhouse with a vine in it and a vegetable garden. The garden was in the charge of a full-time gardener, Mr. Fishpool."[472] In addition to the gardener, the West Road house was serviced by a cook, three maids, and a nanny, Hilda, who had been with the family since their time in Manchester. With Margaret and Patience both in school, Alice had time to spare for more volunteer activities. .She was assigned by Lady Reading to start a Cambridge branch of the Women's Voluntary Service.[473]

The Cavendish Laboratory that Bragg took over had three main research areas: Nuclear physics, which was run by Philip Dee and Norman Feather; the Mond low-temperature laboratory, run by John Cockcroft; and atmospheric research, run by Appleton, since 1936 the Jacksonian Professor of Natural Philosophy. Another important figure was Ralph Fowler, Plummer Professor of Mathematical Physics. More along Bragg's line was Paul Ewald, who had left Stuttgart in 1937 in disgust at the Nazi prohibition on teaching "Jewish physics." Bragg helped Ewald, part-Jewish himself and married to a Jew, to get a temporary position in Cambridge. Bragg also found that George Crowe, the lab boy of his undergraduate days, was still at the Cavendish—somewhat damaged, as he had lost a finger and part of his hearing as a result of exposure to X-rays and radioactivity.[474] Crowe found it hard to think of Bragg as other than a student, continuing to refer to Rutherford as "The Prof."[475]

Shortly after Bragg arrived, Appleton left to take over the Department of Scientific and Industrial Research, vacating the Jacksonian chair. Cockcroft was elected, and Jack Ratcliffe took over the atmospheric research. Bragg brought with him from the NPL (and before that from Manchester) Albert Bradley and Henry Lipson. Bradley became Assistant Director of Research at the Cavendish, a post formerly held by Bernal. Egon Orowan, another metal physicist, soon joined Bragg's research group on X-ray crystallography and metal physics.

Bernal, who had run the crystallography laboratory, had left to take up the chair of physics at Birkbeck College London, vacated when Patrick Blackett went to Manchester. Most of Bernal's group went with him to Birkbeck, but he could not find the money to support a Ph.D. student, Max Perutz. Bragg agreed to take over the supervision of his project. Perutz, who was working on the X-ray analysis of the blood protein hemoglobin, waited 6 weeks for Bragg to visit the crystallography lab. Finally, he took his X-ray photographs of hemoglobin to the Professor's office.

It would not have been entirely surprising if Bragg had regarded the X-ray analysis of hemoglobin as an act of colossal folly. Not counting weakly diffracting hydrogen atoms, the largest molecule whose structure had been solved—the plant pigment phthalocyanine, $C_{32}N_8H_{18}$—has 40 atoms, described by 30 parameters. Hemoglobin has *5000* atoms with *7500* parameters. And phthalocyanine was, as Bragg put it, "a very special case." In fact it was an ideal case—it is a four-fold symmetrical molecule, hence the low number of parameters; a nickel atom can be added to the molecule without affecting the structure

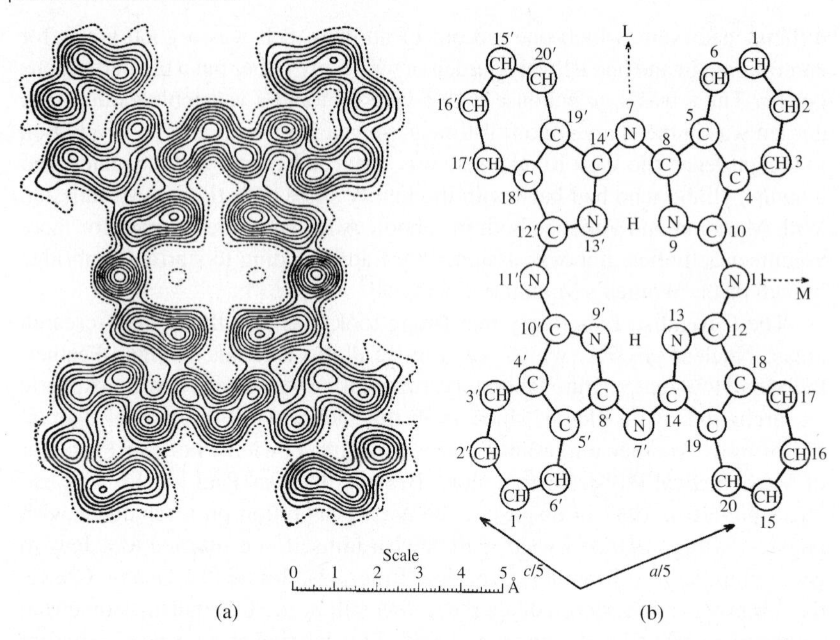

Fig. 6.1 Structure of phthalocyanine. (a) Electron-density map, based on Robertson's Fourier analysis. (b) Atomic locations inferred from (a). Reproduced, with permission, from Figure 2 of Robertson, J. M. (1936). An X-ray study of the phthalocyanines. Part II. Quantitative structure determination of the metal-free compound. *Journal of the Chemical Society,* 1195–209

(the two forms are isomorphous); and, best of all, the nickel atom lies at the center of symmetry of the unit cell (Figure 6.1). This means that in the nickel-substituted form of phthalocyanine almost all F values are positive, thereby solving the phase problem and making the structure amenable to Fourier analysis. Such luck could not be expected for hemoglobin. To begin with, protein crystals could not contain centers of symmetry. Bragg's trial-and-error method would be completely hopeless for a giant molecule in which every atom lay in a general position, and the phase problem presented an insuperable barrier to the application of Fourier methods. As Perutz later recalled: "At that time all my colleagues in the Crystallographic Laboratory thought me mad for choosing so unpromising a subject for my research."[476]

Not only was the project wildly ambitious, but there was some doubt as to Perutz's abilities. In the spring of 1938, Bernal had consulted Bragg about a paper of Perutz's that had been rejected by the *Proceedings of the Royal Society*. Bragg wrote back: "I was rather disappointed in it; it seemed to me that it was too discursive and superficial."[477]

Nonetheless, on seeing Perutz's photographs of hemoglobin, Bragg was hooked: "Some fortunate intuition made me feel that this line of research must be pursued, although it seemed absolutely hopeless to think of getting out the structure of so vast a molecule."[478] According to Perutz, "He realized at once the challenge of extending X-ray analysis to the giant molecules of the living cell."[479] The X-ray diffraction pattern of hemoglobin contained hundreds of spots—including those at low d-spacings, corresponding to atomic planes only an Ångstrom unit or so apart. As Bernal put it in a January 1939 Royal Institution Discourse on "Structure of Proteins," "This indicated that not only were the molecules of the proteins substantially identical in shape and size, but that they had identical and regular internal structures reaching right down to atomic dimensions."[480] Bragg knew that Perutz's photographs contained all the information needed to solve the structure at atomic resolution, if only that information could be interpreted correctly. He had been seeking a new research challenge, and this was to be it; as he told Robert Olby in 1967, "I found that Perutz had taken these wonderful photographs with Bernal of protein crystals; I was thrilled about them and formed the ambition to get out as a final act in my X-ray analyst's life something as complicated as a protein."[481]

Bragg's reaction to Perutz's project illustrates an aspect of his approach to research that is perhaps not obvious—his willingness to take chances on projects that seemed hopeless. In the early 1950s, Bragg told André Guinier and others that he used to give newcomers a set of "golden rules." Guinier remembered two of these as being "Never follow the fashion" and "Never be afraid to carry on an experiment which is declared stupid by the theoreticists [sic] of the laboratory."[482] As Aaron Klug put it, "Bragg, although being, in a sense, conventionally Edwardian . . . was a man who took bold decisions."[483] Francis Crick told Horace Judson: "Boldness? I would have said that Bragg and Pauling were the people who influenced me, and both had that characteristic."[484] In supporting Perutz's X-ray analysis of hemoglobin, Bragg was going further out on a limb than he ever would in his career.

Perutz was an Austrian Jew who had been educated at the University of Vienna before joining Bernal's laboratory in 1936. He became interested in hemoglobin the following year during a visit to Prague, where the husband of his cousin showed him how crystals of the protein changed from trigonal to monoclinic when oxygen was added. Perutz's stay in England had been supported by his parents, but, following the *Anschluss* with Germany in 1938, they had become refugees and Perutz's funding had ceased. As a foreign national, he was not eligible for jobs for which a qualified Briton was available. Bragg suggested that Perutz draw up a research plan for the Rockefeller Foundation, which had become a major supporter of biological research during the 1930s.[485] In his covering letter, dated November 28, 1938, and addressed to Wilbur Tisdale at the Rockefeller International Educational Board in Paris, Bragg wrote: "I wish to take a part in this research myself. My research team at Manchester was largely responsible for making the first extension of X-ray analysis from very simple to complex crystalline patterns, and I should now like to have a share in extending it still further to the very complex patterns of proteins." He requested £275 per year for 3 years for Perutz's salary and £100 for a new X-ray tube.[486]

Perutz's "Programme of Research" stated that differences in the unit cell and space group between hemoglobins of different species would be determined; comparison of methemoglobin, oxyhemoglobin, and reduced hemoglobin[z] used to reveal structural changes accompanying oxygen binding; and that the intensities of all reflections from the three principal zones would be measured and used for Fourier syntheses.[487] Quite a bargain for less than £1000. As Bragg, at least, knew, a Fourier analysis of hemoglobin was going to take a lot longer than 3 years—if indeed it could be done at all.

Whether or not the Rockefeller Foundation realized the gigantic scope of what Bragg and Perutz were proposing, it agreed to provide the requested funding starting January 1, 1939. Perutz had officially embarked upon a 30-year odyssey to find the structure of hemoglobin; like the hero of Homer's epic, he would need both guile and endurance to succeed. In the end, the crystallography of macromolecules was to prove one of the great achievements of what Warren Weaver, head of the natural sciences division of the Rockefeller Foundation, was to call "molecular biology."

In the autumn of 1938, as the support of the Rockefeller Foundation was being requested, Bragg's attention was also drawn to proteins by his involvement in a controversy over the cyclol structure of proteins. Dorothy Hodgkin (née Crowfoot), Bernal's former student and now at Oxford University, had earlier that year published an X-ray analysis of the protein hormone insulin.[488] This included a Patterson analysis, from which Hodgkin drew no dramatic conclusions. Nonetheless, Dorothy Wrinch, an Oxford mathematician who was, like Hodgkin and Bernal, a member of an informal scientific/political discussion group known as the "Biotheoretical Gathering," felt that the Patterson analysis of insulin supported her "cyclol" theory of protein structure. This theory stated that proteins are two-dimensional sheets formed by the folding of the polypeptide chain into rings containing various numbers of amino acids.[489] On at least two occasions, Wrinch sent Bragg information about the cyclol theory,[490] but he does not seem to have responded. Both Bernal and Hodgkin were extremely worried about Wrinch's overinterpretation of the insulin data, and lobbied Bragg to use his influence.[491]

In January 1939, *Nature* published letters critical of the cyclol theory by Bragg, Bernal, and John Monteath Robertson. Understandably—as he knew little about protein structure—Bragg was more critical of Wrinch's grasp of the principles of X-ray analysis than of the inherent implausibility of her ideas. As he pointed out, the Patterson map of even a moderately complex crystal such as an aromatic molecule could not be interpreted—even if the structure were known. "Exaggerated claims as to the novelty of the geometrical method of approach and the certainty with which a proposed detailed

[z] Oxidation of normal (reduced) hemoglobin converts the iron atoms from the Fe^{2+} to the Fe^{3+} oxidation state. The latter form of the protein is called methemoglobin and cannot bind oxygen. Oxyhemoglobin is the oxygen-containing form of reduced hemoglobin.

model is confirmed are only too likely, at this stage, to bring discredit upon the patient work which has placed the analysis of simpler structures upon a sure foundation."[492]

Wrinch must have been upset to be publicly disowned by the Biotheoretical Gathering, but what did more harm to the cyclol theory was a devastating critique published in 1939 by Linus Pauling and Carl Niemann.[493] For Pauling, like Bragg, had become interested in protein structure. In the early 1930s, Warren Weaver told Pauling that the Foundation had no interest in the sulfide minerals he was studying, but was investing heavily in biochemistry. Pauling submitted a proposal for studies on the magnetic properties of hemoglobin, which the Foundation agreed to fund. In April 1939, Pauling wrote to Bragg: "We have structure investigations of crystals of several amino acids and peptides under way in the laboratory, and hope to give this field a thorough examination during the next few years."[494] The "we" included Robert Corey, who had come to Caltech in 1937 from Ralph Wyckoff's laboratory at the Rockefeller Institute for Medical Research. Pauling and Corey decided that the best way of determining the structure of proteins was to use the "bottom-up" approach of measuring bond lengths and angles in crystals of amino acids and simple peptides. Fate had brought Bragg once more into competition with the world's leading chemist.

Bragg also faced some competition closer to home. In May 1939, he wrote to Hodgkin about the intention of her student, Dennis Riley, to perform a Patterson analysis of Perutz's hemoglobin diffraction patterns: "As you may know, Perutz's work is being financed by the Rockefeller Foundation in Paris, and I am responsible to them for its direction. Both because of this responsibility and because I take a great interest in the work, I should like to keep the direction of it in my own hands."[495] Hodgkin agreed with Bragg's suggestion that Riley should do the Patterson analysis only as a collaboration with Perutz.[496]

In his letter to *Nature* about the cyclol hypothesis, Bragg had stated that although the structure of insulin is "far beyond anything as yet analysed by X-rays" its analysis is not "a hopeless task"; in fact "there is every hope of ultimate success."[497] In his accompanying letter, Robertson, who had in 1936 used Fourier methods to solve the structure of phthalocyanine, noted that "The [insulin] molecule does, however, contain a few zinc atoms, and if these could be replaced by mercury, as has been suggested, a very profitable study might ensue."[498] In other words, if the zinc- and (hypothetical) mercury-containing forms of insulin were isomorphous, the phases of the reflected X-rays might be determined. This is because the effect of a heavy atom like zinc (atomic number 30) or mercury (atomic number 80) might be sufficient to make the phases of all, or almost all, reflections positive—as in the case of the superimposed Ca and Mg atoms in the (010) projection of diopside. A similar idea was expressed by Bernal in his above-mentioned Discourse. In the X-ray diffraction pattern of a protein, "we can never know the phases of the reflections corresponding to the different spots. The ambiguity introduced in this way can only be removed by some physical artifice, such as the introduction of a heavy atom, or the observation of intensity changes on dehydration, which have not hitherto been carried out in practice."[499]

Few people in Britain now doubted that war with Germany was only a matter of time. Cockcroft realized that scientific know-how would be a crucial factor in Britain's war effort, and circulated to the heads of armed services research a Central Register of Scientists that had been prepared by the Royal Society. This listed about 12,000 scientists, one-tenth of whom were physicists, together with their areas of expertise. The response to this initiative was discouraging, as the services estimated that only about a dozen scientists would be needed. Undeterred, Cockcroft recruited Bragg and embarked upon a tour of military research establishments equipped with a list of about 100 outstanding young university physicists. The only establishment interested in the list was the radar research station at Bawdsey, near Felixstowe. About 80 young physicists spent the month of September 1939 working at coastal radar stations. When war came, most of them ended up doing radar research—notably John Randall, who worked under Mark Oliphant, Poynting Professor of Physics at the University of Birmingham. Bragg believed that his and Cockcroft's efforts "gave us at least a year's start."[500]

The Cavendish Laboratory was severely overcrowded but help was in sight. In 1936, the car manufacturer Herbert Austin had donated £250,000 to the Laboratory—some of this money was intended for a high-tension laboratory and cyclotron, some for renovation of the Cavendish block, some for an endowment fund; the remainder, £80,000, for the construction of a new wing. By 1938, the Austin Wing had been planned but not built. With war on the horizon, Bragg only got permission to go ahead with the project by agreeing that the wing would be used for military purposes during hostilities (it was used by an Army ballistics unit and a Navy signals unit). When the building was well under way, Bragg was very embarrassed to get a letter from Austin asking when he could lay the foundation stone. The architect, Charles Holden, suggested that the "foundation" stone be in a yet-to-be-built wall flanking the entrance steps. After the ceremony, in May 1939, Austin complimented him on this original concept, but Holden let the cat out of the bag by saying "Oh, you heard about that, did you? We forgot all about the bally thing."[501]

The Austin Wing was completed in June 1940. It was 115 feet long by 45 feet wide, with four floors and a basement—a total floor area of 34,000 square feet. A central corridor ran the length of each floor, with research rooms typically 15 feet by 17 feet opening off the corridor. The external and corridor walls supported the floors, so that internal walls could easily be moved. To facilitate this, the services came up a shaft next to the elevators and ran along the corridor above the doors. The second floor contained offices, a museum bay, tea-room, and library. The colloquium room seated 70 and there were various workshops and storerooms, but no teaching facilities. The building cost £77,000, equipment £10,000. Bragg was allowed to spend £4500 on "magnificent" furniture and fittings; these were stored in the library for the duration of the war.[502]

7

He will have to be Sir Lawrence: World War Two

In June 1939, Bragg and Alice went to the French Alps, hiking and looking for wild flowers. Later that summer, the family spent a holiday in Wales. When war broke out in September, fearing that Cambridge would be bombed, Bragg and Alice sent the children and Nanny Hilda to stay with family friends in Herefordshire.[503] Bragg was on the Commonwealth Fund Fellowship Committee, by which young people spent a year or two in the United States, and were encouraged to tour the country. Committee members took turns to go to the United States to report on the fellows. Bragg's turn came in 1939 and he and Alice had made all arrangements to go when war broke out and the trip was canceled.[504]

Most Cavendish researchers immediately left for war work. The physics departments of Queen Mary College, Bedford College, and St Bartholomew's Hospital Medical School were evacuated from London to Cambridge, so the four institutions pooled their depleted teaching resources. Harold Robinson, head of physics at Queen Mary and a former First World War sound-ranger, "billeted" with the Braggs. George Searle, now almost 80, was recalled from retirement to teach the practical classes in physics, replacing Jack Ratcliffe. A radar lab was set up, and the Mond was used for research for the Ministry of Supply. Brian Pippard and David Schoenberg came up with a laboratory system for simulating the firing of artillery shells, which proved "a real winner."[505] The Cavendish housed special electronics classes for radar operators, and a top-secret research unit on nuclear energy, outwith Bragg's control. John Nye, a research student with Orowan during the Second World War, recalled Bragg saying: "Don't tell me any secrets, then I don't have to remember what is secret and what is not."[506]

Despite the failure of Bragg's critique of sound-ranging at the NPL, he was again called upon as a consultant. Just before the German invasion, Bragg went to France to inspect a sound-ranging station. A small inn near the section was "festooned with beribboned bottles of wine, dedicated by young men who had left for the front to the celebration they would have on their return. Two weeks later all this country had been overrun by the Germans." As Bragg's plane was about to take off on the return journey, mechanics ran out to stop it. It turned out that a colonel returning from India had loaded the luggage compartment with all his belongings while no one was looking, and this extra weight had damaged the rear wheel.

Alice now became very busy. The maids and gardener were called up. The Cambridge WVS was given responsibility for running a canteen for service personnel, collecting salvage, billeting evacuated children, and other war-related activities, and Alice soon "had an organizer in every street of any importance in Cambridge." After the fall of France, a friend in Canada offered to take in the Bragg children. Alice was opposed because she feared that the children would become strangers to them and it would be bad for morale if the head of the WVS took advantage of an opportunity not available to others.[507]

Although he did not know it, the situation was one of some danger for Bragg. His name appeared on a Gestapo list of 3000 Britons to be arrested in the event of a German invasion.[508] As a Jew, Perutz would doubtless have fared worse than Bragg in the event of a German occupation. Because of his Austrian citizenship, however, he was considered an enemy alien by the British authorities. In May 1940, Perutz was taken to a detention center and then shipped to Canada. It was January 1941 before he was returned to Britain and released.

Bragg became directly involved in war-work in the autumn of 1940, when he was asked by the Ministry of Supply for help in detecting landmines and buried bombs. A magnetic device produced at the Cavendish was found to be inferior to designs from the NPL and the Electrical Research Association.[509] However, the Rifle Range at Cambridge proved useful for testing the ability of these probes to detect bombs.

In the New Year's Honors List of 1941, Bragg was awarded a knighthood. WHB wrote to his sister-in-law Lorna Todd: "Isn't that fine? . . . He will have to be Sir Lawrence; we can't have confusion worse than ever. I am so very glad for his sake. In spite of all care, people mix us up and are apt to give me a first credit on occasions when he should have it: I think he does not worry about that at all now, and will never anyhow have cause to do so now. I think I am more relieved about that than he is."[510] WHB was no doubt correct in saying that his son no longer worried about credit for their joint work in 1912–14; his successes with the silicates and the order–disorder transition had been completely independent of WHB, and he had now achieved a job more prestigious than his father's.

That year, Bragg got a real war-job. Edward Appleton, Secretary of the Department of Scientific and Industrial Research and his former student, asked Bragg to head the British scientific liaison office in Canada. Charles Darwin, Bragg's successor as Director of the NPL, was going to Washington as Director of the British Central Scientific Office, so Bragg traveled with him and his family. They sailed from Liverpool on a Baltic trader which had half a dozen passenger cabins and the hold converted to a number of other cabins, and barely enough bunker space for the transatlantic trip; part of a convoy accompanied by destroyers. Bragg and Darwin occupied the grand deck cabin, which had one of only two bathrooms aboard. The other passengers included some others on official missions, aircrews who had flown bombers across to Britain, civil servants going to work in the United States and a party of 17 "mannequins" who were trying to get clothing orders from South America for the war effort. "Their clothes were all being sent in another ship, which seemed to us odd because this was clearly a case where it was desirable to have all the eggs in one basket." The convoy had no problems with submarines, but was bombed on several occasions.

In Ottawa, Bragg was based at the headquarters of the National Research Council, which was responsible for almost all military research in Canada. He stayed in the Rochester Hotel and walked a mile and a half to NRC along the Ottawa River. Behind the "C.P.R. hotel" (Chateau Laurier), and near where the Rideau Canal enters the river, he found an excellent spot for bird-watching. He shared an office with Allen Shenstone, a Canadian who was Professor of Physics at Princeton, and was attached to the NRC to liaise between Canada and the United States. Shenstone and his wife Molly often entertained Bragg in the evenings.[511] Bragg also worked closely with Chalmers Mackenzie, Director of Scientific Research at NRC; during the previous war, the two men had met briefly at a sound-ranging course on Vimy Ridge.[512]

Bragg regularly visited military research facilities in Toronto and Montreal, and made occasional trips to smaller centers in eastern and central Canada. At most stops he gave talks about the role of scientists in wartime and in the post-war period. He also gave several radio broadcasts. Bragg made regular visits to Washington to consult with Darwin and also visited Boston, New York, Fort Bragg in North Carolina, and Fort Monmouth in New Jersey. In May 1941, Bragg and Shenstone visited the United States Naval Research Laboratory, which led to a report on "Theory of Armour Plate Penetration."[513]

The following month, Bragg traveled by rail to Vancouver via Winnipeg and Edmonton: "The big towns, each with its large railway hotel, are like liners in the ocean, compact settled places with miles of open country around them." Bragg was disappointed by the Rockies: "The mountains in the distance are grand, but there is no exciting foreground, just pine trees growing out of gravel." Beyond the Rockies, he liked the "cowboy country" and the coastal range: "The country is wet and lush, the woods full of flowers and ferns—a kind of Devonshire with a backdrop of snow mountains." In Vancouver's Stanley Park, "There seems to be as much wood as air as one walks through the forest threading a way through gigantic trunks of hemlock and Douglas fir." He took the boat over to Vancouver Island, which was "a paradise for birds and flowers" and visited the bishop, who was an old schoolmate. Bragg returned to Ottawa via Lake Louise and Calgary. He took a "charabanc" from Lake Louise to Banff and was impressed by the abundance of roadside wildlife—until he found out that the park authorities put out salt to attract the animals.[514]

On his return to Ottawa, Bragg wrote to Appleton with his "general impressions of the work here." In the Physics section of the NRC, the important areas were radio, naval research, and optics. The main radio project involved designing a mobile unit that would use radar echoes from planes to direct the fire of anti-aircraft guns. The naval research included the development of a device to check ships for demagnetization. Overall, Bragg felt that this was an unnecessary duplication of work being done in Britain. The main achievement of the NRC's Biology Section was a method for non-refrigerated transport of bacon. In Toronto, work was being done on a proximity radio fuse and on aviation medicine, including development of oxygen masks and studies on the physiological effects of altitude, cold, and *g* forces. The aviation research was being hindered rather than helped by the Nobel Prize-winning physiologist, Charles Best, who Bragg considered a menace: "The trouble appears to be that he honestly believes himself to be the world's greatest physiologist, and that he is justified in pulling any strings or doing anything behind the scenes to effect schemes which he judges to be right." Personality problems also bedevilled radio research at McGill University. The west was "a virgin field and one that ought to be exploited" by moving their best men to Ottawa, but "It has not yet been possible to make the Treasury realize that the money spent on research is very small compared with the money which may be saved on production." In general, "I have concentrated on a side of the work which I felt best equipped to do, what one might call the diplomatic side of my functions." Six months was enough and it was now time to bring someone else over. For the future, it would be better either to build up a large organization and use it as a channel of communication between the countries, or else have experts traveling back and forth to handle technical issues while the liaison officer acted as a "diplomatic representative." Bragg favored the latter, as "the liaison officer is a good safety valve."[515]

The letters that Alice wrote to Bragg during the 5 months he spent in Canada illustrate the difficulties she faced in running the household alone in wartime. Of course, these difficulties were only relative—as a wealthy woman, Alice was able to keep her children in some comfort. However, the war was causing servant problems of quality and quantity. Alice was down to two maids from the pre-war figure of three,[516] and those two were not giving satisfaction: "I agree with Bell [the gardener] that maids as a class don't know there is a war on at all."[517] Nellie the cook was "lazy" and "*so* cross with the others."[518] Alice also discovered that the nanny had not been doing as good a job with the children as she had believed: "Honestly I don't think Hilda has brought them up very well (I accept full responsibility for that); they have been allowed to chuck everything about and not take any responsibility for things . . . I am not half as good as you with children I have come to this conclusion. I am much too impatient and get worried that their things will get lost."[519] Being Lady Bragg was proving a mixed blessing: by June, Alice was serving on 23 different committees.[520] There were food shortages and Alice billeted military personnel, including the former Olympic runner Guy Butler.

The 1941 correspondence also illustrates the depth of feeling between Bragg and Alice, now separated for the longest period of their marriage. From afar,

Plate 1 Willie (left), Gwen, and Bob Bragg in Adelaide (undated, c.1895). The Royal Institution, London, UK/Bridgeman Art Library (TRI205278)

Plate 2 Bragg (front row, second from right) at St Omer, France, in 1917. Reginald James is standing behind Bragg. The Royal Institution, London, UK/Bridgeman Art Library (TRI205272)

Plate 3 Bragg at the Solvay Conference on Chemistry in Brussels in April 1925. The Royal Institution, London, UK/Bridgeman Art Library (TRI205283)

Plate 4 Paul Ewald, Charles Darwin, Heinrich Ott, Bragg, and Reginald James (left to right) at Holzhausen, Bavaria in September 1925. AIP Emilio Segré Visual Archives, Fankuchen Collection

Plate 5 Gwendy Caroe (right), Bragg, and Alban Caroe at Watlands in April 1932. The Royal Institution, London, UK/Bridgeman Art Library (TRI205294)

Plate 6 Bragg and Alice (undated). The Royal Institution, London, UK/Bridgeman Art Library (TRI205293)

Plate 7 Stephen, Olga Hopkinson, David, Albert Hopkinson, Patience, Bragg, and Margaret at Windy Howe in the summer of 1937. The Royal Institution, London, UK/Bridgeman Art Library (TRI205292)

Plate 8 Margaret, David, Stephen, WHB, Patience, and Bragg at 3 West Road in 1939. The Royal Institution, London, UK/Bridgeman Art Library (TRI205290)

Plate 9 Bragg in his office at the Cavendish Laboratory (undated). The Royal Institution, London, UK/Bridgeman Art Library (TRI205299)

Plate 10 Conference on "Structure of Proteins" at Pasadena in September 1953. Bragg is at the center of the front row, Linus Pauling is third from the left. The Royal Institution, London, UK/Bridgeman Art Library (TRI205273)

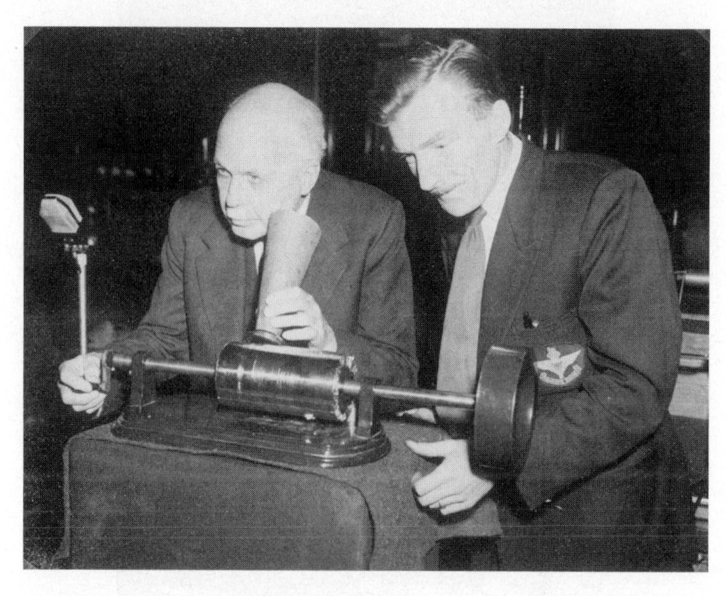

Plate 11 Bragg and Bill Coates (undated). The Royal Institution, London, UK/Bridgeman Art Library (TRI205279)

Plate 12 Members of the Davy–Faraday Laboratory in July 1954. The Royal Institution, London, UK/Bridgeman Art Library (TRI205285)

Plate 13 Bragg and Alice at Camp Pretoriuskop, South Africa in September 1955. The Royal Institution, London, UK/Bridgeman Art Library (TRI205289)

Plate 14 Bragg (center) and Prince Philip (second from left) at the Royal Society tercentenary exhibition in July 1960. The Royal Institution, London, UK/Bridgeman Art Library (TRI205277)

Plate 15 Bragg's sketch of Max Perutz. Archives of the MRC Laboratory of Molecular Biology, Cambridge

Plate 16 Bragg's sketch of John Kendrew. Archives of the MRC Laboratory of Molecular Biology, Cambridge

Plate 17 Bragg birdwatching in Suffolk in 1965. The Royal Institution, London, UK/Bridgeman Art Library (TRI205284)

Plate 18 Bragg in the garden at Quietways in 1965. The Royal Institution, London, UK/Bridgeman Art Library (TRI205297)

Plate 19 The celebration of the 50th anniversary of Bragg's Nobel Prize held at the Royal Institution on October 15 1965. Alice and Bragg are in the front row. Sir George Thomson is on the left of the second row, with Dorothy Hodgkin beside him. The Royal Institution, London, UK/Bridgeman Art Library (TRI205298)

Plate 20 The Nobel Prize ceremonies in December 1965, at which Bragg gave a special lecture commemorating the 50th anniversary of his award. Front row (left to right): Julian Schwinger (physics), Mikhail Sholokhov (literature), Bragg. Rear row (left to right): Richard Feynman (physics), Robert Woodward (chemistry), Jacques Monod, André Lwoff, and François Jacob (physiology or medicine). Sin-Itiro Tomonaga, who shared the 1965 physics Prize with Schwinger and Feynman, did not attend the ceremonies. The peace Prize was awarded to the United Nations Children's Fund. The Royal Institution, London, UK/Bridgeman Art Library (TRI205280)

Alice tried to provide the emotional support that she must have done when Bragg was home. On April 27, 1941, she wrote: "I do hope you are all right and not worrying or spinning. If only you did not spin *what* more could you not do I often wonder."[521] Characteristically, Bragg was plagued with doubts about his ability to succeed in his mission. His predecessor, Ralph Fowler, had been much admired by the Canadians, and Bragg thought himself "a very pale person in comparison." On May 11, Alice wrote: "I was sorry to hear you had lain in bed thinking how complicated the work was going to be and could you do it. Why *do* you work like that? If you'd believe in your self you could do anything."[522]

To Margaret and Patience, Bragg sent amusing illustrated letters. According to Patience, "He was . . . very good at painting and drawing and the letters had glorious little pictures of beavers and even little bits of gnawed beaver wood, I mean, really extraordinarily imaginative. He had this knack of writing a letter that had the things you really wanted to know about, amusing and chit-chatty as if you were there. It's not 'I've done this and I've done that', but 'you would have been thrilled to see . . .' and then these lovely pictures of animals and birds and amusing little anecdotes."[523]

In August, Alice took the children on holiday to Devon. Her letters to Bragg from there are more contemplative than those written amid the hectic domestic routine of Cambridge. The uncertainties of war had made Alice more appreciative of her lot in life: "Even if something parts us for ever I would always feel I had been given more than most women—I think of funny things—watching birds with you at Turnberry, frying sausages on that wet first night on the Broads when I was nearly doubled up with Stephen inside me, you helping me be sick in that funny boat going to Russia . . . and the first time I saw the Alps on our honeymoon and hundreds more, a great long string of happiness and adventure."[524] She had also developed a social conscience: "After the war I'm sure everyone will have equally decent houses with proper lavatories and hot water supplies and food. Altho' we complain that we have not enough food the poor have for ages had to give their children far less. It is all very unfair."[525]

At the end of September, Bragg returned to England by plane from Gander, Newfoundland. His first trans-Atlantic flight was in a bomber fitted out to carry six passengers, two in the tail (including him) and the rest in the bomb-bay. The plane was unheated, so the passengers wore a one-piece flannel garment with a breathing slit. As Bragg later wrote to Mackenzie: "We flew at 21,000 feet, and by an oversight the crew forgot to supply us with oxygen. I therefore had an opportunity of studying at first hand the interesting effects I had been told about when visiting aviation medical research centres. It was too much for two of the passengers, who passed into a state of coma. However, directly we came down again recovery was very rapid, and except for a headache which passed off after an hour or two I felt quite fit."[526] Air travel was still a novelty: When Bragg mentioned to a fellow passenger on the train to Cambridge that he had been in Canada the day before, it created a sensation.[527]

It seems that Bragg was relieved to feel that he had "done his bit." Shortly after his return to England, he wrote to Mackenzie of the NRC: "I feel most grateful for having been given this war job. I could not have asked for anything which was more interesting, and more worth putting all one's best efforts into."[528]

Back at the Cavendish, however, Bragg faced a number of problems. Albert Bradley had developed mental illness and had to be replaced as head of the crystallography section by Henry Lipson. Egon Orowan was also still there. Perutz, released from detention, was back at work on hemoglobin. Bragg wrote to Reg James in December 1941: "Perutz has also done a very pretty piece of work on proteins. The crystal shrinks with alterations in the liquid that surrounds it without distorting the molecules. He has the rotation patterns of successive stages of shrinking, and from the variations in intensity we hope to deduce the signs of the Fourier components. If this is realized, we ought to be able to get some direct evidence of the form of the molecule."[529] Other than this, however, pure research had ground to a halt. The work on metals had been redirected to meet the needs of industry, and with larger classes and a staff reduced by three-quarters, there was little time for anything but teaching.[530]

Physics undergraduates of the wartime and post-war periods exhibited a wide range of reactions to Bragg's lecturing. Uli Arndt, who was at Cambridge in 1942–4, wrote: "I well remember the stamping of the young sophisticates which greeted Bragg when, in a Part II lecture, he said 'we shall take a little bit of x, we shall call it dx and we shall think of it as delta x.' "[531] Pippard attended Bragg's course on optics the first year it was given and "hated (or, more properly, despised) his avuncular style of talking over the rim of his glasses as to a group of nephews."[532] Although the lecture material (strongly influenced by C. T. R. Wilson) "excelled any textbook for clarity and insight," many students absented themselves.[533] David Wilson, a student of the immediate post-war period, felt that the physics teaching was "not of the highest class nor likely to arouse enthusiasm."[534]

However, others had quite a different impression. Antony Hewish, who was in the optics course in 1947, though that Bragg "was a good teacher, if a trifle slow on occasions."[535] John Nye attended Bragg's lectures on optics for Part II of the Natural Sciences Tripos in 1942–3. "It is perhaps significant that I remember them better than I remember all the other lectures. They were practically based and accompanied by demonstrations. He had a gift for the arresting phrase. 'For a telescope we have to make an eye that is as wide as possible; to make a microscope we have to make an eye that can see through 180 degrees.' "[536] Ten years later, Michael Whelan found Bragg "an inspiring lecturer."[537] Peter Hirsch, who graduated in 1946, thought that Bragg's optics lectures were "clear and simple to understand—I remember that I thought that he had a 'pictorial mind.' "[538]

Clearly, optics was Bragg's strong suit. However, David Blow, who did Part II Physics at Cambridge in 1953–4, wrote: "I remember being impressed that the Cavendish professor gave the introductory lecture to quantum mechanics, although I knew it was well outside his own field of research. He described the solutions to the Schrödinger equation for an atom in terms of a cinema whose seating broadened out towards the back. Everyone wanted to be near the screen, so the first two to arrive would fill the front row. The next row had 8 seats in it. Further back, where there were 18 seats, the middle of the row would fill up first and then people might prefer to sit in the middle of the row behind, than at the edges. It was a very clever lecture because at one level it was very easy to understand, but there was obviously another level that we didn't really know anything about, and it tantalised you into finding out more."[539]

It is probably fair to say that Bragg's teaching, like his research, did not appeal greatly to the mathematically minded. The analogies which some students found insightful struck others as simplistic; the demonstrations which illuminated physical phenomena for some were mere parlor tricks to those whose understanding lay at a different level. It is easy to see why he would be more successful in lecturing to the intelligent but non-scientific audience of Royal Institution Discourses, and the bright children who attended the Schools' Lectures, than he was to Cambridge undergraduates.

Bragg's contributions to the war effort included service on several military-related committees. For the Ministry of Supply, he served as Chairman of the General Physics Committee and a member of the Metallurgy Committee. He was also a member of the Council of the Gas Research Board and the Advisory Council to the Committee of the Privy Council for Scientific and Industrial Research.[540]

Apart from that, and his trip to Canada, Bragg's main war work was in sound-ranging and Asdic. He frequently visited the sound-ranging headquarters on Salisbury Plain; it was run by Atkins, his "great friend" from the First World War, and Harold Hemming was also there. Sound-ranging was to have its uses in the war. A square of recording posts, stretching from the south coast to the Wash, was used to locate the launch sites of V2 rockets. Asdic development had been started under WHB in the First World War, but did not reach operational use. It involved using piezoelectricity, the ability of certain types of crystal to produce an electric current when compressed, to generate a pulse of short-wavelength sound waves. The Chief Scientist in charge of Asdic in peacetime was fired during the war because he was felt to be too set in his ways. Bragg was offered the position, "but felt uncertain of my powers as an administrator of so large a centre." However, he became an adviser to the Asdic station, which was located at Fairlie on the Clyde, and made regular visits. Bragg liked working with the practical-minded naval men, but felt the main benefit he provided was in raising their morale. After the war, Bragg continued to advise the Admiralty for a total of 15 years.[541]

Cambridge in wartime was "strangely unreal" with numerous alerts but little actual bombing. Food was rationed, and Patience, hearing her parents discussing the food they used to get, cried "Oh, how I wish it were prewar again!"[542] One way to supplement the food ration was to grow your own.

The gardener was elderly and often in a drunken stupor; Bragg took over the garden himself and found the task rather enjoyable. Wartime restrictions on the activities of the Cavendish Laboratory meant that Bragg had adequate time for garden and family. According to Patience, "He'd come home early sometimes and play with me. He went into the Cavendish early in the morning. He was a very early riser. In the summer, he used to take his scythe and cut the grass before breakfast. At 7 o'clock in the morning, you'd hear the swish, swish, swish in the long grass . . . Then he'd have an early breakfast, a very good cooked breakfast and off he'd bicycle to his nine o'clock lecture. He used to come home and play with me after tea about five-ish. My mother was often out and he would do the most enchanting things with me, like making me beautiful dollshouse furniture, in perfect proportion, and that kind of thing."[543]

WHB was also having a quiet time of it. His son-in-law, Alban Caroe, was on military service, so Gwendy and her children, Martin, Lucy, and Robert, were living at Watlands. For his grandchildren, WHB organized a weekly film showing.[544] He was spending about half the week at the Royal Institution; most of the staff had gone to war work, but Alexander Müller and Kathleen Lonsdale were still there. In June, he wrote to Bragg: "Some people still sleep in the shelter every night, though the number is much less now that bombing has been less frequent . . . My movements are restricted, but with that exception I do not feel old . . . I have some spare time and I amuse myself by applying the simple diffraction formula to various diffuse spot diagrams."[545]

WHB died on March 12, 1942. His heart had been failing for years, and he had become unable to manage the short walk from the Athenaeum to the RI. Gwendy and Bragg had gone to see him and, when he seemed to recover, Bragg went back to Cambridge. He arrived to find a telegram informing him of his father's death.[546] A memorial service was held in Westminster Abbey.

Bragg and his father had never been close. In view of WHBs unusual upbringing—the early death of his mother and his abandonment by his father—it is difficult to criticize his inability to form close emotional bonds with his children. In Bragg's adult life, this emotional distance manifested itself in the communication problems that have been described above—notably the inability of WHB and his son ever to discuss the events surrounding their application of Max von Laue's discovery.

Bragg was no hypocrite. His autobiography does not feign any greater grief over WHBs death than he actually felt. Of the biographical memoir of WHB he and Gwendy co-authored in 1962, Bragg wrote: "I put into that all my dearest and most vivid impressions of him." These impressions, however, were something less than heart-felt or personal; he refers to WHBs "simplicity and a cool wisdom and gentleness . . . diffidence about his own achievements . . . He always thought the best of others . . ."[547]

In his biographical memoir of Bragg, David Phillips describes his relationship with WHB as "distant and guarded." Gwendy disagreed with this, describing the relationship as "difficult but I don't think cold. Warm affection, if somewhat anxious flowed from father to son always, and from son to father not affection but still great *respect* . . ."[548]

None of these problems clouded WHBs relationships with the young people whose careers he had guided. In his obituary of WHB, Bill Astbury wrote: "To many, like myself, he was a scientific father. Simply and affectionately we called him 'The Old Man.' "[549] Bernal wrote to Bragg: "He was in a way, scientifically, my father too . . . His name will live in the structure of matter as sure as Galilei in the heavens or Faraday in electricity."[550]

Bragg not only had to deal with his personal feelings about his father's death—he also had to decide whether or not to succeed him. In the summer of 1942, the Board of Managers of the Royal Institution offered him the directorship of the Davy–Faraday Laboratory. Having only recently reached the top of the greasy pole of British physics, Bragg does not seem to have been greatly tempted, and declined the offer. The physiologist Henry Dale, who was also President of the Royal Society, was then appointed.[551]

A. J. Philpot, director of the British Scientific Instrument Research Association, served on a wartime committee with Bragg, and it is probably this association that led to a research collaboration between the two men. Throughout his career, Bragg had consistently emphasized the analogy between optical and X-ray diffraction. He always made the point, however, that there was one important difference between the two phenomena—unlike X-rays, diffracted visible light can be focussed to a point at which a sharp image of the diffracting object will appear. If only a way could be found to focus diffracted X-rays the way that a camera or the human eye focusses diffracted light, then a direct image of the atomic lattice of the crystal could be produced. It was the lack of a means of focussing X-rays that necessitated the mathematical reconstruction of the diffracted object by Fourier analysis.

In 1939, Bragg had published a design for an "X-ray microscope."[552] This device was inspired by the mutually reciprocal relationship between a diffraction grating, such as a crystal lattice, and its diffraction pattern. Bragg's idea was that by using the diffraction *pattern* of a crystal as a diffraction *grating*, visible light passing through this grating could be focussed to create an image of the original crystal. A grating with the desired properties would consist of holes in a sheet, the positions and sizes of the holes corresponding to the directions and amplitudes of the diffracted beams, respectively. An unavoidable wrinkle was that the positions of the holes would not be the positions of spots on a Laue photograph, but rather would represent the nodes of the reciprocal lattice of the crystal. The reciprocal lattice had been envisaged by Paul Ewald as a Looking-Glass version of the crystal lattice created by drawing a line at right angles to each atomic plane and placing nodes along it separated by a distance that is the reciprocal of the spacing between the planes (*d* in the Bragg equation). An X-ray diffraction pattern of a crystal is a distorted version of a projection of its reciprocal lattice.

The masks for Bragg's "X-ray microscope" (which, of course, did not involve X-rays) were sheets of brass drilled with holes of appropriate position and size. This "grating" was then illuminated with a beam of monochromatic light from a mercury-vapor lamp. Arthur Schuster, in his optics textbook, had described a grating "manufacturing" light of particular wavelength that resulted in a particular diffraction pattern; Bragg's idea was to turn the process around and let the diffraction pattern manufacture light that would construct an image of the object.

If it were that simple, one could solve the structure of any crystal by making a mask of its reciprocal lattice and using it in the X-ray microscope to visualize the atomic structure. However, it is not that simple—there is still the phase problem. An X-ray diffraction pattern consists of beams differing in phase, and this phase information is required in order to reconstruct the object—either optically (the X-ray microscope) or mathematically (Fourier analysis). In his 1939 paper, Bragg had shown that the X-ray microscope could generate a recognizable image of atomic structure in the special case of the (010) projection of diopside, for which all but one of the reflections has a phase angle of zero.

In X-ray diffraction, the number of situations in which all the phases are zero is very small. In his 1942 version of the X-ray microscope,[553] Bragg came up with a means of using the instrument for all centrosymmetrical projections, in which all phases are 0 or 180°.[aa] Philpot had directed his attention to a "film-less photographic process" in which an image is created directly on a glass plate. Bragg used an exposed glass plate (a photographic negative) instead of a drilled brass plate and, in the key development, placed plates of mica that retard light by half a wavelength ("half-wave plates") over the "holes" corresponding to reflections of negative phase. Such glass plates could be used in the X-ray microscope to generate images of the structure of any centrosymmetrical projection of a crystal structure.

This was certainly a step forward, but one still had to know the phases of the reflections in order to know on which holes to place the half-wave plates. However, Bragg had another use for the X-ray microscope that did not require phase information. If one made the holes in the mask proportional to the *intensities* of the reflections rather than to their *amplitudes*, the reconstructed image would be a Patterson map. There was a problem in that Patterson maps typically have a very large peak at the origin, representing the "self-vectors" of all atoms present in the crystal; to simulate this large origin peak, the photographic plate used in the X-ray microscope had to have a large central "hole." Nonetheless, Bragg was able to generate a Patterson map of hemoglobin that looked very similar to one that had been calculated by Perutz.

The "X-ray microscope" was a theoretical means of generating an image of a crystal structure by using its X-ray diffraction pattern as an optical diffraction grating, and a practical means of generating Patterson maps of a structure without calculating the Fourier series. Later in the war, Bragg also explored the

[aa] A centrosymmetrical projection is not necessarily a projection of a crystal having a center of symmetry, but can also represent a combination of symmetry elements that render the projection itself centrosymmetrical. This is the case for the (010) projection of horse methemoglobin studied by Perutz (see below).

opposite possibility—using visible light to produce diffraction patterns from models of crystal structures. In March 1944, he presented his findings in a Friday Evening Discourse at the Royal Institution.[554] Using the photographic process that had been developed for the X-ray microscope, it was easy enough to produce an array of spots corresponding to a projection of a crystal structure onto one face of its unit cell. To obtain high-order reflections, however, it was necessary to have many of these arrays. In order to do this, Bragg used a plate drilled with holes representing the positions of atoms in the projection to make multiple exposures on a photographic plate. When used as a cross-grating, the developed plate, which Bragg referred to as a "fly's eye," gave an optical diffraction pattern corresponding to the X-ray diffraction pattern of the crystal.

It might be objected that a device to generate diffraction patterns from atomic lattices was about as useful as a mean of turning gold into base metal—what was really required was a means of generating atomic lattices from diffraction patterns. However, Bragg argued that the "fly's eye" could be a useful short-cut in X-ray analysis; if it were easier to produce an optical diffraction pattern with the "fly's eye" than to produce an X-ray diffraction pattern, then possible crystal structures could quickly be eliminated. Indeed, the "fly's eye" was to prove of more use in X-ray analysis than the X-ray microscope, which turned out to be little more than an ingenious gadget.

A more sophisticated version of the "fly's eye" was published in 1945 by Bragg and Alexander Stokes.[555] An array of tiny lenses was created by making a regular pattern of indentations with a steel ball on a sheet of copper and then using this as a mould for a sheet of Perspex. A paper mask with corresponding holes was used to cut out light falling between the lenses. The lens array was exposed to an illuminated ground glass screen with black circles placed on it to simulate the positions of atoms in a single unit of pattern, creating a cross-grating. The cross-grating was then placed in front of a lens and the spectra photographed at the focus of the lens.

In a March 1943 Royal Institution Discourse on "Seeing Ever-Smaller Worlds," Bragg, echoing Arthur Lindo Patterson, described X-ray analysis of crystals as being like "solving a cross-word puzzle." He illustrated the phase problem by describing a cross-grating pattern illuminated with monochromatic visible light and viewed with a microscope. At every position of the objective, the intensities and directions of the diffracted waves remain the same, but their relative phases change, resulting in a different image for every position. (This is the case because if the incident light is monochromatic, all the diffracted beams will be of the same wavelength.) "Every image is realistic, and we cannot focus the microscope unless we know what we ought to be seeing." Things are simpler if there is a center of symmetry in the crystal or its projection, in which case "the wave scattered by the unit of pattern as a whole must by symmetry be either in the same phase as, or opposite phase to, that scattered by a point at its centre." As a crossword puzzler himself, Bragg must have realized that the analogy between crosswords and X-ray analysis was limited. In a cryptic crossword puzzle, a solution is obviously correct, because it is defined in two different ways by the clue. For a complex molecule like a protein, an essentially infinite number of combinations of phases gave rise to plausible structures.

As described above, Bernal, in his Royal Institution Discourse of 1939, had mentioned two possible ways in which the phases of reflections from protein crystals could be determined. One was to add to the crystal atoms heavy enough to make the phases of all reflections positive. The other was to determine the phases by observing how the intensities of reflections change as the crystal is dried, a phenomenon that Bernal had first observed with crystals of the enzyme chymotrypsin. Perutz tried the latter approach, leading to a paper on the "X-Ray Analysis of Haemoglobin," published in May 1942.[556] Horse methemoglobin forms monoclinic crystals, in which two of the angles between the crystallographic axes are 90° and two of the axes are equal in length. Low symmetry was emerging as a characteristic feature of protein crystals, presumably reflecting the fact that the protein molecule itself was asymmetrical. Perutz showed that horse methemoglobin crystals are of space group $C2$, which has an axis of two-fold rotational symmetry parallel to the b axis, and contain two molecules per unit cell.

In 1926, determination of the (highly symmetrical) space group of beryl had allowed Bragg and Guy Brown to solve the structure in 15 minutes. For hemoglobin, however, the low symmetry of the crystal meant that every atom might be in a general position. Space-group theory was going to be of little help in solving the structure of proteins.

Perutz was able to arrest drying of horse methemoglobin crystals long enough to record X-ray data at two stages between wet and air-dried. The b-axis remained constant in length throughout drying, the a-axis only changed at the final stage, while the c-axis contracted by a total of 29%. The intensity changes on drying were greatest for $h0l$ reflections, as these were most affected by contraction along the a- and c-axes. Perutz concluded that the hemoglobin molecules must be linked along the b-axis, as the crystal contracts least in that direction. The dried hemoglobin molecule is a "platelet" 36 Å along c, 64 Å along b and somewhat shorter along a. To satisfy space-group symmetry, the molecule must consist of two sheets of protein 18 Å thick related by a two-fold axis, or else one rigid sheet 36 Å thick, itself possessing two-fold symmetry.

This paper also contained the first Patterson maps of hemoglobin, projections on the (010) plane for all four stages of hydration. Not surprisingly, these were difficult to interpret. For a unit cell with N atoms, the number of independent peaks in its Patterson map will be $(N^2 - N)/2$. In the case of hemoglobin, which has 2500 non-symmetry-related atoms (not counting hydrogen), the number of peaks is over 3 million.

In the acknowledgements section, Perutz thanked Bragg for "discussions, help and encouragement." Much as he was interested in the hemoglobin work, Bragg was not prepared to accept coauthorship at this point.

At this time, Bragg was more interested in another research problem—dislocation in metals. Orowan, who "had a most fertile imagination and ingenious mind," was a proponent of the view that metals deform easily because they have abundant dislocations (local disruptions of the crystal lattice) and therefore can change shape by movement of dislocations rather than by a rigid motion of the whole. This idea made a lot of sense to Bragg; in a characteristic analogy, he pointed out that one does not move a carpet by dragging the whole thing at once, but rather by making a "ruck" in one corner and then pulling on the ruck while shuffling one's feet.[557] Bragg was, like any good scientist, a keen observer, and a chance observation led to his model for dislocations; as he later wrote: "I remember I hit on the idea of using these little bubbles when I was mixing the petrol and oil together for my lawn mower. The stirring made small rafts of bubbles of nearly equal size and it was interesting to see them adjust themselves into a structure. I asked one of my assistants, Lipson, to make a mass of bubbles and see whether we could witness dislocations and he came back to report that the idea did not work. I had a look at what he was doing and found his bubbles were far too big. When we made them about $\frac{3}{4}$ of a millimetre in diameter the bubble model worked like a charm."[558]

John Nye's version of the story is slightly different: "Coming in to the lab on Monday he asked his assistant Crowe to set up a small glass nozzle in a soap solution and blow air through it to reproduce the effect. I later heard from Orowan that he was passing at the time and found Crowe puzzled because the bubbles were coming out all different sizes. Orowan suggested to Crowe that he turn the nozzle so that it was pointing upwards; the glass tube being straight, the nozzle was naturally pointing down and the bubbles were colliding with each other as they emerged. Crowe put a bend in the tube so that the nozzle pointed upwards and there was never any problem after that."[559]

Bubbles of the desired size were produced by blowing air though a 0.01-cm orifice 1 cm below the surface of a soap solution. Allowing several "rafts" of bubbles to join resulted in a "polycrystalline mass," in which the individual "crystals" were separated by dislocations along the boundary. To simulate deformation in metals, the bubbles were anchored to springs on opposite sides of the raft, with the pitch of the springs corresponding to the spacing of the bubbles. When one spring was moved laterally, the raft sheared by a row of bubbles slipping past another, one by one across the raft. Bragg and his colleagues measured the stress-strain relationship by attaching two vertical glass fibres to the non-moving spring, and viewing their displacement with a "tele-microscope."

The bubble raft model could also be used to simulate other properties of metals. An irregular raft of bubbles could be "annealed" (made more perfect) by moving the spring back and forth. Stirring the bubbles to promote "recrystallization" was seen as analogous to cold-working a metal. The bubble raft shared many features with a metal crystal. Like atoms, the bubbles were essentially spherical, uniform, and frictionless. The analogy only broke down because the bubbles were essentially without mass.

In Bragg's first paper on the bubble raft,[560] one of the people thanked was his son Stephen. Manpower was scarce, so Stephen was recruited for "a little bit of sub-technician-level help." "I think my father really wanted the pleasure of being able to put a little footnote referring to his son helping with the experiment but I don't think I made a big contribution."[561]

In October 1939, Bragg had taken office as President of the Institute of Physics, a post he held until September 1943. In this capacity, he was involved in planning for post-war needs. This led to a lecture on "Physicists After the War," which he gave at the Royal Institution in March 1942. Bragg well knew that scientific breakthroughs could neither be predicted nor planned—he later estimated the time for fundamental research to produce results was 40–50 years.[562] Nonetheless, he accepted the responsibility that scientists had to benefit society, whether this was the wartime goal of defeating the Axis or the peacetime one of making the country "a better one than it has been during the last twenty years." The Royal Society's central register of scientists provided Bragg the opportunity of conducting some pioneering research in the sociology of science. He noted that few leading physicists came from public schools: "among the heads of physical laboratories in Great Britain, only one in a hundred received his education as a boarder at a public school. The secondary schools and grammar schools are the source from which the body of British scientists is almost entirely drawn."[bb]

Bragg was also critical of the science training university students received in Britain. He found that they could answer harder questions than their American counterparts, but were poorer at handling apparatus. Bragg suggested that physics students work as industrial apprentices for 6 months before entering university or take courses from industrial scientists.[563] (His son did exactly that; on graduating from Rugby, Stephen spent 4 months working as a fitter's mate at the British Thomson Houston factory before going to university in 1942. "Coming straight from school, he was treated as a mate by the men and not as someone of another class from the University."[564])

Another question for the future was how government funds for research were to be allocated. Bragg was not a fan of centralized research institutions that do not have a teaching function. These lacked the flow of bright young students necessary to avoid intellectual stagnation and were ill-suited to react to developments that arose from other branches of science. Not surprisingly, Bragg felt that universities were efficient users of government resources: "The ideal research unit is one of six to twelve scientific men and a few assistants, together with one or more first-class mechanics and a work-shop in which the general run of apparatus can be constructed." Also required was a good stock of "junk." He was more critical of government-funded laboratories. No doubt thinking of the National Physical Laboratory, Bragg expressed "a widespread opinion that the flow from the research associations of Great Britain of fundamental ideas which would profoundly affect our industries has not yet reached

[bb] At a conference on "Science and the Citizen: the Public Understanding of Science" in March 1943, Bragg claimed that the heads of physical laboratories in Britain included single representatives from Harrow, Clifton, Malvern, and Westminster; all others were from secondary schools. [Anonymous (1943). Science after the war. *The Times*, March 22, p. 2.]

the volume which had been hoped for and anticipated." Rather than laboratories conducting research on behalf of industry, Bragg favored them providing facilities for industrial scientists to work undisturbed—something he had done on a small scale in Manchester.

Given his findings about the poor science education in public schools—presumably excepting Rugby—it is not surprising that Bragg reacted to a letter in *The Times* from the Headmaster of Winchester School that lamented the suspension of university arts courses because of the war. In his published response, Bragg, writing as President of the Institute of Physics, took exception to the Headmaster's implication that science courses are a lesser part of a balanced education than arts courses because the former are "technical" and the latter are "general." In fact, he argued, the most significant changes in civilization in the last century have been the result of scientific advances. "The fact that a 'general' education has so often in the past failed to include science has been responsible for many of the weaknesses in our national structure which the war has revealed. It is a tribute to the teaching of the literary subjects to admit that most scientists know something of the arts and wish they knew more, whereas most of their arts colleagues know nothing of science and are often rather proud of it . . ."[565] The indispensability of both sciences and arts was Bragg's position throughout life and he practised what he preached. He was proud of his classical education and had a very active interest in visual art and, to some extent, literature. His impression about the attitude towards science of "arts colleagues" would be proved all too correct when he later offered science courses for humanists at Cambridge.

Another task that Bragg undertook as president of the Institute of Physics was to host a conference on "X-ray Analysis in Industry" in Cambridge in April 1942. He gave a historical review to this conference, which was attended by about 280 people. Further meetings, chaired by Bragg, were held in September 1942 at the Royal Institution and in April 1943 in Cambridge. In July 1943, the X-Ray Analysis Group of the Institute of Physics was established, with Bragg as chair and Lipson as secretary; the inaugural meeting was held in Manchester in October. Biannual meetings were held from that point on; Bragg remained chairman until April 1947, and vice-chairman until his death.[566]

Bragg, as described above, had mixed success in lecturing to university students. However, his Discourses and Christmas Lectures at the Royal Institution had gained him a considerable reputation for conveying scientific ideas to lay audiences. His contribution to a 1943 book entitled *The Exposition of Science* outlined his thinking on the importance of scientific literacy and the craft of talking about science. Bragg's contacts with civil servants and politicians must have left him unimpressed, because he included "most of our leaders" in that great majority of society to whom science is "a sealed book." This ignorance was the result of deficient science education and poor coverage of scientific issues by the press. He took the opportunity to promote once again the equality of what his friend Charles Snow would later call "The Two Cultures": "Manual dexterity, craftsmanship, and science, and the methods of thought which they engender, must rank with the humanities in a balanced education."

Because of general scientific illiteracy, the scientist addressing a lay audience was inevitably told to avoid technical terms. Bragg ridiculed this attitude with a telling metaphor: "What could be more technical than a description of a football match? To picture the scientist's plight, one must imagine the sports expert being told to describe the match using as little as possible such technical terms as 'ball' or 'goal', and of course avoiding the very complex ideas involved in 'try' or 'offside.' "

Bragg had some general advice for those giving popular lectures. One point was that "It is as inartistic to have more than one main theme as it is to have more than one centre of interest in a picture." It was, of course, typical of Bragg to take his example from the visual arts. Other key points were: "The art of giving a good popular lecture, in short, is largely the art of what to leave out" and "It should be a point of honour never to read a lecture."[567]

Bragg may have worried that science was a sealed book to most of the country's leaders, but he was equally wary of a scientific-technological priesthood. In June 1943, Bragg coauthored a letter to *The Times* with Robert Pickard, Chairman of the Joint Council of Professional Scientists, and Alexander Findlay, President of the Royal Institute of Chemistry. This read in part: "The claim that the scientist, as scientist, is entitled to some position of exceptional authority in deciding the policies of Governments, is one which cannot and should not be accepted in a democratic community . . . No social problem can be solved solely by the methods of science; not only material but other values are involved; and it is for the community, of which the scientist is a member, to weigh the different factors and make a decision. A scientific and soulless technocracy would be the worst form of despotism."[568] Having visited the Soviet Union, the home of "scientific socialism," Bragg had seen this for himself.

Quite apart from the wartime limitations on research, Bragg, now in his mid-50s, was becoming more and more an elder statesman and public spokesman for science. In May 1944, he was a member of a committee that recommended changes in the examinations for Cambridge scholarships to avoid over-specialization.[569] Also that year, Bragg wrote a paper entitled "Organization and Finance of Science in Universities."[570] In this, he noted that the university population of Britain was very small per capita compared to the United States—just over 9000 graduates were produced annually. Rather than increasing enrolment, which he felt would make the gap between best and worst students unmanageable, Bragg suggested that all children of ability have the chance of university education and that the best science graduates go on to positions of responsibility and leadership: "scientists of high quality are limited in number, like brilliant musicians or artists or star football players." He again made the point that university research was a bargain. The total cost of the university physics laboratories in Great Britain for 1938–9 was £218,200—"a very modest bill." This included £30,000 for supporting 250 research students,

£15,000 for supporting 50 senior researchers, £117,000 for 180 members of staff, £24,000 for 120 technicians and £4,200 for 70 lab boys (plus apparatus and materials). The total cost of all university scientific research was about £800,000 per year. In the pre-war period, British scientists were better paid than their continental counterparts but their equipment was much inferior—"This is false economy." Scientist who had learned in war-work how effective they could be when provided with adequate resources would not kindly return to "the old regime" now that they had "tasted blood."

In the spring of 1943, with the tide of war now turned in the Allies' favor, Bragg was given another "war-job"—he was sent as a scientific emissary to Sweden. He flew to Stockholm from Leuchars, in Fife: "Dressed up in life-belt, parachute harness, electric light and given a whistle to blow if we fell into the sea, rather alarming preliminaries." A fellow-passenger accidentally inflated the life-jacket under his coat, "so that he swelled to twice the size under our astonished eyes." In Stockholm, Bragg stayed for a while at the British Embassy. Even in neutral Sweden, there were shortages. Cars ran on "transverse sections of birch trunk, quartered like hot-cross buns," with a gadget the size of a dustbin attached to the back of the car feeding gas from the burning wood into the carburetor. Bragg visited the Nobel Institute for Physics: "As usual, felt very depressed at the contrast between beautifully organized, tidy, and clever laboratory installations as compared with ours, and made resolution to tidy up Cavendish." That evening at dinner he was reunited with his old friend Arne Westgren, who had just been appointed Secretary to the Swedish Academy of Science. Westgren told Bragg he was giving up research because he had calculated many crystal structures and wanted a change: "I like W. very much, he has a keen sense of humour and less stiffness than many Swedes."

In Uppsala, Bragg visited Theodor Svedberg's "famous" laboratory, where he was very interested in Arne Tiselius' system for separating proteins by electrophoresis. Bragg and Tiselius "had tea at a rather consciously 'period' restaurant, with waitresses in national dress, and drank mead. I was asked to sign my name, to be engraved in due course on the silver band of one of the huge drinking horns, which form a kind of visitors' book."

After a week in Sweden, Bragg spent several days at Edeby, the sixteenth-century estate of Harry von Eckermann, a prominent industrialist and amateur mineralogist. Located about 60 miles south of Stockholm, the estate had 17,000 acres of forest and lakes, inhabited by osprey, kite and buzzard—a landscape that reminded Bragg of the Laurentian Mountains in Canada. There he feasted on types of foods long unavailable in Britain: "elk tongue, roebuck liver, smoked goose-breast, pike-perch from the lake, eggs and honey ad lib. and many kinds of bread." On his return to Stockholm, he visited the *Teknolog Vereningen*, an organization that promoted links between basic and applied science. Its mandate was similar to that of the NPL, but its *modus operandi* more like that of the Royal Institution: "Each branch has regular meetings, *followed by a supper*, a pleasant blending of science and social festivity." He also had dinner with the physicist Lise Meitner, an Austrian Jew who had taken refuge in Sweden, and found her "A most intelligent woman."

Bragg gave a total of 14 lectures in Sweden. His destinations included Lund, where he "got the impression that Uppsala corresponds to Cambridge and Lund to Oxford." In Göteborg, the links to Tyneside were so strong that the Swedish shipping magnates spoke English with Geordie accents. Signs of war were not hard to find. At Helsingborg, Germans went to and from the ferry to Denmark; from Malmö, Bragg could see fires in Copenhagen from Allied bombing. Although he was safe enough in Sweden, getting back to Britain was a dangerous business. The Mosquito fighter-bomber that had been used because it was too fast to be intercepted having been shot down, Bragg had to take his chances on a flight in a bomber on a night with "very thick weather." The pilot flew such a circuitous route that it took 9 hours to reach Scotland. Bragg's fellow passengers were a group of British airmen who had been shot down in Denmark and smuggled into Sweden by the Danes.

Bragg's report on his Swedish visit recommended more scientific exchanges between Britain and Sweden and a better supply of English-language scientific journals. "In conclusion, I cannot exaggerate the warmth of my welcome. As the first representative of the physical scientists in this country to visit Sweden during the war, I had a royal time. My visit was a continuous series of entertainments, eager discussions, appreciative audiences, and commissions to execute in England. The Swedes want closer contacts with English science, and now is the time to do all we can to establish them, the atmosphere could not be more favourable."[571]

Sending scientists on goodwill visits to neutral countries could not have been a high priority for a nation at war. It seems more likely that Bragg's Swedish visit was some kind of reward for his contributions to the war effort. One important consequence was that his renewed friendship with Westgren and the new one with Tiselius would allow Bragg to influence the awarding of Nobel Prizes when both men later became important figures in the Nobel Foundation.

Back in Cambridge, Bragg resumed his family and professional activities. In October 1943, he wrote to James: "Stephen is now in his second year at Trinity, where he is studying engineering . . . David is still at Rugby. He has to leave next March, when he will be called up . . . Margaret is at school here in Cambridge . . . Patience is also at the Perse School, and is a very lively little person. We are having a tremendous time at the Cavendish because we have such enormous numbers of students and so many of our staff have left for war work. Our first-year class this year is about 350; however, we are coping somehow, though we have had to duplicate all our practicals and lectures . . . The proteins are going quite well, though unfortunately Perutz has been drawn away for war work . . . [John] Randall is now working here with me on electrical sheet steel. We still have about sixty researchers in the Cavendish on various war jobs."[572]

Perutz's secondary research interest, glaciology, which he had initiated before the war as an excuse for mountaineering trips to the Alps, had involved him in a top-secret project, Habakkuk. This was a plan to construct "bergships," giant landing platforms for military aircraft, from wood pulp-reinforced ice. Before Habakkuk was cancelled, Perutz, the former internee, had to be given expedited British citizenship in order to liaise with his American counterparts.[573]

Bradley, whose mental state had been a matter of concern for some years, was now examined by a Harley Street psychiatrist. Diagnosed as suffering from paranoia, he was not considered well enough to carry out his administrative duties but was given the all-clear to do research![574] As a result, Feather was appointed assistant director of the Laboratory. Apart from him, Bragg only had three other teaching staff, and even they spent most of their time on war work. The enrolment in physics was 1200—a six-fold increase from the pre-war figure.[575] Bragg wrote to Henry Dale: "The course we offer now is more 'ersatz' than I care to contemplate."[576] In November 1944, Bragg asked Dale to be excused his Royal Institution lectures for 1945 as he was "desperately short of material. All our work during the last five years has been war work, which we are not allowed to talk about, and I have not been able to do any of the pure research which provides the ammunition for these talks, or think about pure science at all."[577]

Stephen Bragg graduated from his two-year engineering degree with first-class honors in 1944. His decision not to follow the Bragg family tradition of mathematics and physics does not seem to have perturbed his father. There was also an engineering tradition on both sides of the family (Bob Bragg, John Hopkinson, and Bertie Hopkinson). Because of the problems between Bragg and his father, Alice tried to steer Stephen away from physics.[578] Over dinner at Trinity College one night, Bragg discussed Stephen's situation with Sir Roy Fedden, Chief Engineer of the Bristol Engine Company. Engineering was a reserved occupation, and Fedden recommended that Stephen, if given the choice, should work for Rolls Royce. Bragg passed this advice on to his son, who in fact was given the choice between De Havilland or Rolls Royce, and duly opted for the latter. Stephen spent 25 years with Rolls Royce, eventually becoming Chief Research Engineer.[cc]

David Bragg's life was not unfolding so smoothly. According to Bragg's autobiography, "When the time came for him to be called up, he volunteered to go into the mines rather than into the armed forces. He had a grim experience in the coalmine, and this precipitated a severe nervous breakdown which gave us great concern for many years."[579] Bragg's own history of psychological problems could only have increased his concern about David—particularly as depression ran in the Hopkinson family.[580]

[cc] Stephen Bragg was initially posted to the Rolls Royce jet-engine facility in Barnoldswick, Yorkshire. Bragg and Alice visited him there and were "somewhat aghast" to find that he had booked them into the Railway Hotel, "between the gas works and Dewhurst's mill", but realized the wisdom of his choice when they saw the off-ration food that was available: Game, eggs, cream, meat, and poultry. Bragg was amazed when he bought some cakes there and there were grease marks on the paper bag! He took Alice to see Bolton Abbey and WHBs former cottage, Deerstones (RI MS WLB87, p. 100–111).

Alice's volunteer work—and her title—was opening many doors for her. By early 1944, she was running all the War Workers Clubs in the Eastern Area. In 1943, she had joined the Town Council as an independent member for Newnham (no elections were held during the war).[581] According to Stephen Bragg, "At that time, and this, of course, has now completely changed, the local councils were very largely composed of independents. I think there were only two or three Labour members from the wards in the town which were largely composed of working people. Many married women, or professional people who could control their working time, felt that local government was the type of service that they could usefully do . . . So she became involved in local activities. One of these was the Women's Voluntary Service, an organization that did a lot of jobs that needed doing in war time, some connected with Civil Defence and some connected with Air Raid Precautions. So my mother had an office in the town and was, therefore, quite prominent in the town, as it was then, and her work balanced my father's, which was more confined to the university and scientific war-work. She was always very pleased that a letter addressed just to Sir Lawrence Bragg, Cambridge, was delivered by the postman to the Women's Voluntary Services headquarters, because that was the only Bragg the postman had heard of."[582]

8
A message in code which we cannot yet decipher: Cambridge, 1945–53

Bragg's tenure of the Cavendish Professorship began in earnest only when the war ended, his staff returned from war-work and the Austin Wing was made available for civilian research. Although not as bad as Manchester in 1919, it was "a very difficult time indeed" for Bragg. As late as 1947, John Kendrew found that "on the whole everybody thought it was absolutely terrible, the great days of the Cavendish had ended, that they had appointed this man who knew nothing about the main subject the Cavendish did, the worst appointment in the whole history of the place . . ."[583] When James Watson arrived in Cambridge five years later, "everyone" at the Cavendish thought the laboratory had declined under Bragg.[584] Although these statements almost certainly exaggerate the extent of the resentment, there is no reason to believe that the nuclear physicists were ever reconciled to Bragg's appointment.

One of the major problems was filling two vacant chairs. The Plummer Professor of Mathematical Physics, Ralph Fowler, had died in 1944 and the Jacksonian Professor of Natural Philosophy, John Cockcroft, left Cambridge in 1946 to become head of the new Atomic Energy Establishment at Harwell. Nevill Mott, Bragg's choice to replace Fowler, felt that there was "no-one in authority" at the Cavendish, and decided to stay at Bristol. Rudolf Peierls also declined the position.[585] The Electors then dithered between two Dutch physicists, Hendrik Casimir and Hendrik Kramers, in the end obtaining neither man.[586] Eventually, Douglas Hartree, Bragg's old collaborator from Manchester, was appointed to the Plummer. Finding a new Jacksonian Professor was also difficult, as the outstanding candidates had already accepted other positions. Otto Frisch, who was elected, was perceived by some members of the Cavendish as a disappointing choice.[587] As Perutz put it, Frisch was "Certainly not the sort of man to build up a great school of nuclear physics."[588]

Bragg had predicted that the experience of military research would make scientists intolerant of the poor facilities available in the universities.[589] This proved all too true when the physicists returning to the Cavendish demanded resources that they had never had before, such as offices, secretaries, and telephones. Bragg, who recognized the "false economy" of depriving scientists of the infrastructure they required, did his best to meet the raised expectations.

In addition to the new professorial appointments, there were changes in Bragg's crystallography division, which now moved to the Austin Wing from its previous location in the University Zoology Department. Albert Bradley had left in 1944, leaving the metals group to be run by Egon Orowan. Will Taylor was brought back to become head of the division. Henry Lipson, who had wanted this position for himself, replaced Taylor as head of the Physics Department at Manchester College of Technology.

Norman Feather, the assistant director of the Laboratory, left for Edinburgh in 1945, but Bragg managed to find an excellent replacement in the atmospheric physicist Jack Ratcliffe. Like Bragg himself, "Ratcliffe was a grand team-leader, who gave his people excellent ideas and never took the credit for them himself."[590] He reorganized the teaching and proved to be a very effective committee man. Bragg had written in 1944: "nothing scares off the muse of inspiration more than the immediate claims of a host of small responsibilities. A professor should be allowed to be absent-minded, it is part of his defense mechanism behind which he can concentrate on his real work."[591] On committees, Bragg quickly came to depend on Ratcliffe's prompts of "We think, don't we, Professor . . ."[592] By now well aware of his need for administrative support, Bragg appointed E. H. K. Dibden as General Secretary of the Cavendish in 1948.

Bragg's biggest headache was the future direction of the Cavendish Laboratory and in particular the fate of nuclear physics. Although nuclear research was far from abandoned—as late as 1953, it was still the largest research group—Bragg made it clear that he did not intend to compete with the Americans in a field of research which now required such huge machines as cyclotrons and linear accelerators. As Perutz later wrote: "This was an inevitable consequence of the war and the transformation of atomic physics to 'Big Science,' to which the tradition and structure of Cambridge University were ill-adapted."[593] Bragg's decision undoubtedly created resentment, particularly as he was prepared to give research space—admittedly not very good space—to a group of biologists. The fact that Bragg's research area was very far from the mainstream of physics no doubt made it more difficult for him to carry out his reorientation of the Cavendish Laboratory. According to Aaron Klug, who went there as a research student with Hartree in 1949, "The other physicists there were a bit scornful of him because he wasn't in the mainstream physics, he didn't belong in any category."[594]

However, the other research areas that had managed to survive the Rutherford era were encouraged that the Cavendish was to be a more diversified enterprise under its new management. Brian Pippard, who eventually became Cavendish Professor himself, wrote: "Bragg performed a notably excellent job in decentralizing the work of the Cavendish, and thus effectively breaking away from what would have ultimately become the dead hand of the Rutherford tradition. His decision to give each research section as near as possible autonomy, consistent only with very general central principles and of course financial control, has played a significant part in the subsequent developments. Ever since then, the Cavendish has been notable among Cambridge departments for the democratic way in which it conducts its business."[595]

One research area that benefitted from the new regime was radio-astronomy. Bragg managed to convince Martin Ryle to continue in this area rather than move into nuclear physics. According to Crowther's history of the Cavendish Laboratory, Bragg "perceived that the problem of deducing the size and position of radio objects in the sky by interferometric methods was an exact analogue of the determination of the arrangement of the atoms in a biological molecule."[596] Ryle set up two widely spaced aerials on the Rifle Range to study interference patterns produced by radio-stars; "Each aerial was like a wire mattress standing a foot or two above the ground and covering a space some 50 yards by 20 feet." The Trinity beagle pack once caused extensive damage to one aerial in pursuit of a hare hiding beneath it, necessitating the posting of guards any time a hunt was held.[597] In time, the radio-astronomers justified Bragg's faith in them by the discovery of pulsars and quasars.

A new technique in which Bragg placed great hopes was electron microscopy. In his 1943 Royal Institution Discourse, he had stated: "The electron microscope . . . fills in the wide gap between analysis by light and by X-rays . . . it will open up fascinating new fields in the future."[598] The wavelength of electrons depends upon their energy, so illuminating a specimen with an electron beam of sufficiently high energy can produce a spatial resolution far greater than that possible with visible light. Because electrons, unlike X-rays, can be focussed, an image of the specimen can be produced. An added advantage of the electron microscope was that, as had been shown by George Thomson and others in the 1930s, electrons could be diffracted by crystals in the same way as X-rays. The Department of Scientific and Industrial Research provided the Cavendish with one of the first RCA electron microscopes in Britain. Uli Arndt, then a research student, was told that Bragg would not accept a better Siemens electron microscope liberated from a German laboratory because he did not want to give post-war German visitors the satisfaction of finding one of their instruments in the Cavendish.[599]

In general, though, Bragg's feelings about the defeated Axis powers were in stark contrast to his jingoistic attitude in the First World War. As early as 1943, Bragg wrote to J. P. Lawrie: "There is no doubt in my mind that science should remain what it always has been, completely international. We should resume relations with our colleagues in other countries as soon as possible after the war."[600] Shortly after the end of the war in Europe, he wrote to "The Times" stating that scientists previously involved in wartime research should be free to travel to other countries.[601]

Acting on this advice, perhaps, Bragg travelled to Paris in July 1945, to give lectures at the Sorbonne on "X-Ray Analysis: Past, Present and Future," "The Strength of Metals" and "X-Ray Optics." It was presumably during the first of these that Bragg's talent for simplification got him in trouble with a more literal-minded colleague: "My chairman was Professor [Charles] Mauguin, holder of the chair in mineralogy, and he was deeply shocked to hear me begin my lecture with the statement that there were only six minerals in the earth's crust if one neglected tiresome details. When the lecture was over, he led me firmly to a great collection near the theatre where endless specimens were displayed in their cases, and sweeping his arm dramatically said to me 'Les six minéraux!' "[602]

The Paris trip was supported by the British Council, who also funded a trip to Portugal that Bragg made in November 1945. The occasion was the 50th anniversary of the discovery of X-rays. Bragg's first impressions of Lisbon were: "houses in pastel shades, heavy tiled roofs, palms, steep cobbled streets in which everyone stands regardless of traffic, donkeys with panniers, noise." On his first full day in Portugal, he visited the British Ambassador and the Portuguese Minister of State. That evening, he went to a concert by Guilhermina Suggia, a cellist and "the most famous woman in Portugal." The tone-deaf Bragg was not impressed with Suggia's playing or her appearance: "V. like [Augustus] John portrait, most dominating personality although her face, as someone said to me in Oporto, is like an old boot. Wish I were musical enough to appreciate how good her cello playing was." The following day, he gave a lecture on "X-Ray Optics" at Lisbon University. This was given in French, in which Bragg had been fluent since the First World War. However, he lost his notes in the dark and wished that he had chosen to speak in English.

The following week, Bragg gave talks to the British Club about the atomic bomb and the position of science, one in French on "Some Problems of the Metallic State" at the Lisbon Academy and another at the British Institute. He then travelled to Belem, on the Tagus estuary, where he was awarded an honorary degree. His robe consisted of a "garment most like a clerical soutane . . . then a kind of huge villain's black cloak, then a marvellous light blue hood of highly decorated silk, then a cap like a charlotte russe with a spike." His next stop was Oporto, where he gave a lecture and an interview at the British Institute and a lecture at the university. This was a repeat of his French-language talk on the metallic state, but Bragg's jokes did not go over well: "I fear my Portuguese colleagues thought they were not getting their money's worth because it was not recondite and unintelligible." On the train to Oporto, Bragg had met "Graham, of Graham's Port," who invited him for a tour of Factory House, the private club of the port producers. There he saw the visitors' book containing signatures of officers from the Duke of Wellington's army in the Peninsular War. His tour ended at "the holy of holies where a glass of nectar was reverently drawn off for me from the very middle of a cask by a tube." Bragg's next stop was Coimbra, which has the oldest university in Portugal (founded 1290). Bragg had been billed to give a lecture there, but this had to be cancelled in order that he could get back to Cambridge in time to attend the awarding of honorary degrees to Dwight Eisenhower and Bernard Montgomery. The professor of physics in Coimbra was João de Almeida Santos, who had worked in Bragg's department in Manchester. "The old apparatus in the Physics laboratory which Santos showed me was extraordinary. Apparently in the heyday of Portugal's wealth and importance, the Marquis [sic] de Pombal [prime minister of Portugal from 1756 to 1777] decided that its University should be worthy of so great a nation. Everything was to be of the best. The apparatus was made of rare woods and gilt where possible. The steel yard was beautifully engraved. In order to demonstrate that sieves with a range of holes let through particles of different sizes, a rather simple observation, the sieves were held aloft by gilt angels on a walnut stand with lion's feet. The library was an amazing place. Enormously heavy tables with legs as thick as a man's thigh were made of precious woods brought from Portugal's overseas possessions. The ladders to reach the bookshelves were rococo and gilt. A wonderful place."

Back in Lisbon, Bragg concluded his trip by being interviewed on radio about the scientific consequences of Röntgen's discovery, his impressions of Portugal and scientific relations between Portugal and Britain.[603]

In his autobiography, Bragg wrote of Portugal: "I was told that it was a very poor country and run on a shoestring, but the people seemed to have a gay time in spite of their poverty." Historians present a view of post-war Portugal very different from Bragg's Acadian land of merry peasants. The dictatorial regime of António de Oliveira Salazar censored magazines, radio, television, and theater; the secret police killed hundreds and jailed and tortured thousands; child mortality was 126 per thousand in 1940; as late as 1950, 45% of the population was illiterate; factory workers received the lowest wages in Europe.[604]

Bragg's comments about Portugal are not uncharacteristic of his worldview. He was a small-l liberal and not unsympathetic to left-wing causes, such as nationalization and the anti-apartheid movement, but he lived in a privileged cocoon of servants, country houses, High-Table dinners, sailing holidays, and gentlemen's clubs. Despite having socialist friends such as Patrick Blackett, Charles Snow, and Desmond Bernal, Bragg seems to have only dimly understood the lives of those less talented and fortunate than himself.

The British Council offered to send Alice to Portugal, but she had just taken office as Mayor of Cambridge, only the third woman to occupy that office. She had only been a councillor for 2 years, but was following a family tradition, as her paternal grandfather had been Mayor of Manchester. Bragg urged her to take the position; as an independent, she was acceptable to the badly polarized Labour and Conservative factions. He also felt that councillors tended to behave better with a female mayor. In his opinion, Alice was "a great success, beloved by all."[605] However, she had to be both mayor and mayoress, as well as running her household. Bragg pulled some strings with the Ministry of Labour and National Service to get an "Irish maid,"[606] but she turned out to be "a broken reed."

During her year as mayor, Alice also became a magistrate. Perhaps because of her legal training, she enjoyed this aspect of the job and managed to remain on the Cambridge bench until she was 70. According to her elder son, "She, I think, was very good on personal relations generally and could see how people thought about things. This is an important skill for a Magistrate and, of course, as your experience accumulates, you get more and more feel for the way people think and what makes them tick."[607] Alice's experience in local government also led to her becoming the first chairman of the Cambridge branch of the National Marriage Guidance Council.[608]

Bragg had always been deeply involved in the lives of his children. Patience wrote: "My father carved us boats that zoomed down the bath propelled by twisted rubber bands, perfectly crafted doll's furniture, and devised a telephone made of tin cans and string to link us with a neighbouring friend. He constructed a log cabin in the garden, complete with windows, a door, and an outside barbecue."[609]

During Alice's term as mayor, Bragg spent more time with his daughters. He bought a puppy—half fox-terrier, quarter foxhound, and quarter Irish terrier—for Patience. Scrap was "a mixture of amazing vitality and intelligence" and as feisty as his name would suggest, fighting other dogs and chasing the neighborhood cats who overran the Braggs' garden.[610] On one occasion when Bragg was walking him along the Coton footpath, Scrap killed a chicken on an adjoining farm. The irate farmer demanded compensation, and Bragg had to borrow some money from Antony Hewish at the nearby radioastronomy laboratory.[611] For Margaret, Bragg bought a half share in a pony, Joey, from a man who sold cockles and winkles at the Cambridge market, "but while leading it home in triumph from the other side of Cambridge it escaped on Midsummer Common and we had, as my father said, 'one hell of a job' to recapture him."[612] David, still in poor health, was meantime attending art school in London.

Patience provided the following account of life at 3 West Road in the immediate post-war period: "We used to entertain undergraduates to tea every Sunday. We used to play old-fashioned Victorian parlour games and he [Bragg] would join in enthusiastically . . . He was the life and soul of every party because he was a great raconteur and good at telling jokes. He had that sense of drama and timing. But he was terribly vague, too. Once we had two Trinity students after the war when there weren't enough beds in the college and we had some extra room, because my brothers had left home. Dad was reading 'The Times' one morning, when the students had been with us for about two weeks. All at once he put the paper down and said 'Who the hell are you?' The students were dreadfully embarrassed. My father hadn't observed them before."[613]

This may be an example of Bragg's famous absent-mindedness. Peter Hirsch, who was a research fellow in the Cavendish in the early 1950s, wrote: "I remember him bringing visitors to our lab in crystallography. He could not always put names to faces, and he would read the names on the door before coming in, hoping that might help him to identify those inside!"[614] Many years later, Patience claimed that her father once chained his bicycle to an undergraduate student who was bending over to tie his shoelace.[615]

The Braggs were happy to entertain students at West Road but found them sometimes hard to get rid of. The first occasion on which a group of honors students was invited over for Sunday tea, Alice resorted to telling them that she and her husband had to go out to evening service. " 'That's all right', said one of the students, 'We'll be quite happy till you get back. I see you've got a piano!' "[616] Stephen Bragg recalled the solution to this problem: "Students used to be invited to tea at 3 West Road, my mother presiding with an enormous brown teapot, provided scones and jam and cakes, and so on. Then at some point after tea, my father would say "I'd like to show you the garden," and took them round. The tour finished, he'd say 'I think you left your bicycles here' . . . and so off they would go."[617] (Curiously enough, Crowther tells the same story about Lady Rutherford.[618]) Tea parties were also held for the research staff of the Cavendish. According to Hewish, "He [Bragg] was a genial host, but it was a somewhat formal occasion."[619]

Unlike most highly successful scientists, Bragg had many interests outside of the laboratory. The artistic talent he had inherited from his mother, and which very much influenced his approach to science, was expressed through sketches of friends and watercolors of places he visited. Like Vasari's Giotto, he could draw a perfect circle freehand.[620] In Cambridge, Bragg often took Patience to the Fitzwilliam Museum on Sunday afternoons. Later, when they had moved to the Royal Institution, he would take her to the Victoria and Albert Museum. The opening of the Royal Academy also became an important date in his calendar. Bragg's artistic preferences included English eighteenth- and nineteenth-century watercolorists and the Impressionists, whose bright colors and use of light reminded him of South Australia.[621] He was no devotee of abstract art, referring in 1966 to "some modern picture, consisting of an otherwise blank canvas with one button or other object sewn on it at a place which I suppose has enormous aesthetic significance."[622] From a lover of symmetry, this attitude is not surprising.

Bragg read widely in a variety of genres, generally tending to the middlebrow. When Patience read French at Cambridge, she found that Bragg was familiar with many of the novels she was assigned. "He would say, 'Oh yes, Racine, Corneille, turgid stuff. I don't think much of it.' "[623] He was much more comfortable with the classics of the English canon, his favorite works including the plays of Shakespeare, the novels of Austen, Trollope, and Dickens, and the poetry of Browning and Tennyson. In fact, he had a personal canon of ten works that he re-read every decade, including Austen's *Emma*, Tolstoy's *Anna Karenina*, Darwin's *Voyage of the Beagle*, and Homer's *Odyssey*, which he read in the original Greek.[624] Bragg loved detective stories, including the Sherlock Holmes novels of Arthur Conan Doyle and the works of Ngaio Marsh and Agatha Christie.[625]

Bragg was also keen on crossword puzzles. John Nye recalled: "One morning I was in a train from Cambridge to London sitting in a compartment with Bragg. We both had our copies of *The Times* open at the crossword page. He did not seem to be making any progress. 'It's a bit difficult this morning, isn't it', I said. 'No, I don't think so,' he replied; 'I have finished it all right'. He explained that he did the crossword at home each morning, but in his head without writing in any of the answers. This was so that, when his wife came down, she too could do it for herself. He was normally very modest and I do not think he was exaggerating."[626]

Bragg managed to drag his wife away from her mayoral duties for a couple of holidays in 1946. In June, they stayed at the Pleasure Boat Inn on Hickling Broad, a tiny place with only three bedrooms and a tin shed for a dining room. Using a small sailing boat, they were able to reach parts of the Broads inaccessible to the cabin yachts that they had previously used. The family holiday was spent at Blakeney, where they rented a boat from "the fish and winkle shop" and Bragg watched birds on the shore and at a nearby nature reserve.[627]

In July 1946, the Royal Society celebrated the tercentenary of the death of Isaac Newton—delayed for four years because of the war. King George VI visited the Cambridge University Library and various dons were asked to explain the significance of Newton's work to the royal visitor. Bragg later recalled: "I had 'to do gravity' and I remember the King saying 'What's all this about an apple, hadn't many people seen an apple fall before?' 'The point, Sir, was that he realized that the law governing the fall of the apple was the same as the law which kept the moon continually falling towards the earth'. 'Moon falling towards the earth, that's the first I've heard of it' and the King passed on to the next exhibit leaving me feeling how badly I had put it."[628]

Notwithstanding his failure to explain science to the King, Bragg was awarded a Royal Medal in November 1946. Two of these medals are awarded annually by the Royal Society, "for the two most important contributions to the advancement of Natural Knowledge, published originally in His Majesty's dominions." The award citation stated in part: "The implications and applications of the principles and methods of X-ray spectroscopy and X-ray structure analysis are one of the wonders of modern science, and with this manifold triumph the name of Sir Lawrence Bragg is inseparably associated."[629]

Also that summer, a meeting on "X-Ray Analysis During the War Years" was held at the Royal Institution under the auspices of the X-Ray Analysis Group of the Institute of Physics. With about 250 British participants and about 75 from 15 other countries, it was the first post-war occasion for crystallographers to meet. Max von Laue, now Director of the Max Planck Institute for Physics in Göttingen, was one of the foreign attendees. Perhaps in recognition of his courageous anti-Nazism, Laue was excused the travel restrictions on German scientists, but a British scientific liaison officer accompanied him to the RI as a "minder."

One of the main results of the X-ray analysis meeting was the creation of a new body for coordinating crystallographic research. *Zeitschrift für Kristallographie* had ceased publication, and it was decided to initiate a new journal. Bragg suggested that this journal be published by some parent organization, and he was delegated to approach Frederick Stratton, Professor of Astrophysics at Cambridge, who was General Secretary of the International Council of Scientific Unions, about setting up a crystallography union. The resulting International Union of Crystallography held its first meeting in Cambridge, Massachusetts, in July and August 1948. Bragg was not present but was elected President, serving until 1951. He also became an editorial board member of the new journal, *Acta Crystallographica*.[630]

In January 1946, there had been a key addition to the crystallography group at the Cavendish Laboratory. John Kendrew was a chemist by training, but "on the whole I thought chemistry was getting a bit dull, and I wanted to get into biology."[631] During the war, he had met the charismatic Desmond Bernal in North Africa and been persuaded that Bragg's department was the place to go.[632] On hearing that he was a chemist interested in biology, Bragg suggested that Kendrew work with Perutz on proteins. Despite Bernal's warning that crystallography was "a very tedious subject," Kendrew accepted. "Bragg was the *only* crystallographer in Cambridge—apart from Max Perutz—who did not believe we were wasting our time on a project much more complicated than had previously been attempted by the methods of X-ray crystallography."[633] However, if Kendrew had been a crystallographer, he would probably have agreed with the majority opinion: "my own almost total ignorance of this method was fortunate, in that it concealed from me the extent to which contemporary x-ray crystallographic techniques fell short of what was needed to solve the structures of molecules containing thousands of atoms; it was indeed a case of ignorance being bliss."[634]

Kendrew's Ph.D. project was on the differences in gross structure between fetal and adult hemoglobins of the sheep. For his subsequent post-doctoral studies, he decided that he needed to find a different protein to analyze. He came up with myoglobin, an oxygen-storage protein of muscle, which was known to resemble one of the four subunits of hemoglobin. Horse myoglobin could be crystallized, but Kendrew spent several years trying to grow crystals large enough for X-ray analysis. When that failed, he tried the myoglobins of porpoise, seal, dolphin, penguin, tortoise, and carp—all without success.[635] Whale meat was widely available in Britain in the post-war period, a time when other meats were scarce, and Kendrew eventually found that sperm whale myoglobin formed "beautiful monoclinic crystals."[636]

In the meantime, Perutz, together with two assistants, Joy Boyes-Watson and Edna Davidson, had completed a major study on horse methemoglobin.[637] Bragg was not an author on this long paper, but was thanked for "contributing many vital suggestions to this work" and for "active encouragement and interest." As Perutz had reported in 1942, horse methemoglobin crystals contain two molecules of 66,700 daltons in a face-centered monoclinic unit cell of space group $C2$. Although there is no center of symmetry in this space group, a projection of the hemoglobin crystal on the (010) face of the unit cell is centrosymmetrical, because the two-fold rotation axis is perpendicular to that plane. It was therefore much easier to determine the phase angles of the $(h0l)$ reflections that contribute to the (010) projection, as these would be 0 or 180°, than to determine the phases of the reflections that contribute to other projections, which could have any values. From now on, Perutz would concentrate on the (010) projection.

Perutz's new approach relied upon Paul Ewald's concept of the reciprocal lattice. Like the crystal lattice, the reciprocal lattice is a three-dimensional entity, described by reciprocal axes a^*, b^* and c^*. Also like the crystal lattice, the reciprocal lattice can be represented in a two-dimensional projection; the diffraction pattern of a crystal is a distorted projection of its reciprocal lattice.

The $(h0l)$ reflections which contribute to the (010) projection of hemoglobin fall on the reciprocal lattice plane with axes a^* and c^*. Reflections of the type $(00l)$, such as those from planes with Miller indices (001), (002), (003), etc., lie along the a^*-axis of the reciprocal lattice. Reflections of the type $(h00)$, such as those from planes with Miller indices (100), (200), (300), etc., lie along the c^*-axis of the reciprocal lattice. Reflections such as (201), (204) and (403) fall elsewhere on the a^*c^* plane. From intensity measurements, Perutz could calculate the modulus of the amplitude for each spot on the a^*c^* plane. (Because he was not using absolute measurements of intensity, the amplitude measurements he calculated were likewise only relative.) Perutz plotted these amplitude values as lines projecting above and below the corresponding layer lines ($a^* = 0$, $a^* = 2$, etc.) of the a^*c^* plane, the line above the layer line corresponding to the possibility that the reflection had a positive amplitude, the line below the layer line to the possibility that the reflection had a negative amplitude—both possibilities being equally likely. This kind of diagram was referred to as the "molecular transform" of the crystal. The actual amplitudes must vary sinusoidally along the layer line as a series of peaks and troughs; the diagram must also be symmetrical across the c^* axis. However, given that at each point on the reciprocal lattice the amplitude could be positive or negative, there were a large number of possible wave functions by which the maxima/minima could be connected.

This is where Bernal's idea of using the shrinkage stages came in. Since dried horse methemoglobin crystals shrink along the a-and c-axes, the amplitudes measured at different stages of drying would fall at places along the a^*-and c^*-axes of the molecular transform different from those of the amplitudes measured from the hydrated crystals. If there was a change of sign between two adjacent points on the reciprocal lattice of the hydrated crystal—that is, if one was in fact a maximum and the other a minimum—amplitudes obtained from the dried forms that happened to lie between these points should be lower in magnitude. Using this technique, Perutz and co-workers tried to resolve the molecular transform of hemoglobin into "loops" (maxima and minima) and "nodes" (points where the curve crosses the layer line).

Even knowing the location of the nodes, one would not know which of the peaks on either side of it was the maximum and which the minimum. However, one would only need to determine the sign of one reflection on a layer line for all the others to fall into place. To do this, Perutz and co-workers took advantage of Bernal's other idea, isomorphous replacement. As described above, this method had been of great help in John Robertson's solution of phthalocyanine. Because a nickel atom could be added to the phthalocyanine crystal without changing its lattice structure—the substituted and unsubstituted crystals are isomorphous—amplitudes of positive sign were increased and those of negative sign decreased. In like fashion, heavy atoms could be diffused into hemoglobin crystals, where they were thought to occupy aqueous layers between the protein molecules. By comparing the amplitudes obtained from mercury-containing and mercury-free crystals, Boyes-Watson et al. were able to assign signs to the amplitudes of the first four orders of the (00*l*) reflections. Using another approach, the "nodal-point method," they were able to extend this analysis to the (007) reflection.

Patterson maps of the (001), (010) and (100) projections of horse methemoglobin were also prepared. Some of these had been calculated by Dennis Riley, but the Oxford group had otherwise been squeezed out of the hemoglobin work. The total number of intramolecular vectors, about 10^8, was very large, but Perutz hoped that periodic patterns of folding of the polypeptide chain would result in prominent peaks in the Patterson. Lines of high density, spaced about 9–11 Å apart, were prominent in the (010) projection and could also be seen in the other two projections (Figure 8.1). This was similar to the strong 10-Å reflections seen in fibrous and other globular proteins.

The Patterson map was, as usual, ambiguous and the resolution of the molecular transform analysis was very low. About the most that could be concluded was that the hemoglobin molecules were cylinders 57 Å in diameter and 34 Å high, with their vertical axes parallel to the *c*-axis of the crystal—roughly the shape of a hatbox (Figure 8.2). Along the length of the cylinder, there appeared to be four layers of polypeptide: "it is tempting to propose a four-layered structure with the backbones of the polypeptide chains in the plane of the layers and the side chains protruding above and below."

This paper represented many years of work by Perutz, his students, and assistants. How much had been accomplished? The "loops-and-nodes" method of assigning signs to amplitude had apparently been successful, although it was impossible to be sure. However, the phases of almost 8000 reflections would have to be determined in order to generate a structure at atomic resolution. Perutz had now found a grand total of seven. This did not mean that he was even one-thousandth of the way towards a structure of hemoglobin, because the loops-and-nodes method could not be used for the non-centrosymmetrical (001) and (100) projections. It may also have started to occur to Perutz that the Fourier synthesis of hemoglobin, involving a gigantic unit cell and thousands of terms, would require a huge amount of computation and that the projection of a protein crystal, representing the superimposition of thousands of atoms, might be very hard to interpret. Ithaca seemed as far away as ever.

A more immediate problem was how the X-ray analysis of proteins was to be funded. The Imperial Chemical Industries Research Fellowship that now provided Perutz's salary was coming to an end. Towards the end of the war, Bragg had recommended him for a University Lectureship, but only in October 1953, on the eve of Bragg's departure from the Cavendish, would Perutz become a University Lecturer in Biophysics. An April 1946 application to the Royal Society for modest operating funding was unsuccessful, although these funds were subsequently obtained from the Department of Scientific and Industrial Research.[638] David Keilin, who as head of the Molteno Institute of Parasitology

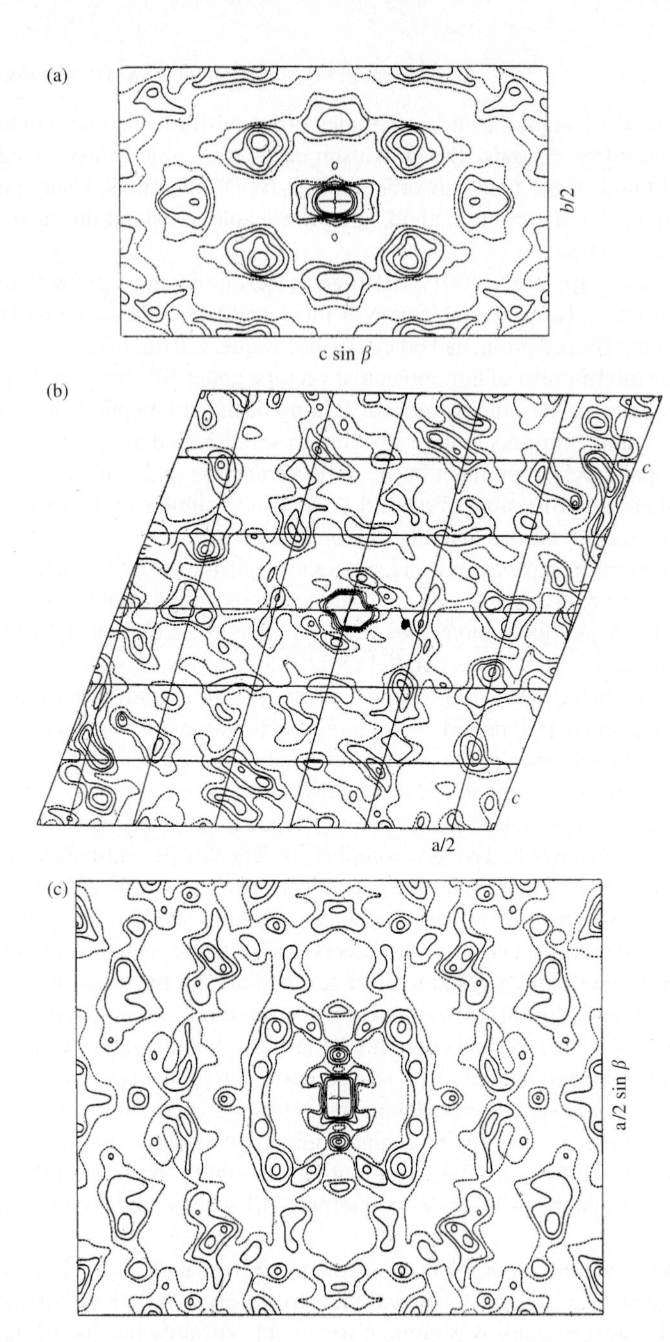

Fig. 8.1 Patterson maps of hemoglobin. (a) Projection on the (100) plane. The prominent peaks are on lines parallel to the *b*-axis and 9 Å apart. (b) Projection on the (010) plane. The most prominent peaks fall on the nodes of a 9 × 9-Å grid. (c) Projection on the (001) plane. Rings of density occur at 10–11 and 20–22 Å from the origin. Reproduced, with permission, from Figure 3 of Boyes-Watson, J., Davidson, E., and Perutz, M. F. (1947). An X-ray study of horse methaemoglobin. I. *Proceedings of the Royal Society of London A* **191**, 83–132. Published by the Royal Society

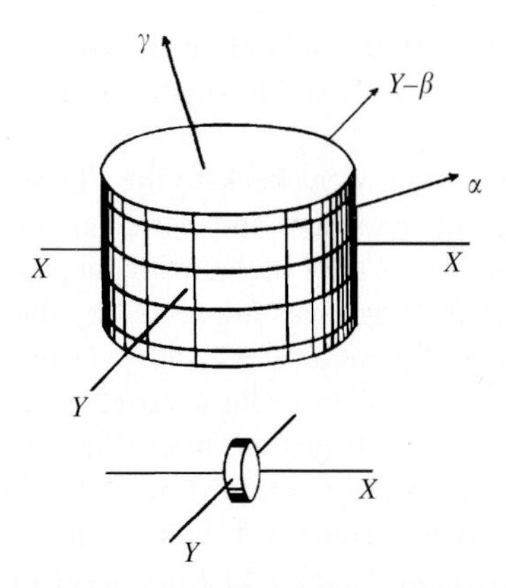

Fig. 8.2 Perutz's "hatbox" structure of hemoglobin. *Y* is the axis of two-fold rotational symmetry. The four lines around the cylinder represent concentrations of scattering matter suggested by the Fourier analysis. Reproduced, with permission, from Figure 12 of Boyes-Watson, J., Davidson, E., and Perutz, M. F. (1947). An X-ray study of horse methaemoglobin. I. *Proceedings of the Royal Society of London A* **191**, 83–132. Published by the Royal Society

had given Perutz and Kendrew laboratory space for crystal preparation, suggested that Bragg contact Sir Edward Mellanby, Secretary of the Medical Research Council (MRC).

In May 1947, Bragg wrote to Mellanby: "We thought it a great triumph to analyse quite simple organic salts by X-ray methods in the early days, and a complex organic molecule then seemed almost as far beyond our reach as the proteins might seem now. Yet, a patient accumulation of clues, and improved techniques, have made it possible to enter the organic field. If the structure of a few molecules of a new type can be analysed, a rich harvest is then reaped, because the structures of many others will then be clear by analogy. I foresee the same happening in the protein field . . ." Bragg clearly believed that proteins shared a common structural principle. The two men met at the Athenaeum, of which Bragg was a member, and Mellanby agreed that an application to the MRC was feasible. On October 20, Mellanby wrote to Bragg: "Rather to my surprise, your project for the establishment by the M.R.C. of a Research Unit at the Cavendish Laboratory, on molecular structure of biological systems, was adopted by the Council at the meeting on Friday October 17, although I had put it forward only for a preliminary run." The official title of the new entity was "Research Unit for the Study of the Molecular Structure of Biological Systems."[639]

Bragg's home life continued to be very full. The winter of 1946–7 being cold and snowy, the Braggs skied and tobogganned on the "Gogs" (hills south of Cambridge) and converted a former sand-pit at the back of the garden into "a kind of inverted Matterhorn and centre for winter sports." The Backs flooded and water came up West Road, though not as far as their house.[640] Margaret was now attending Downe House school, in Berkshire. Unimpressed with the science teaching there, Bragg offered to supply some used apparatus to equip a laboratory.[641]

In April, Bragg and Alice went back to the Pleasure Boat Inn, this time accompanied by Margaret, Patience, and Scrap. In June, Bragg, Alice, and Stephen visited Sweden, where they stayed with the physicist Manne Siegbahn in Stockholm. As in 1943, Bragg was impressed by the quantities and variety of food available: "Plates of great lobsters, smoked salmon, raw fish of various kinds, bear-meat, 'sandwiches' in endless variety, dainties with lashings of cream, cheeses and cakes, it was just too marvellous after the severe and dull rationing which was still in full force in England." Stephen did some train spotting and they also visited Harry von Eckermann's estate at Edeby. After Stephen returned to England, Bragg and Alice went to Gothenburg and then stayed with Johan Hedvall, professor of chemistry at Gothenburg Technical College, and his wife on the island of Orust off the Swedish coast. It was an idyllic spot: "We ate our meals in the open air under the pines . . . The Hedvall's had a motor boat in which we cruised amongst the archipelagoes." Despite his stuffy image, Bragg was unfazed by the Scandinavian habit of nude bathing. After that they went to Fjällbacka, near the Norwegian border, where their hosts were the Westgrens.[642]

By 1948, Bragg's reorganization of the Cavendish Laboratory was complete. There were now 160 researchers (including 110 research students), compared to 40–5 before the war, and only a third of applicants were accepted. In addition to the workshops of each research group, there was a "students' workshop" and a "boys' workshop" for the training of research students and laboratory assistants, respectively, and also "special techniques" and electronics workshops. Bragg wrote: "I regard the special-techniques workshop as a kind of library of dodges; any one of us who hears about some special device, of whatever nature, files it there, so to speak, for future reference." The clerical staff now numbered nine. "Each group . . . has its own secretary, a most important individual nowadays when paper-work assumes such amazing proportions." There were five practical laboratories—most members of the teaching staff had a college appointment and were responsible for a group of about a dozen students from their college. The number of Part II Physics students at Cambridge had increased from about 70 before the Second World War to about 125 in 1947.

The research groups at the Cavendish were described in Bragg's 1948 departmental report as: "nuclear, radio and low-temperature physics, crystallography, metal physics and mathematical physics, with some minor groupings." The radio group was studying the upper atmosphere using radio waves and also radio waves from space and from the sun. At the Mond Low-Temperature Laboratory, research on superconductivity and the properties of liquid helium was being performed. Orowan's metals group was studying various phenomena, including creep, brittle fracture, plastic flow, and rolling. The Laboratory had become much more diversified since Rutherford's time, which, even from Bragg's point of view, had some disadvantages: "The days are past when most researches in a department were on closely related lines and a joint colloquium was possible."[643]

The Crystallography Laboratory directed by Will Taylor now included Helen Megaw, a former student of Bernal's and now the first female staff member of the Cavendish. Another new member of the crystallography division was William Cochran, a post-doctoral fellow from Edinburgh. Cochran provided the following picture of life in the crystallography laboratory in the late 1940s: "He [Bragg] visited the crystallography group every two or three weeks to see what was going on ... W.H. Taylor managed the group in a very diplomatic style but those of my age scarcely appreciated the part he played in holding together a group which contained a number of prima donnas ... Social life centred on coffee parties in the late evenings, occasional dinners and sometimes a sherry party in a college, but we were an abstemious lot." At a Bragg Christmas party, a game of charades was played, and Bragg enacted hemoglobin ("hay-mow-glow-bin").[644]

One of the problems Taylor had to resolve concerned a large rotating-anode X-ray tube completed in 1948. This was intended to produce a high-intensity microbeam for studying metals, but Bragg put the cat among the pigeons by proposing that it be used instead for proteins—a telling insight into how his personal research priorities had changed, as well as into his limitations as an administrator. Taylor managed to negotiate a compromise between the two groups.[645]

Bragg would have been the first to admit that he was not a natural administrator. John Nye recalled: "He would sometimes appear by himself in the late afternoon, obviously bored by administrative matters, and wanting to do some science. 'You know,' he said, 'you dictate a lot of letters and Miss [Brenda] Smith [his secretary] makes them look beautiful, and you think you have done a job of work'. He meant that it wasn't real work."[646] Even Perutz, who admired Bragg greatly as a person and a scientist, found it hard to be positive about his administrative performance: "He's not a powerful persuader of men, and a person who sits on committees and gets his own way, and gets the money... I think he was rather ineffective in his dealings with the university, and in getting anything for the Laboratory ..."[647] Alice agreed, referring to "University politics which he hated."[648]

Ratcliffe, who was probably best-placed to judge, wrote: "A Cavendish Professor plays at least four parts. He must be a scientist, run the laboratory, uphold the interests of the department in the University, and act as an Elder Statesman of Science outside. Bragg was pre-eminently the active scientist, and he ran the laboratory extremely well. I do not think he played the part that some others have done in the University itself, and I am not sure that his part as Elder Statesman was quite as large as theirs would have been. I found him extremely helpful and, and above all a real gentleman in every way. He was quite open and straightforward and ready to help anyone who had the good of the laboratory at heart. I think there was an extremely good feeling in the laboratory during his time and all liked him."[649]

Certainly Bragg encouraged a devolved administrative structure and a collegial style of decision-making. Pippard wrote: "Bragg was no autocrat in the Rutherford mould—there were indeed those who felt he was too readily swayed by his closest counsellors."[650] Like his father, Bragg hated confrontation, which caused him sometimes to neglect personality problems that should have been addressed. His daughter Patience thought that Bragg was "a bit soft-shelled . . .. He did not like sorting out political differences within the lab."[651] Ronald King, who later worked closely with Bragg at the Royal Institution, noted that "He hated any situation where a rebuke was necessary."[652]

Bragg's ambivalent attitude to administration was rooted in his conviction that it was incompatible with research. Clearly speaking from experience, he wrote in 1969: "When one is trying to work out some knotty problem a process goes on in one's head like the Titans piling Pelion on Olympus and Ossa on Pelion in their attempt to scale the heavens. The structure tumbles down each time it is disturbed and has to be started again."[653] His feelings on the subject are neatly encapsulated in the following epigram: "The Muse of Science . . . flees from the busy man."[654] That researchers in his department should not be distracted by engagement books, intrays, telephones, typewriters, streams of callers and other "deadly enemies of scientific work"[655] was, according to Gwendy, "a lesson printed like Calais on his heart."[656]

Thirteen years after Bragg died, Lipson recorded a pungent criticism of Bragg as a team-leader: "I believe that he did not appreciate other people's problems, possibly because success had come to him so easily in his formative years. Evidence of this is that he did not support his staff in their applications for higher posts. I believe that Arthur Wilson and I were the only ones of his research group to obtain professorships in this country, and that was only some time after our period in the Cavendish." One of those who missed out was Reg James, who, according to Lipson, should not have had to go to South Africa to obtain a chair. Another was Ewald, who had come to Cambridge with Bragg's help in 1937, and moved to Queen's University in Belfast two years later—"W.L. was, he said, concerned about his English, but it was much better than most Englishmen's." Lipson also blamed Bragg for Alexander Stokes' failure to get a fellowship at Trinity. However, he regarded Bragg's appointment of Bradley as assistant director of the Cavendish to be "his greatest mistake"; Bradley was a "genius," but "hopeless as an organizer," and William Kay, the Manchester lab steward, had said that Bradley should not have been put in charge of anything. Lipson was also indignant that Bragg had made a point of telling him that he did not have anything to do with Bradley's FRS, and believed that good men did not need any "pushing"—"I only wish life were like that!" At the end of the war, Lipson wanted to be Director of the Crystallography Laboratory, but Bragg told him he preferred Will Taylor, who Lipson felt had become "too conservative."[657]

There is some truth in Lipson's criticism. Surely Ewald, a brilliant crystallographer and good friend of Bragg's, deserved his support in finding a permanent position in Britain. Bragg must also take some blame for appointing Bradley, whose madness was long in the making, as assistant director of the Cavendish. However, it is not entirely true that Bragg did not believe in "pushing" for his subordinates; as will be seen below, he supported Max Perutz's and John Kendrew's candidacies for the Royal Society and nominated Perutz, Kendrew, Jim Watson, Francis Crick, and Martin Ryle for Nobel Prizes. Certainly Bragg did not believe in recommending individuals merely on the basis of their association with him. Although John Randall admitted that Bragg's support was crucial at two points later in his career, he resented the fact that Bragg had advised him to pursue a career in industry rather than a Ph.D. in physics.[658] In the end, however, one must agree with Lipson to the extent that the trainees Bragg directly supervised did not go on to achieve great scientific eminence—the exception, Perutz, was inherited from Bernal. In this respect, he was quite unable to live up to his father.

As noted above, the only hope for the X-ray analysis of proteins seemed to be that the molecules had some periodic pattern of folding such as the "hexagonal fold" that Astbury had proposed for α-keratin or, for that matter, Wrinch's cyclol structure of insulin. An important paper had been published in 1943 by Maurice Huggins of the Kodak Research Laboratories in Rochester, New York.[659] For the first time, Huggins proposed that the polypeptide chains of proteins could fold into a spiral, or helical, conformation. As he pointed out, a helical polypeptide chain would have a screw axis of symmetry (a rotation combined with a translation along the rotation axis). Screw axes were common in the mineral crystals that Bragg had studied—for example, iron pyrites has a two-fold screw axis. In a protein, the asymmetric unit (fundamental unit of pattern) is one amino acid, so the number of amino acids per turn of the helix would be equivalent to the rotational symmetry.[dd] Thus, if the polypeptide chain rotated 180° between adjacent amino acids, the helix would have a two-fold screw axis of symmetry; if the chain rotated 120°, the helix would have a three-fold screw axis. The two-fold helix would actually be a "ribbon chain," in which the side-chain groups of adjacent amino acids project from opposite sides of the helix, and would have no "handedness." A helix with three-fold symmetry, however, could be either left- or right-handed. Huggins produced models with either two- or three-fold screw axes. The latter was a right-handed helix, stabilized by hydrogen bonds between amino acids three positions apart in the polypeptide chain.

One researcher who was very interested in Huggins' work was Linus Pauling. With Robert Corey, Pauling had now solved the structures of a variety of amino acids and small peptides, allowing them to make accurate measurements of bond lengths and angles. One important principle had emerged. The polypeptide "backbone" of proteins consists of three different bonds alternating

[dd] Since there were known to be about 20 different amino acids in proteins, and these seemed to occur in many different permutations, an amino acid was not strictly speaking equivalent to the asymmetric unit of a crystal. However, it had been accepted that the folding of a polypeptide chain such that amino acids were equivalent in space but not necessarily equivalent in structure would result in an X-ray fiber diffraction pattern.

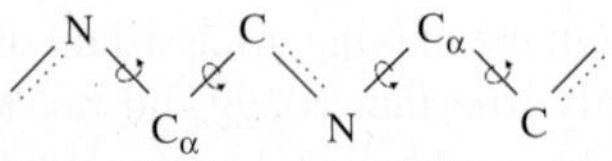

Fig. 8.3 Bonds present in the polypeptide "backbone" of proteins. Rotation of the chain occurs around the N–C_α and C_α–C single bonds, but not around the C–N bonds, which have partial double-bond character

along the chain: N—C_α, C_α—C and C—N, where C_α is the carbon atom to which the side-chain attaches (Figure 8.3). There seemed to be no restrictions on rotation of the chain around the N—C_α and C_α—C bonds, but Pauling and Corey had realized that the C—N bond—the "peptide" bond which connects adjacent amino acids—has partial double-bond character and will not allow rotation. The ability of polypeptides to rotate around only two of the three types of bonds in their backbones did not restrict the number of possible configurations in a general way—the number of ways a protein could be folded was still essentially infinite—but it did significantly limit the number of periodic structures, such as helices, that could occur.

Pauling spent 6 months of 1948 as Eastman Professor at Oxford University. Jack Dunitz, who was studying the structures of vitamin D derivatives and related compounds with Dorothy Hodgkin, found that Pauling was a "superb lecturer" but "made me feel stupid"; in contrast, Bragg, a frequent visitor to Oxford, was "courtesy itself."[660] While in England, Pauling visited Bragg. The relationship between the two men was friendly, if not warm, and Pauling left his son Crellin, then nine, with the Braggs for a week.[ee] As in 1930, Pauling found Bragg unwilling to talk shop: "At that time, too, Bragg and I did not have any serious discussions about science . . . Some years later I was told that Bragg resented my having intruded into the fields of crystallography and mineralogy in which he was working, and that he considered me to be a competitor. This information came as a shock to me."[661] It could not have been that much of a shock—Pauling certainly viewed Bragg in the same way. From England he wrote to Corey: "I am beginning to feel a bit uncomfortable about the English competition. They have a gift for driving straight at the heart of a problem, and getting its solution by hook or crook."[662] He was starting to think that it was time to start work on proteins or proteolytic fragments rather than amino acids and dipeptides. Pauling was also dismayed to find that the X-ray facilities at the Cavendish were far more extensive than those he had at Caltech.[663]

In March, confined to bed with a cold, Pauling decided to try to work out plausible structures of the polypeptide chain using nothing more sophisticated than pen and paper. His basic assumption was that all amino acids were structurally equivalent, which meant that the chain had to be helical. By folding a

[ee] Patience Bragg was delegated to take Crellin Pauling to a cricket match which featured the great Don Bradman. She was embarrassed when Crellin compared cricket unfavorably with baseball.

piece of paper along lines drawn through the α-carbons, Pauling tried to bring the C=O and N–H groups that participate in hydrogen bonds between amino acids into positions about 2.8 Å apart, which he knew to be the average length of a hydrogen bond in other substances. He also tried to ensure that the hydrogen bonds were roughly parallel to the rotation axis. It took him only "a couple of hours" to come up with two structures which satisfied these criteria. One of these had 3.7 amino acids per turn of the helix and seemed as if it would result in an X-ray diffraction pattern like that of α-keratin, so Pauling subsequently called it the α-helix (Figure 8.4). The other, which had 5.1 amino acids per turn, he called the γ-helix.[664]

When Pauling got back to Caltech and showed his helix structures to Corey, he was disappointed to find that there was in fact a discrepancy between the α-helix and α-keratin. The α-helix had a "pitch" (the distance along the rotation axis corresponding to one full turn of the spiral) of 5.4 Å, while the most prominent X-ray reflection of α-keratin was 5.1 Å. Pauling decided to put his helical structures away until he could resolve the discrepancy or find evidence for a strong 5.4-Å reflection in other proteins.

Like Pauling, Bragg was away from home a great deal in 1948. The Easter holiday, with Alice and the girls, was with Hopkinson relatives at Freshwater Bay, Isle of Wight, where they walked over the downs and along the cliffs. The Braggs then revisited the Pleasure Boat Inn. In July, Bragg gave lectures at a number of Dutch universities and visited the Philips company in Eindhoven. The latter offered him a weekend holiday of his choice, so he requested bird-watching, now one of the great passions of Bragg's life. Philips arranged for the president of the local natural history society to take him to the island of Schouwen, which was protected by a double set of dikes. Bragg found that numerous waders, including avocets and the Kentish plover, inhabited the shoreland between the dikes. In the sand dunes that line the North Sea coast of the island, Bragg saw other rare species of bird. He and his host also visited the medieval town of Zierikzee.[665]

In September, Bragg attended the British Association meeting in Brighton, where he gave the presidential address, on "Recent Advances in the Study of the Crystalline State," to section A (physics and mathematics). In this talk, he reiterated the central problem of X-ray crystallography: "Hidden in the pattern of diffracted spots lies all the information required to establish the structure, if only one can find the 'open sesame' which would reveal the treasure." Although the art of X-ray analysis had advanced to the point where structures with a hundred parameters could be solved, this still left the proteins far out of reach: "The photographs which Perutz and others have taken are like a message in code which we cannot yet decipher."[666]

The following month Bragg also went to Brussels to fulfil another presidential duty. He had been elected president of the Solvay Conference on Physics, and was responsible for organizing the 1948 conference on "Elementary Particles." Ernest Solvay, son of the founder, was "strongly pro-royalist and right-wing," whereas the conference secretary was the opposite. Bragg's letters

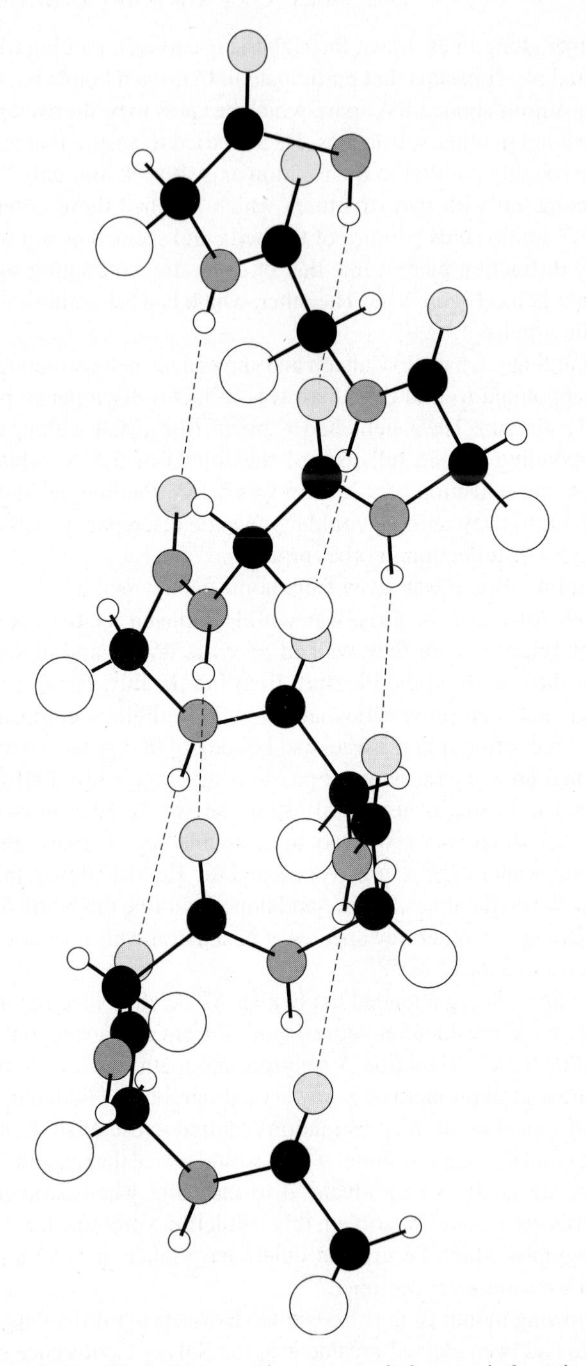

Fig. 8.4 The α-helix. Black circles: carbon atoms; dark grey circles: nitrogen atoms; light grey circles: oxygen atoms; small white circles: hydrogen atoms; large white circles: side-chains. Note that all C = O and N–H groups are involved in hydrogen bonds (dashed lines) and that the side-chains (which contain many atoms) are on the outside of the helix

to Brussels were often not answered, and he had to do most of the organizing himself.

The final trip of 1948 was the most ambitious—a five-week lecture tour of the United States and Canada. Alice accompanied Bragg, but not as a tourist this time—now a public figure in her own right, she gave 12 lectures to his twenty. They sailed in mid-October to Boston, where they stayed with the physicist Karl Compton, President of the Massachusetts Institute of Technology, and spent some time with their son Stephen, who was at MIT on a Commonwealth Fund Fellowship. After Boston, they went to Toronto, Rochester, Ann Arbor, Chicago, Pittsburg, and New York. Bragg lectured on "A Dynamical Model of a Metal Structure"—the most recent version of the bubble raft model—and "Recent Research Work in the Cavendish Laboratory." Along the way he met up with several old friends from previous visits to North America. Bragg had always enjoyed the contrast between North America and Britain, whether it was the unique landscapes of the former or the different attitudes of its inhabitants, but this time he found himself being more negative. In Pittsburg, the cars lining suburban streets reminded him of "aphis on a rose bud," and he also resented the prevalence of air-conditioning and consequent lack of fresh air.

Most of the tour was spent in Pittsburg, where Bragg had been invited by Alexander Silverman, Professor of Chemistry. He gave a series of lectures at the Mellon Institute that "drew record numbers." In New York, Bragg received the Roebling Medal, awarded by the American Mineralogical Society for "scientific eminence as represented primarily by scientific publication of outstanding original research in mineralogy."

The journey back to England was difficult. A dock strike meant that Bragg and Alice could not sail from New York, so they had to make a 36-hour train ride to Halifax, Nova Scotia. On the first night of the voyage home, Bragg had the alarming experience of awakening to find the cabin awash with six inches of water. It turned out to be a flood created by the passenger in the next cabin having left a tap running. The boat had no cargo because of the dock strike and "rolled abominably." Alice spent the whole voyage in bed and the stewardess who tended to her broke her arm in a fall. They arrived back in England on December 4 to complete "a record year of stays."[667]

Bragg's visit to Rochester was hosted by Maurice Huggins. Although protein structure was very much on Bragg's mind at this time, it appears that he did not discuss this topic with Huggins—his notes of the tour mention neither Huggins nor proteins.[668] A month after his return to Cambridge, Bragg wrote to Huggins: "I should be grateful for a copy, or two if you can spare them, of your papers on the structures of fibrous proteins which you published in October 1942 [sic]. Perutz and I have been thinking again about these alternative ways of folding the chains since he has got his more accurate results on the globular proteins, and it would be useful to have these copies to work with."[669] Bragg had missed an excellent opportunity to pick the brains of the man who was, excepting only Pauling, the leading expert on the folding of proteins.

Soon after Bragg returned from North America, Perutz published his second major paper on hemoglobin.[670] The Patterson map of the (010) projection had proved sufficiently tantalizing that the labor of generating a three-dimensional Patterson seemed worthwhile. Without a solution to the phase problem that would make a Fourier analysis possible, there was no real alternative to the Patterson method. Even so, Perutz was gambling on there being a periodic pattern of folding: "if the globin molecule consisted of a complex interlocking system of coiled polypeptide chains where inter-atomic vectors occur with equal frequency in all possible directions, the Patterson synthesis would be unlikely to provide a clue to the structure."

The three-dimensional Patterson analysis of hemoglobin, which apparently was performed by Perutz alone, resembled the labors of Hercules more than the trials of Odysseus. All 7840 reflections corresponding to d-spacings of 2.8 Å or greater were used: "The photographing, indexing, measuring, correcting, and correlating some 7000 reflexions was a task whose length and tediousness it will be better not to describe." On top of this, the amount of computation needed to construct the Patterson made Bragg's Fourier analysis of diopside look like a back-of-the-envelope calculation: using intervals of $x/120$, $y/60$, and $z/60$, corresponding to a resolution of approximately 1 Å, meant calculating the Patterson function at 58,621 points. For each of these points, a Fourier series with about 7000 terms had to be summed—a total of almost half a billion operations!

This was too much even for the dogged Perutz. By this time, however, the calculation of Fourier series could be done automatically. As early as 1939, Pauling wrote to Alexander Todd: "The principal activity in the molecular structure field now involves the possibility of getting our calculations made by machine."[671] Dorothy Hodgkin's Fourier analysis of penicillin used a punched-card machine developed for convoy planning in the Second World War. Such primitive computers could easily be adapted to perform the summations otherwise performed using Beevers–Lipson strips.[672]

Perutz's three-dimensional Patterson analysis of hemoglobin was performed by a computing services company using a punched-card device. Final results were plotted as 31 contour maps parallel to the (010) plane. The average value of F^2 being taken as "sea-level," only values above this were plotted. The main features of the Patterson were a "shell" of high density at about 5 Å from the origin and a series of rods parallel to the a-axis and 10–11 Å apart. Along these vector rods were peaks about every 5 Å. In an end-on projection, there appeared to be a central rod with six others arranged in a hexagon around it. Perutz's interpretation was that the polypeptide chain was folded in a zigzag pattern in the (001) plane such that the straight parts of the chain were parallel to the a-axis (Figure 8.5). Probably there were four such layers of protein in the hemoglobin molecule, as had been suggested by the one-dimensional Patterson. However, it was impossible to be sure—the proposed structure "may be regarded as plausible and consistent with all the known evidence, but proof for the time being eludes me."

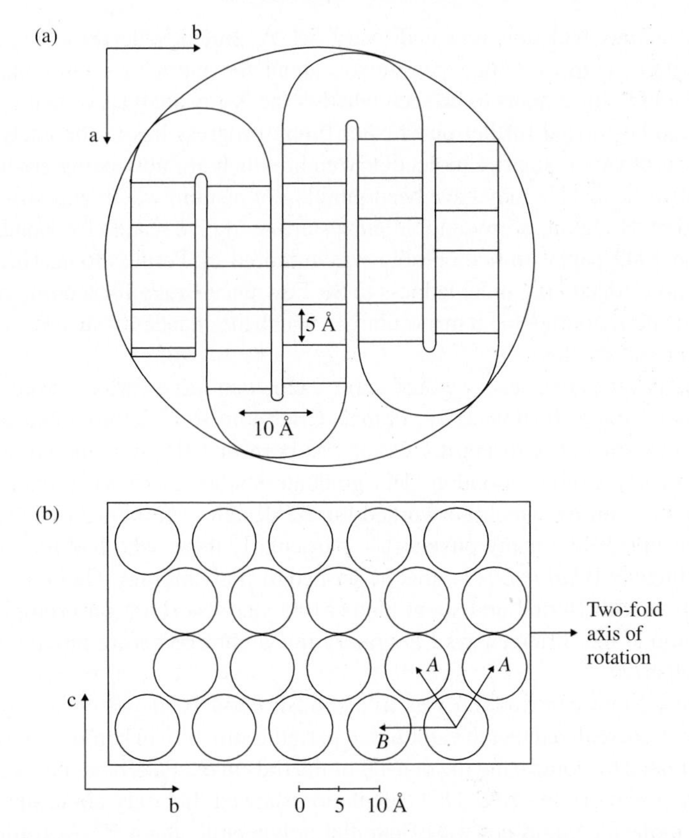

Fig. 8.5 Perutz's four-layer structure of hemoglobin. (a) Projection on the (001) plane. The protein is folded into five parallel segments. The prominent 10-Å peak on the Patterson map arises from the vectors between atoms 10 Å apart in adjacent segments; the 5-Å peak arises from vectors between atoms in amino acids adjacent in the polypeptide chain. (b) Projection on the (100) plane. In each layer, the five segments of the folded protein are seen in cross-section. Four such layers occur in the monoclinic unit cell. *A* and *B* indicate the major inter-atomic vectors in the (100) projection. Adapted, with permission, from Figure 23 of Perutz, M. F. (1949). An X-ray study of horse methaemoglobin. II. *Proceedings of the Royal Society of London A* **195**, 474–99. Published by the Royal Society

If Perutz's interpretation were correct, there may be a structure common to the globular and fibrous proteins. In α-keratin, Astbury had found strong reflections at 5.1 Å along the fiber axis and at 9.7 Å at right angles to it. These were temptingly similar to the 5 Å spacing along the rods and 10–11 Å spacing between them in the Patterson map of hemoglobin. From the density of the hemoglobin crystal, it could be estimated that the 5-Å unit repeating along the rods represented about three amino acids, which meant that the polypeptide chain must be folded in some way. In Astbury's proposed structure for α-keratin, a three-amino-acid unit repeated every 5.1 Å. But despite these intriguing similarities, Perutz seemed pessimistic about the merit of continuing with hemoglobin: "It remains to be seen whether the X-ray analysis of hemoglobin itself can be carried further or whether future progress lies in the analysis of simple proteins of smaller molecular weight which are now being studied by several workers." He may have been thinking of insulin, which was still being studied by Hodgkin, or myoglobin, now starting to give results for Kendrew.

The 1949 paper on hemoglobin was authored by Perutz alone. However, he acknowledged "my indebtedness to Sir Lawrence Bragg for lending me his support and encouragement in a venture in which the chances of success seemed forlorn to most others."

The chances of success would soon seem even more forlorn, even to the optimistic Bragg. In June 1949, Francis Crick joined the hemoglobin project as a Ph.D. student with Perutz. Crick was born in 1916 and studied physics at University College London. His graduate studies there were interrupted by the war, during which he worked on the design of mines. After the war, Crick decided, like many physicists—particularly those who had read Erwin Schrödinger's *What Is Life*?—that he wished to study biology. He failed to get a position with Bernal and spent almost two years working on protoplasmic streaming at the Strangeways Laboratory in Cambridge before moving to the Cavendish.[673]

Crick's first contribution was to demolish Perutz's model of hemoglobin structure. He realized that the amount of periodic structure in hemoglobin could be estimated by comparing the density of the rods in the Patterson map with the density of the origin peak. This calculation showed that only about one-third of the molecule could consist of parallel polypeptide chain.[674] According to Robert Olby, this analysis "marked the end of the belief in regular geometric structures for the globular proteins."[675] In a Royal Institution discourse given in April 1949, Bragg had stated: "One must be prepared to abandon without hesitation any proposed model, however attractive it may appear, if further evidence indicates that it is not valid."[676] This fate now befell the hatbox model of hemoglobin.

As in the previous year, Bragg did a good deal of traveling in 1949, although it was personal and closer to home. Now that Margaret was at boarding school, Patience was the only one of the Bragg children living at West Road. She conceived of a plan to take her father on a sailing holiday in his beloved Norfolk Broads. After two years of saving, the holiday took place at Easter, 1949. Bragg, Patience, and Scrap set sail on the "Wild Rose," "a delightful cabin yacht for two" with calor gas and electric light among the innovations since Bragg had last sailed there. "In the evening, while the bitterns boomed in the background, my father would sketch from the deck while I wrote up the log and my dog, Scrap, splashed muddily through the reeds, returning quite filthy to curl up on my bunk."[677] For Patience, "It was idyllic, it was my idea of absolute heaven. He was a very good teacher and showed me how to sail. But the fun was that we had a boat without a motor, which is quite tricky, and when we went through

Yarmouth harbour we didn't get the tide right, so we drifted backwards. We had all sorts of adventures. We took my dog with us and he kept falling overboard. When we wanted to wash up, we just put everything in a bag and trailed it behind us in the water. We once ate four pounds of strawberries between us for supper. We just enjoyed ourselves."[678]

Frosty in the mornings at first, later in the week there was a heatwave, and they "explored all the rivers above Acle." Once they had just gone under a bridge and Bragg had the difficult task of raising the mast single-handedly. He had just about got it secured when a "boat with four young men in bobble caps" ran into them and brought the mast down on Bragg's head. "Patience said afterwards 'Daddy, I didn't know you knew those words.' " Meat was still on ration, but Alice had given them a small package of bacon. They decided that they would have the bacon for dinner on the last day, but a "sailing distraction" occurred and they found the bacon hanging from Scrap's mouth. They retrieved the meat and cooked it anyway.

Bragg wrote in his autobiography: "The Broads have always fascinated me. I think it is the way in which one moors at night right away from everything, with the vast expanse of sky, the aromatic marsh vegitation [sic], and the night noises of bird and beast, and yet one has all the comforts of a snug home which one carries with one like a snail its shell." He lamented the presence of "despicable motor cruisers" and the disappearance of the wherries, which could outsail a yacht with their great sails "like the inside of mushrooms."

In June, Bragg and Alice spent a week in Spiggie, Shetland. Bragg's first impressions of Shetland were not favorable, as they drove through "dreary moors" set with "hideous houses." However, he and Alice greatly enjoyed walking along the tops of cliffs densely populated with nesting seabirds, and among which "the sheltered clefts in which the little streams ran were brilliant with flowers." Some mornings they arranged to be rowed out to islands and picked up in the evening. Skuas attacked them if they approached their nests, but seals would swim towards them on the beach. Machair, land where sand had drifted over the soil, was "a carpet of thyme, buttercups, orchids, pansies and other dwarf flowers." At the high latitude of Shetland, darkness never fell, and larks sang all night.

Also in the summer of 1949, Bragg and Gwendy had a sketching holiday at Portesham, Dorset. The sketching was poor but "It was a great delight, though, to have her all to myself." In August, Bragg and Alice went to Polzeath, Cornwall, with Margaret and Patience and two exchange students from Switzerland and France. But this holiday was not a success—Alice was ill from the strain of the war and her busy mayoral life, and David showed up with "the absolutely appalling 'Nigel,' " a con-artist.[679]

Bragg was still publishing on the bubble raft model. Otherwise, his role in research was now as a team-leader rather than investigator. He wrote in 1948: "It is, alas, a long time since I last analyzed a structure personally."[680] Perhaps it was his April 1949 Royal Institution discourse on "Giant Molecules" that stimulated Bragg to take up the cudgels one last time. Whatever the reason, Bragg now took a much more active role in the protein work. According to Perutz, "at that stage Bragg thought that the first priority was to determine the nature of the fold that gave rise to these rods in myoglobin and haemoglobin."[681] To that end, he wrote to Astbury in the autumn of 1949, requesting reprints of his papers on keratin. Astbury wrote back: "To put it as simply and briefly as possible, our inference that the 5.1 Å repeat in α-keratin and its analogues corresponds to a group (fold) of three amino acid residues rests on the two following findings: (a) that the β-keratin diagram represents a system of extended polypeptide chains along which the average length per residue is roughly 3.4 Å; and (b) that the β-configuration is approximately twice as long as the α-configuration; this was concluded from our demonstration that that the full range of repeatable, reversible extensibility of wool and other hairs is to approximately twice the normal length—and similar results were later obtained with oriented strips of extracted myosin [a fibrous protein from muscle]. It follows that the average length per residue along the chains of α-keratin is about 1.7 Å, giving three residues in the repeat of 5.1 Å."[682]

That much seemed clear. However, Bragg did not agree with Astbury's structure of α-keratin: "It seemed to me that Astbury's model of a kind of Greek key pattern was extremely improbable, and that a helix was a far more likely structure because it placed each amino-acid residue in the same kind of position in the chain."[683] This was correct—indeed, Pauling had used a similar line of reasoning in producing the α and γ helices[684]—but such classical crystallographic thinking would soon lead Bragg astray.

To illustrate his point, Bragg came to the Crystallography Laboratory with a broomstick into which he had hammered nails in a spiral pattern. He set Perutz and Kendrew to work on analyzing all possible helical arrangements of amino acids in a polypeptide. The resulting analysis was published in a 1950 paper entitled "Polypeptide chain configurations in crystalline proteins."[685]

Bragg, Kendrew, and Perutz made the following assumptions: that the bond lengths and angles in proteins were the same as those measured by Corey and others for amino acids and dipeptides; that proteins fold into structures with screw axes of symmetry; and that the folded structure is stabilized by hydrogen bonding between amino acids at different positions in the polypeptide chain. The structures generated were classified by their screw-axis symmetry (two-, three-, four-fold or higher); the number of "backbone" atoms in the ring formed by hydrogen-bonding the C=O group of one amino acid to the N–H group of another (7, 8, 10, 11, 13, or 14); and the fraction of the total C=O/N–H groups involved in hydrogen bonds. For example, the "$2_{13} \cdot \frac{1}{3}$" helix had two-fold symmetry, 13 atoms in the ring and $\frac{1}{3}$ of the possible hydrogen bonds (Figure 8.6). In most cases, structures could exist in either right-handed or left-handed forms. In rating the plausibility of structures, preference was given to those in which hydrogen bonding was maximal and in which an apparent repeat distance of about 5 Å contained three or four amino acids.

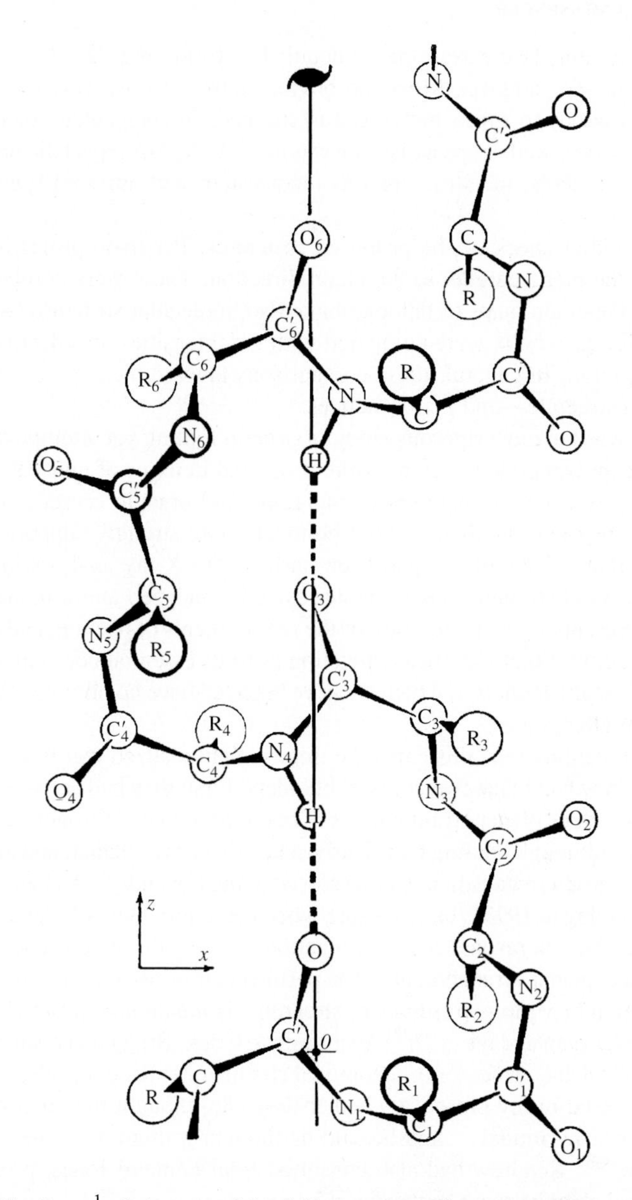

Fig. 8.6 The $2_{13} \cdot \frac{1}{3}$ chain, one of the folded conformations of proteins proposed by Bragg, Kendrew and Perutz in 1950. "2" refers to the presence of a two-fold screw axis of symmetry (vertical line), "13" to the presence of 13 atoms in each of the "loops" formed by the hydrogen bonds (dashed lines) and "$\frac{1}{3}$" to the fact that only one-third of the C=O and N–H groups are involved in hydrogen bonds. Unlike in Pauling's α-helix, there is no requirement that the peptide bond be planar (if there were, atoms O_4, C'_4, N_5 and C_5, for example, would all be in the same plane). Reproduced, with permission, from Figure 8 of Bragg, W. L., Kendrew, J. C., and Perutz, M. F. (1950). Polypeptide chain configurations in crystalline proteins. *Proceedings of the Royal Society of London A* **203**, 321–57. Published by the Royal Society

Ten structures were examined in detail: Two forms of 2_7, 2_8, $2_{13} \cdot \frac{1}{3}$, $2_{14} \cdot \frac{1}{3}$, 3_7, 3_8, 3_{10}, 4_{11}, and 4_{13}. Most had obvious flaws: Atoms were too close for hydrogen bonds to form, the ring was strained, the percentage of hydrogen bonds was too low, the repeat distance was not 5 Å, the 5-Å repeat did not contain three amino acids, the structure was inconsistent with infrared spectroscopy data, etc.

As another check of the proposed structures, Patterson projections were made in the plane parallel to the chain direction. These were compared with Perutz's Patterson maps of hemoglobin. Also, molecular structure factors calculated for each helix were compared with the intensities of reflections found for myoglobin. In general, these comparisons favored the $2_{13} \cdot \frac{1}{3}$—Astbury's α-keratin structure—and $2_{14} \cdot \frac{1}{3}$ helices.

This was the most rigorous study of protein folding yet attempted. For the first time, polypeptide structures were described in terms of unit-cell coordinates, as was done for analyses of inorganic and organic crystals. However, Bragg's hope that one helix would be much more strongly supported by the analysis than all the others had been dashed: "In X-ray analysis in general, when a crystal structure has been successfully analyzed and a model of it is built, it presents so neat a solution of the requirements of packing and interplay of atomic forces that it carries conviction as to its essential correctness. In the present case the models to which we have been led have no obvious advantages over their alternatives."

When Pauling read this paper, he immediately realized that Bragg, Perutz, and Kendrew had made two colossal blunders. First, they had considered tetrahedral as well as planar arrangements of bonds around the nitrogen atom of the backbone, although Pauling had clearly stated that it was planar, and although a dipeptide structure had already been solved in the Cavendish. As Pauling wrote to Robert Olby in 1973, "I am not sure when the requirement of planarity of the atoms involved in resonance or conjugation was first discussed. The existance [sic] of a coplanar arrangement of the pertinent atoms as a requirement for the contribution of various [resonance] structures is mentioned in the 1933 paper by Jack Sherman and me . . ."[686] According to Crick, Bragg had been misled by Charles Coulson, professor of theoretical chemistry at King's College London: "he was asked in my presence whether they [sic] thought the nitrogen would be planar or pyrimidal . . . and he said he thought it might be either, which is nonsense."[687] Kendrew had also consulted John Lennard Jones, professor of theoretical chemistry at Cambridge.[688]

Second, Bragg, Perutz, and Kendrew had failed to consider structures in which the screw symmetry of the helix—corresponding to the number of amino acids per turn—was a non-integral number. Huggins had stated in his 1943 paper that "there is nothing about this [3_{10}] structure which requires exactly three residues [amino acids] per turn of the spiral. In fact, it would seem, from the models that have been made, that the bond distance and angle requirements are best satisfied by a slightly smaller number of residues per turn." Nonetheless, Perutz recalled, "It never occurred to us that it might not be integral."[689]

The first blunder was not critical—it only meant that Bragg et al. considered implausible structures as well as plausible ones—but the second was, as it specifically excluded from consideration Pauling's α- and γ-helices.

In their analyses of mineral crystals, Bragg and Pauling, although having very different scientific training, were quite evenly matched. When it came to proteins, though, Bragg was at a distinct disadvantage. Jack Dunitz thought that he knew little chemistry.[690] Kendrew, "never found him very interested in biology—not even in the chemical structure of the compounds the structures of which he analysed... Basically he was a puzzle-solver; to him the great fascination was to interpret the complicated diffraction pattern, say of a protein crystal, in terms of its three-dimensional structure."[691] Bragg's 1952 paper on the form of the hemoglobin molecule includes the following statements: "the molecular weight of globin is 33,000" and "globin contains three separate chains."[692] (Hemoglobin contains four globin chains of about 16,000 daltons each.) As late as 1967, Bragg wrote: "Some [amino acids] are acidic, some are basic, some polar, some neutral," which fails to recognize that "polar" and "neutral" are not mutually exclusive concepts.[693] According to Crowther, "Physicists do not like chemistry, because there are too many details in it that have to be learned."[694]

Pauling knew that his α-helix was a much better candidate for α-keratin than any structure that Bragg, Perutz, and Kendrew had discussed. However, it did not fit the diffraction data and he thought it was only a matter of time before the Cavendish group "learned enough chemistry" to find the α-helix.[695] If he published the α-and γ-helices, Pauling risked being associated with a wrong structure; if he waited, he might be scooped by Bragg or someone else.

Pauling's state of mind during the period between his discovery of the α-helix and its publication is revealed by a letter he wrote to his former student, David Harker, in March 1951: "I have had some ideas about protein structure that I have not published, nor even mentioned to other workers, except those in the laboratory here, during the past three years... If I had told Perutz about them, or Bragg, they probably would have checked up on them, and might not have published the paper that they did publish in the October 1950, Proceedings of the Royal Society."[696]

As a compromise, Pauling and Corey published a very short description of the α- and γ-helices in the November 1950 issue of the *Journal of the American Chemical Society*.[697] Containing just enough details to unambiguously characterize the structures, the paper was a modern version of the "sealed note" that Sommerfeld had deposited with the Bavarian Academy of Arts and Sciences to establish Laue's priority for the discovery of X-ray diffraction. Pauling and Corey had already begun to adopt Perutz's "top-down" approach of studying intact proteins to determine the pattern of polypeptide folding, using X-ray data on lysozyme provided by Kenneth Palmer of the Western Regional Research Laboratory in Albany, California.

Bragg turned sixty in March 1950. In addition to his newfound enthusiasm for research, he continued to enjoy the life of a grand old man of science.

In April, he took Alice to Portesham, where he had stayed the previous year with Gwendy. In June, there was another lecture tour, this time to Switzerland. The main event was a series of talks at the International School for Training in Management in Geneva. Once more Bragg and Alice were exceptionally well treated—they found the guest flat of the school fully equipped with cocktail cabinet, restaurant coupons, tram tickets, pre-stamped postcards, etc. After Geneva they went to Grindelwald, where they enjoyed the Alpine flower-meadows: The countryside, flowers, "anything to do with water" and architecture being their chief mutual interests.[698] They also visited Zürich, Basel, and Bern, where Bragg gave lectures on the bubble model, protein structure, and the Cavendish Laboratory.[699] As Cavendish Professor, Nobel Laureate, and recipient of many other awards, Bragg was much in demand on the international science circuit. With his stock of well-polished lectures he was able to gratify Alice's love of travel and the high life, and his own interest in nature. Those who wished to attract the great man soon realized that the prospects for success were greater if Alice were also invited—and greater still if bird-watching were part of the itinerary!

The Bragg summer holiday for 1950 was in Blakeney. Bragg entered a local sprint in which the runners were handicapped by age, and won a prize.[700] He and Patience were also successful in a yacht race, using the fish-and-winkle man's boat: "we had a gun all to ourselves as second in, and it was hard to say whether Patience or the winkle man beamed most." They were joined in Blakeney by Stephen, now working for Rolls Royce in Derby, and his fiancée Maureen Roberts. Maureen was the daughter of Dorothy Amos, who had been a childhood friend of Alice's.[701]

In October, Bragg was an invited guest at a celebration of the 50th anniversary of the General Electric research laboratory in Schenectady, New York. Afterwards, Bragg and Alice visited Pocono Manor in Pennsylvania: "we stayed in the off-season in one of those enormous hotels in the wilds which only America produces. There were arrangements for every kind of activity, including specially laid out honeymoon trails, and an old gentleman with a long white beard to take one nature walks and explain the names of the trees."[702]

Another jubilee occurred two months later. To commemorate the 50th anniversary of the Nobel Prizes, all living Nobel Laureates were invited to attend the annual award ceremony. The highlight of the celebration was dinner for 1400 in Stockholm town hall. Each course was brought to the tables by waiters carrying silver trays, marching down the grand staircase accompanied by music. Dessert was delivered on blocks of ice into which electric lamps had been frozen. After dinner, Alice recalled, "Stockholm was so beautiful with deep, hard snow, and all the Christmas decorations shining and twinkling in the dark that dropped soon after lunch."[703]

Bragg's report for 1950 noted that the total number of students in the Cavendish was 513, up from 459 in 1948–9; however, first-year students were down to 204 from 234, as the post-war boom passed. The staff included three professors (The Cavendish, Jacksonian, and Plummer), four readers, eight lecturers in experimental and two in theoretical physics. Research students and staff totalling 161 had produced 114 papers during the year. A new course in theoretical physics had been started, as well as a summer course on electron microscopy. Bragg had also introduced non-credit Arts courses on "The Ancient World," "The Growth of English Literature," "Literature and Thought from the 17th to 19th Century," and "Modern Poetry." In taking this initiative, Bragg was putting into practice his conviction that a balanced education required both arts and sciences. The average attendance at these arts courses was a very respectable 220.[704] In subsequent years, the topics presented included "The Novel, from Conrad to Virginia Woolf," "Science and the Modern Novel," "The History of North America," "Economic and Social Development Since the Middle Ages," "Music," and "Man and His Environment"; attendance remained very healthy. The science courses that Bragg offered for arts students were, however, received with much less enthusiasm.[ff]

A storm that had been brewing over Bragg's unsuspecting head since the publication of the paper on protein helices burst in the spring of 1951. Pauling's ambivalence about the α-helix was resolved when the synthetic polypeptide polybenzylglutamate was shown to give an X-ray reflection corresponding to a 5.4-Å spacing along the helix axis—just like the α-helix. On February 28, 1951—his fiftieth birthday—Pauling sent off a full paper on the α-and γ-helices.[705] In May, *Proceedings of the National Academy of Sciences* published seven papers on protein structure from Pauling's group. In Thomas Hager's words, "It was as though a single composer had debuted seven symphonies on the same day."[706]

Not knowing quite what to make of the α-and γ-helices, Bragg decided to take Pauling's paper to Alexander Todd, professor of organic chemistry at Cambridge. Bragg's group had been collaborating with Todd's on studies of both peptides and nucleotides, but this was apparently the first time Bragg had set foot in the Chemistry Department. Todd shared Pauling's low opinion of Bragg's grasp of chemistry, and told him so in no uncertain terms.[707] (Crick and Kendrew both expressed doubts about this version of events.[708] However, Todd's story is supported by Bragg's later insistence that the double helix model of DNA be checked out by Todd before it was published [see below].)

Perutz's reaction to the α-helix was that it should give a prominent reflection at 1.5 Å, corresponding to the spacing of adjacent amino acids along the helix axis—too far out in "reciprocal space" to appear on the plates normally used at the Cavendish. He photographed a horsehair with a larger plate and immediately found the 1.5-Å spot. It soon became clear that the α-helix also occurred in hemoglobin and myoglobin. A letter from Perutz to Pauling written

[ff] After Bragg had left the Cavendish, Ratcliffe proposed the creation of a "Principles of Science Tripos" that would consist of science courses for arts students. This proposal was rejected by the University, but "History and Philosophy of Science" was introduced as a half subject in Part I of the Natural Science Tripos [Budden, K. G. (1988). John Ashworth Ratcliffe. *Biographical Memoirs of Fellows of the Royal Society of London* **34**, 671–711].

on August 17, 1951, contains the rather fawning remark that "the discovery of this reflexion in haemoglobin has been the most thrilling discovery of my life."[709]

Of the 1950 paper on polypeptide chain folding, Bragg later wrote: "I have always regarded this paper as the most ill-planned and abortive in which I have ever been involved."[710] What was worse was that his come-uppance was at the hands of Pauling, who had scooped him on mica, corrected him on cyanite and embarrassed him on the rules of crystal packing. According to Perutz, "Bragg was annoyed that Pauling should have beaten him to it the second time. He didn't blame Pauling for that, he blamed himself."[711] Bragg was nothing if not magnanimous in defeat, writing to Pauling: "We have been tremendously interested in your broadside of papers on protein structure. Your solution of the α-keratin chain carries conviction. It fits in so beautifully with many facts. I think we were led astray in our review of chains by a feature of the Patterson projection that originates from something else than the chain structure. A spiral pattern has always appealed to me much more strongly. The Astbury chain always seemed such a very artificial one for so universal and fundamental a structure. I do congratulate you most warmly on what I feel is a very real and vital advance towards the understanding of proteins."[712]

It was difficult for Perutz to be so gracious. There was still no solution to the phase problem, the amount of periodic structure in hemoglobin was only about one-third of the total—and now Pauling had solved the structure of that part of the molecule. All that was left was the remaining aperiodic two-thirds with no method to attack it. This was the lowest of many low points on the odyssey.

Things looked so bleak that Bragg was willing to consider a new research direction. One possibility was deoxyribonucleic acid (DNA). Long considered a structural component of the nucleus, this molecule was now coming to be thought of as functioning with proteins in the transmission and expression of hereditary characteristics. Although Bragg would certainly not have known this, and it is doubtful that Perutz would have, a 1944 study performed by Oswald Avery at the Rockefeller Institute for Medical Research in New York had suggested that DNA might itself be the hereditary material, at least in bacteria.[713]

X-ray analyses performed by Astbury in the 1940s indicated that DNA was helical. At the Cavendish, students supervised by Cochran had determined the crystal structure of the purine compound guanine, one of the four "bases" of DNA, and were now investigating nucleosides, which consist of a base and the sugar deoxyribose. DNA would have been a logical choice as a sideline, if not an alternative, to the globins, but the decision was made to leave this field of research to John Randall's MRC Biophysics Research Unit at King's College London. Randall had been a physics student in Manchester when Bragg was professor there, and had then done an M.Sc. under the supervision of Reg James. During the Second World War, he and Henry Boot had developed the cavity magnetron, a critical component of British radar technology, which Franklin Roosevelt described as "the most valuable cargo ever brought to the shores of America." Randall was now Wheatstone Professor of Physics and director of the biophysics unit at King's. DNA was being studied by Maurice Wilkins, a New Zealand-born biophysicist doing post-doctoral research under Randall's supervision. He and Crick had been friends since they served together in the Admiralty during the war.

Bragg told Robert Olby in 1967: "Now I remember at that time I was feeling very low that we had so much missed the bus over the alpha helix, and I said to Perutz, 'Now look, are we missing any other buses—what about DNA, ought we to have a shot of that, because we have some elementary knowledge about the purines and pyrimidines?' And Perutz said. 'We musn't have a shot at that; Wilkins has done so much work and got such wonderful results with the material he so painstakingly collected that really nobody else ought to enter in on that field; we must leave that to him."[714] Perutz, however, claimed that it was Bragg who insisted that DNA be left to the King's College group: "Bragg felt very strongly that . . . we should let King's College have their structure, and go ahead on our own structures, a gentleman's agreement."[715]

Biological fibers were of great interest to the King's group, and Wilkins' studies on DNA had significantly added to Astbury's earlier X-ray analyses of this enigmatic macromolecule. In January 1951, the physical chemist Rosalind Franklin joined Randall's group and was assigned to the DNA project over Wilkins' objections.

Bragg or Perutz or both may have decided that the gentlemanly thing was to leave DNA to the King's group, but other members of the Cavendish Laboratory did not necessarily agree. In the autumn of 1951, James Watson joined the MRC Unit as a post-doctoral fellow with Kendrew. Watson had done a Ph.D. on bacteriophage (a type of virus that infects bacteria) with Salvador Luria in Indiana and had spent about a year working on nucleic acid chemistry in Copenhagen. He was convinced that DNA was the genetic material and that the solution of its structure would be of the greatest importance. Watson persuaded Crick, easily distracted from his thesis work on hemoglobin, that they should work on the structure of DNA in their spare time.

Around this time, Bragg made an unwitting contribution to Watson's and Crick's informal DNA project. Following Pauling's publication of the α-helix, Bragg asked Cochran, who "acted as consultant sometimes to the protein-crystallography group," to work out the Fourier transform of a helix. Busy with other work, Cochran did not immediately respond, but was finally prodded into action when Bragg gave him an unpublished manuscript on the same subject written by Vladimir Vand of the University of Glasgow. Cochran concluded that Vand's theory was correct for a continuous helix but not for helical arrangements of discrete units such as atoms. In the darkroom one day, Cochran mentioned to Crick that he was working on a helical diffraction theory. Crick, who had the habit, as Bragg put it, of "doing someone else's crossword," decided to solve it himself.

The solutions produced by the two theorists were formally identical, although Cochran's approach, which involved Bessel functions, was undeniably more elegant. According to these theories, a helical polymer will give an

X-ray diffraction pattern consisting of two rows of spots in the shape of an "X." From the slope of the arms of the "X" and the spacing between the spots, the radius and pitch of the helix can be determined. In addition to the "X"-shaped array of spots, there will be spots on the meridian (vertical axis) of the diffraction pattern, the distance of which from the origin depends upon the spacing of the monomeric units along the helix axis. Cochran then realized that the X-ray diffraction patterns he had obtained from the synthetic polypeptide polymethylglutamate, which Bragg had asked him to analyze, were indicative of a helix.[716] Meanwhile, Alex Stokes, Cochran's counterpart in Randall's group, had derived a similar theory.

Bragg probably first became aware that Watson and Crick were working on DNA in late November, when a three-stranded helical structure they had come up with was rubbished by Franklin and Wilkins. According to Watson, Bragg then sent word via Perutz that DNA was to be left to the King's College group. It is commonly assumed that this edict resulted from an agreement between Bragg and his former student Randall. Such an agreement could have been arrived at by telephone or, for example, at the Athenaeum, of which both men were members. However, there is no record of *any* communication between Bragg and Randall on the subject of DNA, and it is quite possible that Bragg, quite independent of King's, decided to continue the "ban" on DNA work that he had earlier agreed upon with Perutz. Bragg's version of the story, given in a May 1965, letter to Warren Weaver of the Rockefeller Foundation was: "I cannot remember urging Crick to stick to protein work but the story may have originated in this way. I must confess that I did at times get exasperated with Crick (between ourselves) instead of sticking to his own experimental results he was always setting himself to interpret the results which other workers had got with blood and tears, and this caused much upset in the laboratory. He meant no harm, but he could be very irritating, and again I must confess this made me at first underestimate his genius."[717]

At the time he told Watson and Crick to leave DNA to Randall's group, Bragg did not realize that Pauling had already entered the field. In a memorandum to Corey dated July 24, 1951, Pauling described his interpretation of Astbury's X-ray photographs of DNA and concluded: "I think that it is likely that a helical structure could be formulated for the thymonucleate [DNA] ion, and that it could be tested by calculating its form factor. May I suggest that one activity to be carried out under the new NFIP [National Foundation for Infant Paralysis] grant, in case that it is made, would be to prepare some oriented specimens of sodium thymonucleate and to photograph them, for comparison with a calculated x-ray pattern corresponding to a helical structure."[718] Fresh from his triumph with proteins, Pauling now set his sights on another macromolecule.

Characteristically, Bragg took to heart his abortive attempt to determine the pattern of protein folding. Uli Arndt wrote: "one sometimes wondered whether he did not brood to much on what he regarded as his failures, such as the fact that he was pipped at the post by Pauling's discovery of the α-helix."[719] Now only a few years from retirement, Bragg was starting to slow down physically;

on holiday in the Lake District in the summer of 1951, he "found I now had to go a bit slow on up-hill work."[720] Under the circumstances, it would have been quite understandable if Bragg had left the globins to Perutz and Kendrew and spent the last spell of his Cavendish Professorship playing the role of a senior scientist. But Bragg was still hooked on the problem of protein structure. In the early 1950s, he published first-author papers at a rate higher than he had done since the 1920s. In 1952, Bragg and Perutz published three papers on hemoglobin structure. With typical Bragg ingenuity, these papers pieced together information from a variety of sources to make deductions about the external and internal form of the hemoglobin molecule.

The key to determining the shape of the hemoglobin molecule was that horse methemoglobin crystals could be prepared with or without ammonium sulfate in the water of crystallization. From the difference in density between the "salt-water" and "salt-free" crystals, it was possible to calculate the fraction of the crystal from which salt was excluded, and therefore the volume of the hemoglobin molecule. This came out to be 116,000 $Å^3$. To determine the dimensions of the hemoglobin ellipsoid, Bragg took advantage of Perutz's molecular-transform approach. He subtracted the amplitudes of the salt-water crystal from the corresponding amplitudes of the salt-free crystal ($|F_{water}| - |F_{salt}|$) and plotted these values on the three reciprocal planes (a^*b^*, a^*c^* and b^*c^*). From the theoretical diffraction behavior of an ellipsoid, Bragg knew that the first zero contour corresponded to the outline of the molecule. He was able to trace this contour well enough to determine that the hemoglobin molecule was a prolate ellipsoid with $a = 65$ Å, $b = 55$ Å and $c = 55$ Å—roughly the same relative dimensions as a hen's egg. So instead of a cylinder 57 Å in diameter and 34 Å high, with the axis of the cylinder parallel to the c axis of the crystal, as in the hatbox model, it now appeared that the hemoglobin molecule was an ellipsoid 55 Å in diameter and 65 Å long, with the long axis parallel to the a-axis of the crystal.[721]

One of the reasons for discarding the hatbox model was that a molecule of those dimensions could not be packed into the unit cells of some hemoglobin crystals. In the second 1952 paper, Bragg and Perutz addressed the packing of their new ellipsoid. For four different forms of hemoglobin crystal, two from human and two from horse, they were able to find ways of arranging ellipsoidal molecules of 55×65 Å such that the space-group symmetry was satisfied and there was no molecular overlap.[722]

The third paper dealt with the internal structure of the hemoglobin molecule—specifically, it addressed the question of how a linear polypeptide chain could be packed into an ellipsoidal molecule.[723] A similar analysis had been performed by Perutz in 1947 with the now-abandoned hatbox model. Pauling's work encouraged a new attempt, as it proved that helical segments of polypeptide chain occurred in proteins and showed what the characteristics of these segments were. Perutz's earlier work suggested that the hemoglobin molecule contained "rods"—presumably α-helices—that were 10.5 Å apart. Hemoglobin consists of about 580 amino acids; assuming that these are in an α-helical configuration, the total length of the folded polypeptide chain is about 870 Å. Bragg's problem was how to fold a polypeptide chain 870 Å in length and 10.5 Å in diameter into an ellipsoid 55 × 65 Å. An additional constraint, arising from Perutz's 1949 three-dimensional Patterson, was that the chains are viewed end-on in hexagonal packing in the projection of the molecule onto the plane defined by the *b*- and *c*-axes (100), meaning that the straight portions of the chain are parallel to the *a*-axis. From the length of the polypeptide chain and the size of the ellipsoid, Bragg calculated that there would be about 15 such segments. The problem thus reduced to how 15 or so circles 10.5 Å in diameter (representing the cross-section through an α-helical chain segment) could be packed in approximately hexagonal fashion into a circle 55 Å in diameter (representing the circular cross-section of the ellipsoidal molecule). Using the signs of amplitudes deduced by swelling and shrinking the crystals, and the "fly's eye" to test possible arrangements of hexagonally packed cylinders, Bragg et al. concluded that a likely structure contained 17 chain segments organized into five layers. The central layer contained five segments, the next layers each contained four and the outermost layers each contained two.

However, there were a couple of inconsistencies. If the five-layer structure were correct, the number of electrons contributing to the (063) reflection, $F(063)$, after correction for temperature, chain folding, etc., would be about 6000; however, the measured value of $F(063)$ was only 2050. There was also the fact that whereas the proposed structure envisaged a single polypeptide chain neatly folded into parallel rows and layers, it was well known that hemoglobin could be dissociated into four subunits. In the end the five-layer 1952 model proved no better than the 1947 four-layer one; Bragg's assumptions of parallel chain segments and 100% α-helix content were unfounded.

Bragg's analysis of the external form of the hemoglobin molecule was much more successful than his analysis of the internal structure, as the former relied upon basic X-ray optics and crystallography rather than dubious assumptions about chain folding. Perutz later wrote of the two papers on the shape of the hemoglobin molecule: "The papers . . . were largely written by Bragg. They mirror his originality, his lucid arguments and his profound understanding of diffraction, but they made me feel that I ought to have thought all this out myself."[724] For Crick, these papers were "a revelation to me as to how to do scientific research."[725]

Bragg's Indian summer of 1952 demonstrated that he was not an "old buffer" blowing bubbles and inventing Heath Robinson-ish gadgets to illustrate diffraction phenomena. His artistic flair and profound understanding of the physics of diffraction allowed him to come up with innovative solutions to problems that had stumped much younger men. But Bragg's papers on the hemoglobin molecule had a more important effect than just determining the size and packing of the hemoglobin molecule—they initiated the unravelling of the phase problem that had hindered X-ray analysis from its beginnings. The studies on both the internal and external form of hemoglobin involved absolute measurements of reflection intensity—the first time these had been carried out on proteins.

When Perutz made these measurements, he was surprised how small the *F* values were. This removed one major barrier to the use of isomorphous replacement to determine the phases of reflections. As described above, Robertson had solved the structure of phthalocyanine by exploiting the fact that a nickel atom could be added to the center of symmetry of the molecule; this had such a strong effect on all reflections that the signs of their amplitudes could be determined. Hemoglobin is a much larger molecule than phthalocyanine—or any other molecule that had been solved by isomorphous replacement—so Perutz and Bragg had assumed that the effect of a heavy atom or two would not be large enough to significantly affect the amplitudes of reflections. In fact, Bragg had written in 1949, "No heavy atom could stand out in such a crowd."[726]

It now appeared that a heavy metal atom might, in fact, stand out in the crowd of light atoms that make up a protein molecule. However—and this was the second major barrier to the use of the isomorphous replacement method—the addition of the heavy atom(s) must not change the structure of the protein. This still seemed highly unlikely.

Lacking any other way of determining the signs of reflections, Bragg and Perutz returned to the loops-and-nodes method. In the molecular transform of the centrosymmetrical (010) projection of horse methemoglobin, each layer line consists of amplitude maxima and minima (loops) separated by points of zero amplitude (nodes). The trick was to find which loops were maxima (positive amplitudes) and which were minima (negative amplitudes). In Perutz's 1947 paper, analysis of different stages of shrinkage and swelling of the hemoglobin crystal had provided enough amplitude values along the zero layer line to determine signs of peaks close to the origin. These correspond to low-order reflections. It is the higher-order reflections, however, that arise from features of the intramolecular structure.

It was again Bragg who was able to take the next step forward. He realized that the minimum distance between maxima and minima along a layer line of the molecular transform was determined by the dimensions of the diffracting object—in this case, the hemoglobin molecule. In a 1952 paper with Perutz, Bragg illustrated what became known as the minimum-wavelength principle with a characteristically striking analogy. He plotted on a line the times that trains from Cambridge arrived at Liverpool Street Station on Sundays—essentially a random distribution of points—and then determined the Fourier transform of this "grating." This he compared with the transform of a grating consisting of the same number of points but distributed uniformly along a line of the same length. The central maximum, first minimum and next maximum of the two transforms, corresponding to low orders of reflection, were very similar. As Bragg pointed out, "Whatever the arrangement of the scattering points in the structure, the maxima and minima succeed each other with a certain minimum distance of separation, or minimal wave number, which depends upon the overall dimensions of the molecule in a corresponding direction."[727]

Armed with the minimum-wavelength principle, Bragg and Perutz made another attempt at determining the signs of the loops in the molecular transform of hemoglobin. They used six shrinkage stages of horse methemoglobin that had identical unit-cell dimensions but in which the value of β, the angle between the a- and c-axes, varied from 84.5° to 127.5° (α and γ, the other angles formed by the axes of the monoclinic unit cell, are 90°). For each of these crystals, the points of the reciprocal lattice lay at the same distances along the a^*-axis but at different distances along the c^*-axis. For example, the c^*-axis distance of the reciprocal-lattice point corresponding to the spacing between (001) planes of the crystal varied from 42.3 to 54.9 Å, depending on the value of β. The presence of numerous $|F|$ values along the zero layer line ($a^* = 0$) allowed Bragg and Perutz to identify where $|F|$ decreased and then increased, indicating that the curve crossed the axis at that point. Knowing where the transform changed sign, and knowing that the origin peak was positive, the signs of the amplitudes could be determined.

So far this was identical to the approach that Perutz had used in his 1947 study. Using this method, Bragg and Perutz could unambiguously follow the transform curve along the zero layer line through the first three groups of F values, which they labeled B, C, and D. Between D and E, however, there were no low F values, so it was not clear whether the curve crossed the axis. However, the minimum-wavelength principle indicated that the curve must cross the axis, as otherwise D and E would both be minima and these would be "far closer than the minimal spacing" (Figure 8.7). Based on the Patterson synthesis previously obtained, "The peaks in the c^* projection would then represent layers of the chains parallel to the ab [(001)] plane."

Despite the elegance of the minimum-wavelength principle, it had allowed Bragg and Perutz to determine only one more sign. But even if it had enabled them to assign signs to every amplitude in the molecular transform of the (010) projection, they would still be far from a molecular structure of hemoglobin. The transform only used reflections greater than or equal to 7 Å, a small fraction of those needed to obtain atomic resolution. Also, the use of projections meant that even if a Fourier synthesis could be achieved at atomic resolution, the electron-density map obtained would represent the superimposition of scores of layers of atoms. It would be akin to taking an intricate three-dimensional structure, squashing it flat, and then trying to determine what it had originally looked like. And there could no longer be any doubt that the hemoglobin molecule was a very intricate structure: "It appears certain that the molecule is a far more complex entity than a simple picture of sheets of parallel chains would suggest . . . The nodes and loops of the complete ($h0l$) transform . . . have a highly complex distribution . . ."

In a lecture on "X-Ray Analysis of Proteins" given in March 1952, Bragg noted that the different swelling and shrinkage forms of hemoglobin and the minimum-wavelength principle allowed one to determine where the molecular transform changed sign along each layer line. The only loop whose sign was "given" was the one at the origin of the reciprocal unit cell (corresponding to

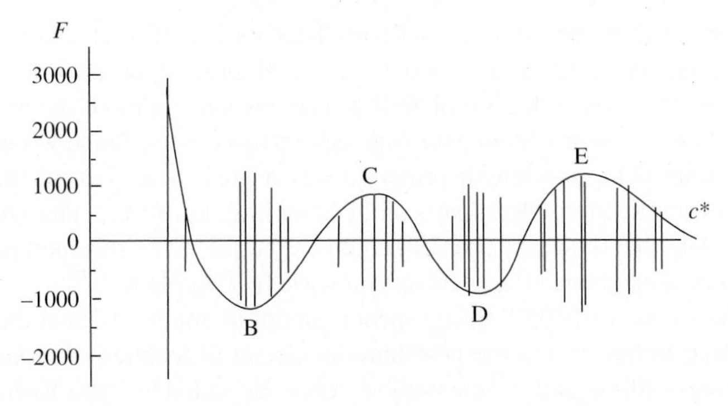

Fig. 8.7 Part of the "molecular transform" of hemoglobin. Vertical lines represent F (amplitude) values of $(00l)$ reflections from different shrinkage stages of the protein. These are drawn above and below the abscissa as both positive and negative values are theoretically possible. The signs of "loops" B (negative), C (positive) and D (negative) were determined by inferring that a "node" occurs where the F values decrease and then increase; the sign of loop E (positive) was determined using the minimum-wavelength principle. Adapted, with permission, from Figure 4 of Bragg, W. L. and Perutz, M. F. (1952). The structure of hemoglobin. *Proceedings of the Royal Society of London A* **213**, 425–35. Published by the Royal Society

the center of symmetry of the projection). This meant that the signs of loops along the zero layer line could be determined, but not those of loops on other layer lines. For the latter, it was necessary to be able to relate the signs on one layer line to the next. Swelling and shrinkage forms were of no use for this purpose, but Bragg and Perutz had been able to tentatively assign signs for the first and second layer-lines by comparing hemoglobin crystals from different species. Despite the low resolution of the transform, Bragg was "optimistic about the possibility of a solution . . . If we can turn this corner . . . I feel that the most difficult pitch has been climbed." However, this optimism was fueled by the realization that abandoning the analysis at this stage would be very difficult: "We have, as it were, sunk much capital in haemoglobin."[728]

In May, Bragg wrote to Pauling: "Perutz and I have been making a frontal attack on the structure of the helioglobin [sic] molecule. I am no biochemist and have felt my best contribution is to see what one can deduce purely by X-Ray Analysis without making any assumptions about the structure of the molecule. We are not there yet, but I think we have got a long way. In the $(h0l)$ projection which is centrosymmetrical the different shrinkage forms enable one to plot many values of $F(h0l)$ along the $h =$ constant layer lines. We have so many points that we are able to establish the nodes and loops along these layer lines and of course we know they alternate $+$ and $-$ in sign. If we can take the further step of relating the signs of each layer line to the next we can then make a Fourier picture of the crystal. We have not made this further step yet but have hopes of doing so. We can already relate layers one and two to the central layer and are working hard to do the rest."[729]

It was another false dawn. As Perutz later wrote: "The absolute signs of the 00*l* reflection were fixed unambiguously and those of the 20*l* reflections in the central region with high probability. The absolute signs of the remaining six layer lines were left open, but many sign relations were found within them. Bragg's minimum wavelength principle was a great help, but all the same, weakness of reflections along parts of the layer lines left several ambiguities. It became clear that the phase problem could not be solved by this method alone, not even in projections in the centrosymmetric [(010)] plane."[730]

In the summer of 1952, Bragg spent a month in South Africa at the invitation of Reg James. It was the now-familiar circuit of lectures, receptions and sightseeing—although this time without Alice. He sailed to Cape Town, where he gave lectures on "General Ways of Attacking Difficult Structures" and "The Atomic Patterns of Matter," attended a meeting of the South African Association for the Advancement of Science, sketched and went bird-watching. In Pretoria, he lectured on "The Application of X-Ray Analysis to Complex Structures" and "Current Researches in the Cavendish Laboratory." In Johannesburg, he lectured on proteins and the bubble model, and descended 4000 feet into the City Deep Mine. Bragg again talked about the bubble model at Rhodes University, where he also gave a public lecture on "Atomic Patterns of Matter." His reward for all this lecturing was to be sent on a safari. After a 300-mile drive across the veldt, Bragg and his host reached a game reserve where they saw hippos, giraffes, warthogs, baboons, monkeys, and jackals, as well as many birds. The return trip featured the bad luck that was now a characteristic feature of Bragg's overseas trips. His plane was damaged while landing in Khartoum and he had to fly to London via Wadi Halfa, Cairo, and Rome.[731]

Alice's absence from Bragg's trip to South Africa was probably due to her extensive committee work, both in local government and various non-governmental organizations. This now extended to the national level. In 1951, she began a three-year term on the Central Advisory Council for Education for England; from 1952 to 1956, she was a member of the Royal Commission on Marriage and Divorce.

These activities, which involved regular trips to London, were now easier, as the Bragg children were grown up. Stephen had married Maureen Roberts in September 1951. David was often home, as his poor health meant that he could work only intermittently. Margaret had surprised her parents by electing to go to Oxford University to read history. Patience had followed her sister to Downe House School, and then went to her mother's old college, Newnham. The house in West Road was now too large, and the Braggs' lease on it was running out. In 1952 they bought 10 Madingley Road—"a pretty house not far away, in a charming garden."[732]

The year 1952 marked the 40th anniversary of the discovery of X-ray diffraction by crystals. To mark the occasion, the X-Ray Analysis Group held a conference at the Royal Institution in October. Bragg gave both the opening and concluding talks. The former was an account of the early days of X-ray analysis, which made the point that the seeds of many of the subsequent developments had been sown during the period 1912–15. Bragg's second talk was an overview of his and Perutz's recent work on hemoglobin. Although leaving the audience with no illusions about the difficulty of the problem, he concluded that "we are now in sight of the promised land." Another participant was Max von Laue, who gave a talk about his own role in the development of the field and delivered a speech at the conference banquet.[733]

The hitherto tranquil atmosphere of the MRC Unit was now being disturbed by Crick's abrasive personality and Bragg's growing irritation with him. According to Perutz, "Crick was not a model of tact and he talked incessantly in a very loud voice and Bragg just couldn't bear it." Crick was present when Bragg told Perutz about the minimum-wavelength principle. Accounts of this incident differ, but Crick either doubted the correctness of Bragg's principle, claimed that he had thought of it first or even insinuated that Bragg had stolen the idea from him![734] Whichever it was, Bragg "blew my top off." He called Crick to his office and told him the sooner he finished his Ph.D. and left the Cavendish, the better. Bragg also sent a letter critical of Crick to Harold Himsworth, who had replaced Mellanby as Secretary of the MRC in 1949. Perutz and Kendrew interceded on Crick's behalf, and by the time Himsworth showed up to investigate Bragg had forgotten the incident.

In the autumn of 1952, Peter Pauling, Linus's second son, came to work at the Cavendish. From him, Watson and Crick learned that the Caltech group had come up with a DNA structure. When they obtained a copy of Pauling's manuscript in January 1953, Watson and Crick saw that the structure was a three-stranded helix very similar to their 1951 model. Now that Pauling was involved, Bragg gave Watson and Crick permission to resume work on DNA. As Perutz later put it, "Bragg felt . . . that we shouldn't encroach on Randall's preserves but, when Pauling published a structure of DNA, he felt that now it was really a free-for-all . . ."[735]

Bragg later described his view of the limits of scientific decorum in his foreword to Watson's book *The Double Helix*: "When competition comes from more than one quarter, there is no need to hold back."[736] That the new player was not only the world's leading chemist but also his old rival made Bragg's decision a lot easier. As Kendrew told Judson, "We were all a bit sore with Pauling over the α-helix; we didn't want the man to scoop us again."[737] Bragg's attitude towards competition was summed up by a comment he made in a 1918 letter to his father: "this 'getting the better of the other fellow' is not bred in us very much, is it? Though I must confess I loath the idea of it far more than I dislike doing it if someone jolly well makes me do so."[738] After all his setbacks at the hands of Pauling, Bragg would be only too happy to see Watson and Crick get the better of him.

Wilkins had shown Watson X-ray diffraction patterns that Franklin had obtained from the hydrated "B" form of DNA; Perutz gave Crick a copy of an MRC report on the King's group that contained information about the unit cell of the B form. From this information, Watson and Crick were able to come up with a double-helical structure that agreed with Franklin's crystallographic data and also explained some curious chemical characteristics of DNA. The helical diffraction theory of Crick, Cochran, and Vand now came into its own, as it could be used to relate the Watson–Crick double helix to Franklin's B-form diffraction pattern.

Bragg had "forgotten all about nucleic acid" and was off work with influenza in early March 1953 when the double helix structure was conceived. Prewarned by a telephone call from Crick, he returned to the Cavendish just in time to see the unveiling of the physical model.[gg] Remembering the fiasco over the helical structures of proteins, Bragg—or possibly the Rockefeller Foundation's Gerard Pomerat—insisted that Alexander Todd approve the DNA structure before it was published.[739] Bragg's pleasure at Watson's and Crick's success must have been enhanced by the fact that the double helix was a last great success for the trial-and-error method of X-ray analysis.

Kendrew was given the delicate task of phoning King's to tell Wilkins that the DNA structure had been solved.[740] Watson, Crick, and Wilkins then came up with a plan for publishing the double helix that gave credit to all concerned. Randall later wrote to Bragg, "I don't think you and I were really closely in touch with each other about all this."[741] On April 25, 1953, *Nature* published a group of three papers on DNA structure: One by Watson and Crick, another by Franklin and her student Ray Gosling, and the third by Wilkins, Alex Stokes, and Herbert Wilson.[742]

Bragg had the more pleasant job of making the first public presentation of the double helix. This occurred at the Ninth Solvay Conference on Chemistry, which was on the topic of "Proteins," and took place in April 1953 in Brussels. Pauling, who visited Cambridge on his way to Belgium and discussed the double helix with Watson and Crick, said at the conference: "Although it is only two months since Professor Corey and I published our proposed structure for nucleic acid, I think that we must admit that it is probably wrong. Although some refinement might be made, I feel that it is very likely that the Watson–Crick structure is essentially correct."[743] Although it was a vicarious victory, Bragg had finally won a round against the Wizard of Pasadena.

Perutz must have felt at least a tinge of envy that Watson and Crick had, after only a year and a half of part-time work, solved the structure of DNA, while he was still a long way from a structure for hemoglobin. According to Bragg's autobiography, "Perutz at times became quite discouraged, very naturally, and wished to change to the other line of research in which he was interested, the flow of glaciers. Why I continued to be optimistic I shall never understand. Certainly our x-ray crystal colleagues thought we were on a wild goose chase, and I think Perutz might have dropped the project had I not spurred him on.

[gg] On November 2, 1962, Gerard Pomerat wrote to Bragg: "When the announcement [of the Nobel Prize] came out today, I looked over some of my old diaries to see if I could find the date on which, one morning, Perutz and Kendrew and Watson and Crick first showed you the model of DNA they had stayed up all night to construct. It was on April 1, 1953, the first day of your return to the laboratory after a week of 'cold.' " (RI MS WLB32E/7). This date is incorrect. The DNA model was completed on Saturday, March 7. Bragg probably saw it the following Monday.

I remember Perutz sending me a long memorandum to say that it really was not worthwhile pushing on with measurements unless some quite new line of attack was discovered . . ."[744]

In July 1953—soon after this memorandum was written—Perutz received from Austin Riggs of Harvard University a set of reprints which claimed that mercury atoms could be attached to hemoglobin without affecting the oxygen-binding characteristics of the protein. If the function were conserved during heavy atom addition, surely the structure must be too. Perutz immediately used Riggs' method to react hemoglobin with parachloromercuribenzoate and compared its X-ray diffraction pattern with that of the unreacted protein. To his immense satisfaction he found that the positions of the spots were identical, indicating that the two crystals were isomorphous. As Bragg later put it, "the molecule takes no more notice of such an insignificant attachment than a maharajah's elephant would of the gold star painted on its forehead."[745] Also, the intensities of the spots differed by exactly the amount Perutz had predicted. "Madly excited, I rushed up to Bragg's room and fetched him down to the basement dark room. Looking at the two pictures in the viewing screen, we were confident that the phase problem was solved."[746] Perutz was "elated beyond measure."[747]

Progress on the X-ray analysis of hemoglobin, previously as slow as one of Perutz's glaciers, now speeded up considerably. On Cochran's advice, Perutz's student David Green subtracted the $h0l$ intensities of the mercury-free form from those of the mercury-containing form and used these to calculate a "difference Patterson" map of the (010) centrosymmetrical projection. A single large peak, corresponding to the vector between the two symmetry-related mercury atoms, was found. Using the center of the vector as origin, a structure factor could be calculated for the mercury atom, and the signs of all $h0l$ reflections affected by it could be established. These, 150 in number, were then plotted in reciprocal space and superimposed on the molecular transform. "It was a triumph to find that there were no inconsistencies. The signs determined by the isomorphous replacement method exactly fitted the loops and nodes which had been so laboriously worked out in the course of the previous two years, and confirmed the great majority of the sign relations established by the transform method."[748] Not only were the old sign assignments vindicated, but the new ones were obviously correct. As Bragg told Horace Judson, "it was thrilling, because it was working, you could see it was self-consistent. That's the key word there. Everything fitted."[749]

Now that the signs of the molecular transform were established, it was possible to use them to make Fourier maps of the (010) projection of hemoglobin—just as Bragg had done for diopside in 1929. Initially, Bragg and Perutz subtracted the amplitudes of salt-containing and salt-free crystals of hemoglobin and used these differences to make a "salt-water" Fourier representing the outline of "ghost" molecules of uniform electron density. The projection on the (010) plane showed a row of molecules parallel to the a-axis and centered on the two-fold rotation axes, with neighboring molecules related by two-fold screw axes. As the thickness of the molecule at the screw axis was approximately 60 Å, about the same as the length of the *b*-axis of the crystal, it was concluded that the molecules must be in contact at that point in the crystal. The shape of the individual molecule conformed to the previous prolate ellipsoid model, with long and short axes of 71 and 54 Å, respectively, with the long axis tilted at 15° to the *a*-axis. The electron density decreased at the rotation axis, indicating the possible presence of a "dimple" on the surface of the molecule, or possibly a channel through it.

The salt-water Fourier was, as noted above, of uniform density, so it revealed only the overall shape of the hemoglobin molecule, but not any of its internal structure. To study the internal structure, Perutz constructed a Fourier map of the salt-free crystal at 6.5 Å resolution (Figure 8.8). The boundaries of the molecule seemed to conform to the one-electron contour. At the screw axes, the electron density was much higher, in agreement with the conclusion from the "salt-water" Fourier that neighboring molecules were in contact along the *b*-axis. Because of the relatively low resolution, the positions of the iron atoms

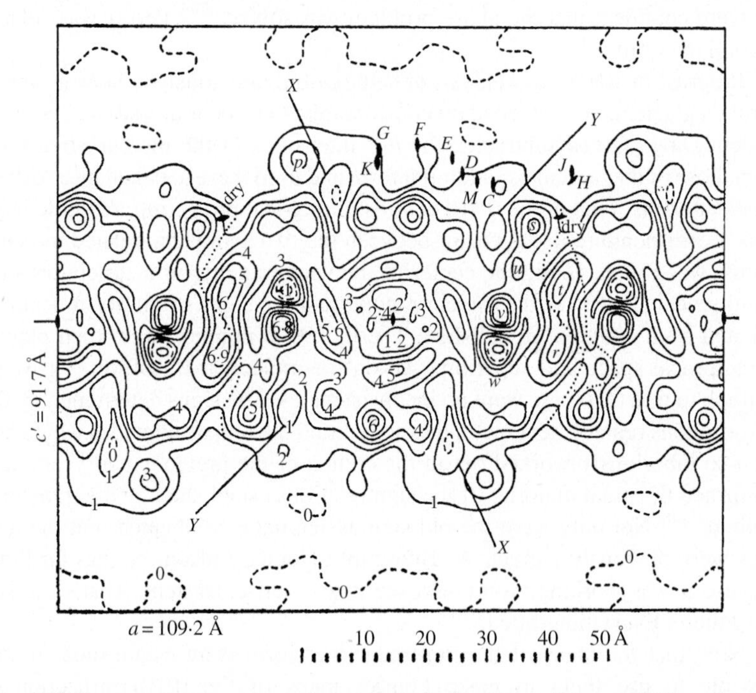

Fig. 8.8 Electron-density map of the "salt-free" form of hemoglobin. Row of hemoglobin molecules projected onto the (010) plane. Contours are drawn at intervals of 1 electron/Å^2 above or below the electron density of water. Lines X–X and Y–Y represent possible planes along which the protein could dissociate into two half-molecules. Reproduced, with permission, from Figure 4 of Bragg, W. L. and Perutz, M. F. (1954). The structure of hemoglobin. VI. Fourier projections on the 010 plane. *Proceedings of the Royal Society of London A* **225**, 315–29. Published by the Royal Society

could not be determined. In general, "The internal structure of the molecule exhibits a striking system of peaks and depressions which have so far defied interpretation."[750] The five-layer structure suggested by the 1952 analysis could, like the earlier hatbox model, be discarded.

The Fourier map of the (010) projection of hemoglobin represented the superimposition of all the atoms along the 63-Å length of the *b*-axis—about 40 atomic diameters. This meant, as Kendrew later wrote, "that the various features of the molecule were superimposed in inextricable confusion"[751] It had become clear that "the amount of structural information which could be derived from a projection was almost nil."[752]

Nonetheless, it was a tremendous thrill for Perutz to finally see an electron-density map of hemoglobin, even one at low resolution and in projection. It was also a wonderful moment for Bragg. Perutz told Judson: "I remember going around to his house with the Fourier projection—and the great thrill of having this, this problem all solved which seemed so *inscrutable* for all these years."[753] He also phoned Dorothy Hodgkin, who drove from Oxford to see the projection.[754] As she had expected, it was uninterpretable. However, Hodgkin was able to give Perutz a useful tip—she drew his attention to a technique by which the phases of reflections in non-centrosymmetrical crystals could be determined using two heavy atoms occupying different positions.[755] The bad news was that this would involve measuring 12 times as many reflections as had been used for the projection.[756] It was to be a long time before this could be accomplished; for Perutz, his "triumph" of 1954 was to be followed by "several years of bitter frustration."[757]

Perutz's initial studies on isomorphous replacement of hemoglobin were presented at a September 1953 conference on protein structure held in Pasadena and organized by Linus Pauling. At this conference, Watson and Crick discussed the DNA double helix and Hugh Huxley, another member of the MRC Unit, presented his new sliding-filament mechanism of muscle contraction. It was a great vindication of Bragg's belief that physics techniques could usefully be applied to biology, and it occurred in the backyard of his great rival. Pauling, still basking in the glory of the α-helix, was in a magnanimous mood, taking Bragg and Alice for a four-day tour of California.[758]

With the DNA structure solved and the structure of hemoglobin seemingly only a matter of time, Bragg realized that his gamble of supporting Perutz had paid off handsomely. As he wrote to Niels Bohr: "When we started this work, it seemed almost unthinkable that one could ever devise a direct method of analysis of molecules containing 10,000 atoms. It has been one of the greatest pleasures in my studies of X-ray analysis that I should have been associated with research in which this problem was solved by straightforward methods of X-ray optics . . . It is rather exciting that out of this little unit at Cambridge under Perutz and Kendrew, which I started with the help of our Medical Research Council, there has come the first analysis of a protein, the analysis of nucleic acid, which holds out such fascinating possibilities of explaining the handing on of the hereditary characteristic, and the brilliant work by young Huxley on the way the muscle fibre works. I am so glad I was able to start a bio-physics section while I was at the Cavendish."[759]

It was also a great note on which to go out. Bragg had already resigned the Cavendish professorship and within a few months would take up a new challenge—the directorship of the Royal Institution.

At his farewell dinner, Bragg spoke of the difficulties of succeeding Rutherford and being "smothered by the gigantic folds of his mantle." Then war had come and by the end of the war the Cavendish had "flown to pieces" and the loss of Fowler had been devastating. But by 1953 "they were living in proud times." The Mond was a leader in superfluidity, the radio research of Ratcliffe and Ryle was a great success and in crystallography they were at the leading edge: Bragg was especially pleased at the application of X-ray methods to biochemistry. His contribution to the Cavendish was not to plan these developments, but rather to recognize the importance of the ideas of others and help them to realize them. Bragg thanked Ratcliffe for his support in "affairs of state," including dealing with boards of examiners and setting up a course on history and philosophy of science.

It was typical of Bragg that, even on an occasion when he was being honored for his successes as Cavendish Professor, he made a point of listing his failures. He apologized for his "shyness of personal problems," "inability to remember names" and discomfort with administrative duties.[760] Bragg's farewell present from the Cavendish Laboratory was a pair of binoculars. He thanked his colleagues by saying: "Whenever I look at a strange bird, I shall think of you."[761] Alice acerbically commented: "So useful for observing the bird-life of Albemarle Street."[762]

9
The art of popular lecturing on scientific subjects: The Royal Institution, 1954–66

The Royal Institution (RI) was founded in 1799 by a group of men headed by Benjamin Thompson, Count Rumford, and including Henry Cavendish, Sir Joseph Banks (President of the Royal Society), William Wilberforce, and the Bishop of Durham. Its goal was "the promotion of science and the diffusion and extension of useful knowledge." Rumford and his colleagues bought a large house, 21 Albemarle Street[hh] in Mayfair, to provide living quarters for a professor, lecture facilities, and laboratory space. Thomas Garnett, from Anderson's Institution in Glasgow—founded in 1796, and after which the Royal Institution was modeled—was hired as the first Resident Professor—so-called because he lived on the premises. The administrative structure of the RI consisted of a President (initially George Finch, 8th Earl of Winchilsea), nine Managers, and a group of "Proprietors," who had each contributed money towards the institution.

Garnett was replaced after 2 years by the physicist Thomas Young, who was in turn succeeded a few years later by the chemist Humphry Davy. Under Michael Faraday, who became Supervisor of the House in 1821 and Director of the Laboratory in 1825, the RI acquired its modern form. In 1826, Faraday started a series of Friday Evening Discourses and the Christmas Lectures "for a juvenile auditory." Faraday retired in 1861, being replaced by John Tyndall.

The RI expanded in 1896 when Ludwig Mond, a German chemist who established the Brunner-Mond Company (later Imperial Chemical Industries), bought the next-door property, 20 Albemarle Street, and converted part of it to a research laboratory named in honor of Davy and Faraday. Initially, the Davy–Faraday Laboratory was used for outside scientists and James Dewar, now Resident Professor, worked in the RI basement. WHB, who succeeded Dewar in 1923, was the first Resident Professor to use the Davy–Faraday. A further expansion of the RI occurred in the 1930s, when 19 Albemarle Street was purchased, but most of this property was leased to others.[763]

[hh] The house was purchased from the executors of John Mellish, who had been shot through the head by a highwayman on Hounslow Heath. [Bragg, W. L. (1958). The contribution of the Royal Institution to the teaching of science. *School Science Review* **40**, 240–5.]

The RI served a public function but was essentially a private club. Among other things, this meant that its funding was always precarious and its constitution rarely logical. The same individuals tended to hold a variety of different titles whose responsibilities were insufficiently defined: Professor of the Royal Institution, Supervisor of the House, and Director of the Laboratory. As described above, the RI was in a state of decline when WHB became Resident Professor in 1923. During his almost 20-year tenure of the position, the RI became an important research facility again, and its Discourses as popular for the social aspect as for the science. WHB and Gwen introduced the practice of entertaining speakers and other guests before and after the Discourse. However, WHBs death in 1942 plunged the RI into another crisis. Sir Henry Dale, a physiologist who had succeeded him as President of the Royal Society in 1940, became "caretaker" of the RI. In 1946, the physical chemist Eric Rideal took over as Resident Professor. John Moore-Brabazon, first Baron Brabazon of Tara, became President in 1948.

Rideal did not care for the entertainment function and suggested to Brabazon that Edward Andrade, who was Vice-President of the RI and Professor of Physics at University College London, become responsible for lectures while he, Rideal, remained Director of the Davy–Faraday Laboratory. This arrangement not being accepted by the Managers—Brabazon said "There cannot be two Kings in Babylon"[764]—Rideal resigned, being replaced, in January 1950, by Andrade.

Andrade's position was Fullerian Professor of Chemistry,[ii] Superintendent of the House, and Director of the Davy–Faraday Laboratory. Although WHB had used the title of "Director in the Royal Institution," it was not an official position and Dale's efforts to make it one had failed. Brabazon, and presumably the Managers in general, felt it inappropriate for a club to have a director. As it was, the resident professor answered to a Board of Managers, consisting of the President, Secretary, Treasurer, and 15 members. There was also a Committee of Visitors which, according to a later Resident Professor, George Porter, "was a second, elected body of members which duplicated most of the work of the managers and challenged almost every decision on principal [sic]."[765] Andrade's ambivalence about this unique organizational structure was expressed in his 1943 obituary of WHB, in which he wrote of the post he would later occupy: "The whole complicated position is one that could only exist in England, but it works, though whether it works better than a simpler and more unified administration would do has not been proved."[766]

Andrade must have seemed an excellent choice to lead the oft-beleaguered RI.[jj] He had been a member since 1924, a Manager on several occasions, Chairman of the Library Committee and Vice-President. A man of wide interests

[ii] The Fullerian Professorship was founded in 1833 by a grateful John Fuller, whose ill-health prevented him from sleeping—except during Royal Institution lectures! [Bragg, W. L. (1958). Scientific apparatus in the Royal Institution. *Nature* **182**, 1541–3.]

[jj] Except where otherwise indicated, the following account of the Andrade affair is based on James, F. A. J. L. and Quirke, V. (2002). l'Affaire Andrade or how not to modernise a traditional

and an excellent lecturer, he had given many Friday Evening Discourses and the Christmas Lectures in 1927–8 ("Engines") and 1943–4 ("Vibrations and Waves").[767] However, Andrade's autocratic personality was a serious liability in his efforts to navigate the treacherous waters of the RI. His replacement of several long-serving staff members with his own people alienated many Members. The Secretary, Alexander Rankine, was a good friend of Andrade's and had argued vehemently for Andrade's appointment, but the two men soon were at loggerheads—particularly when Andrade engineered the dismissal of Thomas Martin, the long-time General Secretary.

Bragg took a great interest in the RI. Not only had his father been Resident Professor there, but Bragg himself had been Professor of Natural Philosophy since 1938, had lectured at the RI almost every year since and had given the Christmas Lectures in 1934–5. It is no surprise that he came down in the anti-Andrade camp. Bragg apparently had no strong opinion about Andrade when they had briefly both been at the Cavendish together in late 1911 or early 1912. However, it was a different story when Andrade served as a sound-ranger in the First World War. In June 1917, Bragg had written to WHB: "Andrade got jolly well kicked out of the show here as he became absolutely the limit. He is a hopeless chap, I am sorry for him too sometimes but he has got a bad kink in him somewhere. There was an awful to-do about it all, the officers in his section refused to work with him any longer and told their colonel so, and as they were A1 chaps and about eight others had one by one begged to leave his section, he departed. He used to tell them off in front of the men!"[768] This poor impression was subsequently confirmed by others. Bragg's former student Guy Brown, who worked in Andrade's department at University College London, wrote to Bragg in December 1942: "Andrade did not allow research students to members of the staff, except in exceptional cases and after a row, and then only Indians whom he despises."[769] Andrade's poor leadership skills were also referred to in a December 1947 letter from Norman Campbell to Bragg: "I hope you *wont* ask Andrade to write your father's life—I am sure he has not the temperamental sympathy with your father that would be essential for the proper performance of the job!"[770] Bragg was also one of those who disapproved of Andrade's personnel changes—particularly the "accelerated retirement" of William Green, principal lecture assistant, in 1950.

At the 1951 Solvay Conference in Brussels, Bragg had a heart-to-heart talk about the RI with George Thomson: "It was becoming apparent that it was headed for disaster." Thomson, who was a Manager, had agreed to Andrade's appointment, but now bitterly regretted it.[771] This conversation may have prompted Bragg to write to Brabazon. The colorful Brabazon replied in March 1951: "I cannot tell you what a lot of worry, being President of the Institution has been for me. Worry, right from the very start, and I gasped with horror when Andrade was chosen by the Managers. Not that I don't like him; I do like

institution. In "*The Common Purposes of Life": Science and Society at the Royal Institution of Great Britain* (F. A. J. L. James, ed.), pp. 273–304. Ashgate Publishing Company, Burlington, Vermont.

him. As you say he has something, but he really is too difficult altogether. This is the morning of Monday and our Meeting is this afternoon. God knows what will happen, but if you were here, I should put my arms round you, and in a continental way kiss you on both cheeks, for you voice exactly my feelings in the matter, and I am going, this afternoon, to fight it out fair and square. If I go down, well I shall have done my best, but I am not going to have complete domination of the Institution by Andrade."[772] At that meeting, the Board of Managers accepted Rankine's terms for withdrawing the letter of resignation he had submitted a few weeks earlier. Bragg wrote back to Brabazon with a list of "good men and true" among RI members, including Dale and Alban Caroe.[773]

In January 1952, Bragg was contacted by a sub-committee of the Board of Managers that was examining the constitution of the RI. He was asked to advise the Managers concerning "the proper administrative structure of the RI." Bragg wrote to Dale, who had also been asked for advice. Dale replied: "I think it *most* important, however, that the Director should be the executive officer... No Director ought to be asked to accept a position in which the Asst. Secretary can 'cock a snoot' at him, and say 'I am not responsible to you, but to the Managers or their House Committee'... I am convinced that the structure in the present Bye-Laws only lasted as long as it did because your father was an angel and because my tenure was wholly in war years."[774] A few days later, Bragg attended a meeting of the sub-committee.

The following month, the crisis at the RI worsened when the Earl of Halsbury, a personal friend of Andrade's who had been nominated for the post of Treasurer, failed to be elected after Rankine had publicly and privately expressed his opposition. A Special General Meeting of Members was called for March to consider a motion requiring Andrade's retirement. Four Managers and three Visitors resigned, apparently in protest. At the meeting, Brabazon sided with Rankine and Andrade lost a vote of confidence by 250 to 136. Six more Managers resigned, leaving only five.

Not only was the RI in the midst of a civil war, but it also found itself in conflict with a sister institution. At the Special General Meeting of March 1952, Brabazon made a remark about the Royal Society that deeply offended some Fellows present. There were no Fellows of the Royal Society among the remaining Managers of the RI, and, despite Bragg's efforts, none was elected to the vacant positions in the Annual General Meeting in May. In a highly symbolic move, Robert Robinson, President of the Royal Society from 1945 to 1950, resigned his RI membership.[775] The rift widened when Edward Salisbury, Biological Secretary of the Royal Society, gave up the Fullerian Professorship of Physiology at the RI.

Andrade resigned the Fullerian Professorship of Chemistry on May 23, and subsequently his other offices, on the understanding that compensation would be arbitrated. An Arbitration Court was set up, to which Bragg gave written evidence unflattering to Andrade. During the arbitration, Brabazon and Rankine, who had resigned as Secretary, offered to support Andrade's nomination to a non-resident Fullerian Professorship in Chemistry. On December 23, 1952, the arbitrator awarded Andrade £7000 plus legal costs. Brabazon duly nominated Andrade to the non-resident professorship at a meeting of the Committee of Managers in January 1953. Rankine, however, objected that the appointment should not be made until a Resident Professor had been selected. At the March meeting of the Managers, Brabazon formally proposed Andrade and the motion was defeated. Andrade, feeling that the Managers had not kept their end of the bargain, immediately sued for damages. Mr Justice Vaisey of the High Court of Justice disagreed, dismissing Andrade's suit with costs on March 25.

Meanwhile, Bragg had been approached about the vacant resident professorship. On January 4, 1953, he wrote to Brabazon: "I would give an invitation to direct the Royal Institution very serious consideration because of my warm affection for the place and appreciation of all it stands for." However, "I could not come if Professor Andrade retains any official connection whatever with the Institution. I say this with no feeling of animosity towards Andrade, with whom I have always had friendly relations. I am certain, however, that in that event my position would be an intolerable one . . . This was my first reaction when you told me of the suggestion to give Andrade a Professorship, and it is confirmed by reflection . . . If I am to be considered, I feel the managers ought to know my stand on this point before they reach any decision about the proposal concerning Andrade which you outlined to me."[776] In a personal letter to Brabazon, Bragg stated that George Thomson agreed with his position on Andrade. As long as Bragg agreed with Andrade, all would be well; but as soon as he had to say no to him, "it would be pure hell."[777]

On April 2, 1953, the new Secretary, Stanley Robson, wrote to Bragg to offer him the positions of Fullerian Professor of Chemistry and Resident Professor: "The salary for the combined posts is £2000 per annum, with an allowance for expenses of £600 per annum, and the occupancy, during the tenure of office, of the Resident Professor's flat, the three historic rooms of which are furnished." Robson hastened to assure Bragg that he need not worry about problems with the Managers: "you would find a cooperative and very sympathetic Committee of Managers and Officers who would be happy indeed to help you in every way all the time."[778]

It is not clear what effect Bragg's January 1953 letter to Brabazon had on the Managers' apparent change of heart about offering Andrade a non-resident professorship. Robson's letter specifically absolved Bragg of any blame: "your letter of January 4th to Brabazon was read to the Managers *after* they had decided that it was inappropriate to give Andrade an official position." However, Bragg had specifically asked that the Managers make his opinion of Andrade a factor in their decision about the non-resident professorship—it would seem quite a coincidence that the Managers, having agreed to Andrade's appointment during the prior arbitration process, had now quite independently of Bragg changed its collective mind. However, even if Brabazon had told the Managers in January that Bragg would not accept the position if Andrade were non-resident professor, and the Managers had then reneged on their arrangement with Andrade, Bragg had done nothing wrong. He had merely stated his conditions for considering the position; if the Managers felt they were legally or morally unable to accept those conditions, they were at liberty not to pursue the matter further.

All ties between Andrade and the RI having been severed, Bragg now had to decide whether or not to accept the position. The arguments in favor were many and strong. He would have to retire from the Cavendish Professorship in a few years, and was not yet ready to devote himself full-time to gardening. There was a family tradition to maintain. The mandate of the RI, to increase public understanding of science, was one to which Bragg was passionately committed and excellently suited. It was, in fact, the perfect job for a distinguished scientist with a genius for explaining science to children and other lay audiences.

Alice's initial reaction to the prospect of moving to the RI was unambiguously negative: "I think it would be dreadful."[779] They would be leaving their many friends and her many jobs in Cambridge to live in central London without a garden. There was also the unpleasant prospect of dealing with the fallout from the Andrade affair. At Alice's suggestion, she and Bragg went to consult Baron Adrian of Cambridge, an old friend of Bragg's who had shared the Nobel Prize in Physiology or Medicine in 1932. As President of the Royal Society as well as Master of Trinity College, Adrian was perfectly placed to read the political tea leaves. In addition, the close links between the two families meant that he could be relied upon to put Bragg's interests first. Alice considered Lady (Hester) Adrian to be "one of my dearest friends;"[780] the Adrians' son Richard would later marry Gwendy's daughter Lucy. Pacing the drawing room of the Master's Lodge, Adrian opined that Bragg should go to the RI—if only because no one else would.

It appears that Adrian proposed what he felt was best for the RI, not what was best for Bragg. As Alban Caroe later learned from Lucy Adrian, "at that time Adrian's personal view was that Willy ought *not* to take the post, but that as PRS [President of the Royal Society] Adrian would never have allowed his personal view to colour any advice he gave to anyone."[781]

At the annual meeting of Members held on May 1, 1953, Bragg was appointed Fullerian Professor and Resident Professor as of January 1, 1954. The normal retirement age of 70 was increased to 75 for him. The notice of Bragg's appointment in *Nature* stated in part: "He will bring to the Royal Institution not only an unrivalled record of scientific achievement and experience, but also those personal attributes of friendship and sympathy so characteristic of his father."[782] Once again, he would have large boots to fill.

As in 1919 and 1938, Bragg had taken up a poisoned chalice. At Manchester, as an inexperienced professor, he had had to overcome the hostility of students and colleagues; at Cambridge, he had had to reorient the Cavendish Laboratory away from Rutherford's highly successful program of nuclear physics. At the RI, he now had to reorganize a dysfunctional and impoverished organization while enduring the wrath of a substantial and influential body of the British scientific establishment.

According to Gwendy, "Willy and Alice found their first couple of years at the RI dreadfully difficult, not just through practical difficulties and 'nerves' within the RI, after the Andrade troubles there, but because so many of their R.S. friends practically 'cut' them."[783] Because so many influential scientists were both Fellows of the Royal Society and Members of the RI, the rift between the two organizations made Bragg's new job much more difficult. There was a personal price to be paid, too—not only were Bragg and Alice deeply hurt by the cold reception they received by many in London, but the Andrade affair may well have cost Bragg the honor of becoming, like his father before him, President of the Royal Society. In 1955, George Thomson refused to let his name go forward for President on the grounds that Bragg was the better candidate. The chemist Cyril Hinshelwood was then elected, violating the informal Royal Society tradition of alternating chemists and physicists, or chemists, physicists, and biologists. To make things worse, Hinshelwood was, in Alice's opinion, "extremely anti the Royal Institution." Robert Robinson, who, as noted above, had resigned his RI membership over the Andrade affair, was also very hostile to Bragg.[784]

Bragg faced a number of other challenges at the RI. The most urgent priority was ensuring its financial well-being. When Bragg took over, the income of the RI was only about half the expenditure and its capital was being drained. There was a small research fund but it was rapidly being depleted to cover overhead costs. Bragg's main innovation was to introduce a category of corporate memberships, which soon brought in more money than the individual ones. After 5 or 6 years, both operating and research accounts were "out of the red." By 1963, member subscriptions were less than 10% of total income.[785] Nonetheless, in December 1964, Bragg wrote to D. A. Oliver: "Do you remember our discussions, at the time I came here, about the Royal Institution's activities and financing them? Plans for this are still my main pre-occupation after eleven years. . ."[786]

The next priority was to reform the constitution. Alban Caroe had told him that "Your father often said to me that the antiquated byelaws of the RI were a great encumbrance, but that too great an upheaval would be needed to get them altered."[787] The Committee of Managers had to approve all expenditures except "petty items," approve requests for use of rooms by other organizations, etc.[788] Despite Robson's promises of support from the Managers and officers, there was considerable resistance to Bragg's proposed reforms. According to Alice, "there was a lack of cooperation; in their fear of WLB taking too much power, managers and officers too often thwarted him, and did not back him up properly."[789]

However, Bragg found some good advisors. Among them was Sir Alfred (Jack) Egerton, former Professor of Chemical Technology at Imperial College London. As secretary of the Royal Society from 1939 on and chairman of a 1951 committee on the finances and constitution of the RI, Egerton was well placed to advise on both internal and external problems. Bragg wrote of him in 1961: "He was the kindest and wisest counsellor and friend when I came to the Royal Institution in 1954. It was a time of great difficulty, when the financial position was at its lowest ebb, and much bitterness and suspicion still existed owing to the troubles which had led to the resignation of my predecessor, Professor Andrade. Among those who helped to save the Royal Institution for posterity, Jack Egerton was one of the foremost."[790] Another was Ronald King, whom Andrade had made Assistant Director of the Davy–Faraday Laboratory. Although he was Andrade's man, Bragg not only reappointed King to this position but also made him Deputy Superintendent of the House. Like Ratcliffe, his counterpart at the Cavendish Laboratory, King found Bragg to be a benevolent leader: "I never heard him speak harshly of anyone critically [sic]. Yes, angrily occasionally. The most damning appellation I remember him applying to anyone was 'Juggins.' "[791] Other key supporters were Bragg's old friends Charles Darwin and George Thomson.[792]

Bragg wrote to Harold Spencer Jones, the outgoing Secretary of the RI, in 1960: "In my opinion, the malaise from which the Royal Institution has been suffering for a considerable time before things came to a head in the time of my predecessor was due to the fact that this public character had largely lapsed. The Royal Institution had come to be considered by the outside world, and to quite an extent by the members themselves, as a place which mainly existed to supply privileges enjoyed by a restricted body of members." He proposed that the key function of the RI is "diffusion of knowledge," and that the title of the Resident Professor should be "The Director, Royal Institution."[793] In a letter probably written some time in 1963, Bragg solicited the views of his old friend Patrick Blackett. Bragg's own ideas included acting as a home for small scientific societies or philanthropic societies or becoming "a university centre." Administratively, there were questions about whether the RI could survive as a private organization—should it instead become a trust with wider representation or seek government support?[794]

The last step in Bragg's constitutional overhaul—the granting of the title "Director of the Royal Institution" to the Resident Professor, was fiercely resisted by the Secretary, Brigadier Harry Hopthrow, and other Managers who clung to the "club" concept. Not until 1965, his last full year in the job, was Bragg given the title of Director. By that time, George Porter, non-resident Professor of Chemistry at the RI, had been appointed as his successor; Porter's insistence on the title of Director was crucial in overcoming Hopthrow's resistance.[795] To ensure a smooth transition, Bragg and Porter both sat on the Board of Managers during the two years preceding the former's retirement. Not until 1984 were the Boards of Managers and Visitors merged into a single Council.

In his contribution to the Bragg memorial volume *Selections and Reflections: The Legacy of Sir Lawrence Bragg*, King told the story of a time Bragg took him sailing on the River Deben. The wind got up when they were returning and they had forgotten the oars. Bragg successfully maneuevered by sail through the moorings in front of the sailing club. "Is it too fanciful of me to be put in mind of the way in which he steered the good ship RI through difficult waters under often very critical eyes?"[796]

King, who had been a teacher in rural Wales, suggested that the RI offer courses for science teachers. Bragg was "not immediately enthusiastic," but then had the idea of inviting school-children as well. Some of the Managers were not keen—one member said "You know, King, we are not going to let this place become a kindergarten."[797] The first Schools' Lectures, a series of three talks on "Electricity," were given four times in 1954–5 to an audience of sixth-formers and teachers. The following year, Bragg presented a lecture on "Famous Experimenters in the Royal Institution," also given four times. A regular routine was established of giving a series of four lectures on Tuesday and Wednesday of consecutive weeks. The series was then repeated a number of times. A lecture series given in a particular year was given again 3 years later, when the school population had changed. The Schools' Lectures were a great success and were subsequently extended to fourth-formers and those from preparatory schools—by 1958, 16,000 tickets were being distributed to schools throughout London and the Home Counties. Far less successful were the traditional afternoon lectures, open to the public for a small fee and free to university students, but nonetheless poorly attended. These were quietly dropped.

Bragg's approach to the Schools' Lectures was the same as that he had adopted to the Christmas Lectures of 1934–5. The key was to show children a scientific phenomenon, not tell them about it. As Bragg wrote in 1957: "The lectures I remember most vividly as a student were the good old-fashioned ones, copiously illustrated by brilliant experiments."[798] The experimental demonstrations were organized by Bragg in cooperation with King; the RI Librarian, Kenneth Vernon; and a Lecture Assistant.

Members of the research staff of the Davy–Faraday Laboratory were also expected to help out. One day soon after Bragg was appointed, he asked the laboratory assistant, Bill Coates, to come up with a demonstration of the contraction of rubber on heating. Coates "produced a model in which the long molecules were meccano chains, and the heat was a large sheet of aluminium vibrated by blows from drumsticks," which Bragg used in a lecture on materials. According to Coates, the Director was also capable of coming up with his own demonstrations: "On one occasion he asked me how far I could throw a wooden dart. I replied that I was not quite sure. He produced a two-foot length of dowel stick with a paper flight attached and invited me to throw it down the long red corridor at the RI. I tried and made a throw of about fifteen feet—the corridor must be at least fifty feet long. WLB's reply was to take the dart and cut a small notch in the dowel stick near the flight. To this he added a piece of string rolled at one end. Hitching the string to the dart and round his hand, he hurled the dart the whole length of the corridor such that it slammed into the far wall, narrowly missing a large oil painting. 'Simple leverage!', said WLB. 'This was used by the Australian spearthrower when hunting.' I agreed, but have often wondered how, if the dart had gone through the oil painting, I would have explained to the Managers of the RI that the damage had occurred while playing darts with the Director."[799] In 1957, Bragg persuaded Coates to leave the research laboratory to become Lecture Assistant.

According to Uli Arndt, a member of Bragg's research group: "His RI lectures to school children were enormously popular; one of the highlights was when he demonstrated the field-free region inside a Faraday cage by climbing into the cage himself, which was then charged by the original Whimshurst [electrostatic generator] machine. He would explain that his hair would stand on end if he cautiously raised his head out of the trap on top, but as he had no hair he would content himself with raising a stick to which paper streamers had been attached."[800] George Porter, Professor of Chemistry at the RI, recalled: "the fluid-bed of sand on which a 'ship of the desert' miraculously surfaced."[801] Some of the most effective demonstrations used only the simplest of apparatus. Bragg would demonstrate the limits of resolution by signing his name about two feet high first with a crayon and then with a distemper brush.

One of Bragg's secrets was shared with the nuclear physicist Thomas Allibone, who gave the Christmas lectures for 1959–60. The first lecture was, Allibone felt, "a bit dull." Bragg suggested that he get an "unexpected" result. Allibone managed to do so for the remainder of the lectures and "the atmosphere was marvellous."[802]

In 1961–2, Bragg gave the Christmas Lectures himself. As in 1934–5, the topic was "Electricity." V. T. Saunders, a secondary-school science teacher who attended with his 16-year-old grandson and 12-year-old granddaughter, wrote an account of the lectures for *Contemporary Physics*.[803] In one lecture, Bragg used a van de Graaff generator charged to 100,000 V to demonstrate lightning: "A doll's house under an artificial thunder cloud was unscathed when protected by a lightning conductor, but when the protector was removed the lightning shattered the roof." In another lecture, Coates swallowed a "radio pill" with a pressure transducer and was poked to give a response. The principle of the magnetron was demonstrated by Bragg focussing 3-cm electromagnetic waves with a glass lens and reflecting them from a piece of tin. Saunders concluded that Bragg "has shown how the bridge over the gap between the scientist and the intelligent non-scientist can be established" and "it is possible to introduce modern [!] physics, without misleading sacrifice of accuracy, to school boys and girls."

In lecturing to school children, Bragg was far more successful than he had ever been to university students. King wrote: "As a scientific expositor he was superb at all levels, but when he spoke to young people his imagery and demonstration technique and above all his obvious enjoyment, established a special rapport."[804] Phillips agreed on the essential features of Bragg's lecture technique: "his gift of illustration by analogy coupled with an infectious enthusiasm."[805] Phillips also wrote: "it was always a delight for those of us who worked in the laboratory to slip into the gallery at 5 o'clock and watch him enthrall, stimulate and occasionally provoke his audience."[806] Porter simply stated: "Bragg's lectures were a 'tour de force'..."[807] In Bragg's own mind, the easy rapport he had with children had a simple explanation: "We enjoy the same things."[808] Perhaps more revealing is another comment: "It is astonishing how many great scientists are unable to project themselves in this way [into the mind of a 17-year-old]."[809]

The other major educational program of the Royal Institution was the Friday Evening Discourses. Because of the varied nature of the audience, the Discourses were more of a challenge than the Schools' or Christmas Lectures. As the chemist William Ramsay wrote: "The RI audience [is] the most ticklish in Britain to lecture before, because the most critical and the most refined, and possessing also in equal shares so much knowledge and so much ignorance."[810] Bragg's approach to these lectures to a (generally) non-scientific audience was similar to that he used in the Schools' Lectures. In a 1958 article on "Interpretation of Science to the Public," Bragg wrote: "To the layman the difference between the description of an experiment and the actual witnessing of it is as great as the difference between looking at a foreign country on the map and visiting it . . . lectures on art or music can only be appreciated when they are related to such first-hand impressions [as paintings or concerts]. The primary way to interest the general public in science is to show experiments and demonstrations, and so let them share in the thrill of understanding how things work, which is after all a good popular definition of science."[811]

The Discourses allowed Bragg full rein of his gifts for simplification and analogy. In a November 1959 Discourse on "Atoms and Molecules," he stated: "One can compare the union of sodium with chlorine to a joining forces of two eightsome reel parties, one of which is short of a partner for the complete dance figure, and one of which has an extra wallflower who cannot take part in the dance."[812] In a February 1964, lecture on "Minerals," he described the structure of mica as being "like two slices of bread with butter in between."[813]

According to his son Stephen, "his Discourses were always lucid and very well planned. You know, they make rather a fetish there of lasting exactly an hour. He used to arrange his material in five-minute tranches and had about 14 of them, so that, if he was running late, he could leave out two or three and, if he was running early, he could put in the extras so it was to be exactly the time . . . he was a great admirer of Faraday's correspondence on how to lecture and he thought Faraday's ideas were very good. He only disagreed in one sense with Faraday, I think, in that Faraday was rather against angling for laughs, whereas my father always liked to bring a little bit of humour into a lecture."[814]

As with many things at the RI, the Friday Evening Discourses were steeped in tradition. According to RI legend, Charles Wheatstone had suffered an attack of stage fright shortly before giving a Discourse in 1846 and bolted from the premises. Faraday, then Resident Professor, entered the auditorium promptly at 9 o'clock and delivered a lecture on Wheatstone's work that was exactly one hour long.[kk] As Bragg wrote in 1957, "Ever since then we have kept up the

[kk] Frank James has shown that this story has no basis in fact. [James, F. A. J. L. (1985). 'The optical mode of investigation': Light and matter in Faraday's natural philosophy. In *Faraday*

practice of immuring our lecturer in a small room before he starts, guarded by the senior lecture assistant, and it is the Resident Professor's duty to take over the victim from his guard and duly deliver him in the lecture theatre."[815] Ralph Wain, who gave a Discourse on "Plant Growth and Man-Made Molecules" in April 1958, recorded that Bragg, "a great traditionalist," came into Faraday's office at 8:55 p.m., pointed him towards Faraday's specimen of rock crystal and said "May your Discourse be as clear as the crystal." He then turned Wain towards a tank of barnacles and said "May you stick to your timing as the barnacles stick to the rock."[816] Bragg and Alice revived WHBs tradition of inviting speakers and audience members to dinner in the Resident Professor's flat—amounting to 120–150 guests a year. They also invited people from professions not well represented among the RI membership, such as politics, law, and industry.[817]

Phillips provided the following account of the Friday evening ritual: "At 8:55 p.m. each Friday evening during term the audience (mainly in evening dress) would quieten as Lady Bragg led in the dinner party guests followed soon afterwards by the President himself escorting the lecturer's wife and supported by a stately procession of Managers. The doors would be closed until the stroke of 9 when the lecturer appeared dramatically through one door while Sir Lawrence soon afterwards came in through the other one . . ."[818]

Bragg was able to continue the family tradition of Friday Evening Discourses to a third generation when, in November 1963, his son Stephen gave a talk on "Oscillation and Noise in Jet Engines." Two years later, Alice delivered a Discourse on "Changing Patterns in Marriage and Divorce." Following her work on the Royal Commission on Marriage and Divorce, she became chair of the National Marriage Guidance Council, a position she held for 20 years.[819]

The educational mandate of the RI extended beyond its famous lecture theater. In 1959, Bragg gave a series of six lectures on "The Nature of Things" which was recorded by the BBC before an invited audience at the RI. After these aired, he wrote to Sir Eric Ashby: "It is surprising how many strangers recognise one and speak about them in shops and trains, and I was even accosted in the street the other day."[820] Max Perutz recorded that "He was intrigued when the greengrocer woman in Soho told him that he was 'the spitting image of a man she saw on the telly last night' and modestly signed the bill for her to keep as a souvenir."[821] This media exposure had its downside, however, as Bragg received letters from crank scientists around the world. Thus he was among the first to learn about the "Ortona Ray," the "Uratome Theory" and the "Electronic Mind Tap," as well as anti-gravity devices and perpetual-motion machines.[822]

Bragg also made a series of films of experimental demonstrations for distribution to schools. One of these films was faulty, and the producer asked him to repeat part of the demonstration wearing the same suit used for the original filming. Unfortunately, Alice had meanwhile given the suit, which she

Rediscovered: Essays on the Life and Work of Michael Faraday, 1791–1867 (D. Gooding and F. A. J. L. James, eds.), pp. 137–161. Stockton Press, New York.]

had long hated, to a jumble sale. Bragg managed to locate the man who had bought the jacket and waistcoat, but the trousers had disappeared. He borrowed the two items and the producer agreed to re-shoot the missing sequence from the waist up.[823]

Bragg once compared the Royal Institution to "one of the ductless glands in our bodies, which have an influence out of all proportion to their size."[824] He was determined to use that influence to promote science literacy among the British public and its leaders. One of his allies in this mission was his old friend from Cambridge, C. P. Snow. In March 1957, Snow published a two-part article in *The Times* bemoaning the state of science education in Britain, particularly compared to the Soviet Union—this was 7 months before the flight of Sputnik shocked the West out of its comfortable assumption of scientific superiority. Snow reported that when he had asked a group of "highly educated and cultivated people" to define the term "conservation of energy" or "machine tool," none of them could. Scientific illiteracy, he felt, was regarded as "one of those inexplicable English phenomena, like hanging, ringing changes on church bells, and county cricket." Snow recommended increasing the scientific and mathematical content of school and university education.[825]

In a response to Snow's letter, also published in *The Times*, Bragg blamed the contempt for science found among humanists on the public schools, as it was the grammar and secondary schools that produced Britain's scientific leaders: "no-one should be regarded as having a sound cultural education unless this has been based on a good grounding in both the humanities and sciences."[826] In an address to the Federation of University Women, Bragg put the point more bluntly: "An ordinary educated person should not so often regard a scientific achievement with about the same extent of understanding as an Australian aborigine or South African Bushman looking at an aeroplane."[827]

One of Bragg's most important efforts to promote the public understanding of science was his presidency of the International Palace of Science at the Brussels World Fair, which ran from April to October 1958. The Palace of Science consisted of displays on "The Molecule," "The Crystal," and "The Living Cell." The latter contained models of protein, nucleic acid, and virus structures. The fourth scientific theme of the World Fair, "The Atom," was housed in a giant model of the unit cell of an iron crystal. The "Atomium" consisted of brilliant aluminum-skinned spheres, 18 m in diameter, connected by steel tubes. As the iron crystal is body-centered cubic, there were nine of these spheres (eight representing cube corners, one representing the body center). For aesthetic reasons, the unit cell rested on a cube corner, so a three-fold rotation axis of the cube (connecting opposite cube corners) was perpendicular to the ground. Six of the spheres were accessible to the public by means of an elevator and several escalators contained within the connecting tubes. The uppermost sphere held a restaurant; others housed replicas of Aston's mass spectrograph and Wilson's cloud chamber.[828]

It must have been a proud moment for Bragg when he looked over the World Fair site, dominated by the spectacular 102-m high Atomium. There, magnified

165 billion times, for all 42 million visitors to see, was a unit-cell structure of the type that he had first shown to be the basis of all crystalline matter. Admittedly it was a body-centered cube, rather than the crystallographically and historically more important face-centered type. However, Bragg could also take pride in the fact that he had made fundamental contributions to the other three areas of science featured at the Fair—chemistry, crystallography, and biology. Ironically—given that such was never the intention—no celebration of Bragg's career, before or since, captured its wide-ranging significance more than the 1958 World Fair.

In addition to restoring the finances of the RI and starting new educational programs, Bragg had to adjust to the lifestyle of living in a flat with no neighbors and where, as Patience put it, "Fortnum & Mason and Harrod's were the two local grocers." Margaret only lived at the RI for a short while before her marriage, after which Patience was the only Bragg child living there. She sometimes accompanied her father when he went to have tea with his research team, and listened to his lucid summaries of the previous night's Discourses over Saturday morning breakfast.

For Bragg, the biggest drawback to his new accommodations was the lack of a garden. He arranged to use a plot of land at the back of a nearby house, 11 Little St James' Street, that had been bombed during the Second World War. Patience liked to accompany her father as, dressed in gardening clothes and an ancient hat, he trundled a wheelbarrow across Piccadilly to his allotment.[829]

Francis Crick has described how Bragg, missing his Cambridge garden, hired himself out as a gardener to an elegant lady living on The Boltons, an exclusive street in South Kensington. She knew him only as "Willie" until one day one of her guests looked out of the window and asked "My dear, whatever is Sir Lawrence Bragg doing in your garden?"[830] Similar accounts are given by Kendrew[831] and Watson.[832] Unfortunately, these stories are apocryphal. According to Patience, the garden in which Bragg was spotted working was actually that of Celia Hensman, an old friend of the family who lived in Hereford Square.

Such part-time gardening activities did not compensate for the lack of a real garden. Partly for this reason, Bragg and Alice purchased a house in the country—Quietways, in the village of Waldringfield, Suffolk. It was located on a farm lane close to a river, with a coppice and a garden. According to Stephen Bragg: "It was not an old house but it was made of old materials from some cottages that had been pulled down... There was an attractive garden with very light soil. They lived there mostly from Easter onwards to September or October, for weekends during the term and then largely during vacations. The people in Waldringfield, which is on the river Deber, live by the tide. Everybody has a boat and, if the tide is coming in, you sail up to Woodbridge, and then back when the tide turns. If the tide is going out, you sail down to Felixstowe Ferry and then back. It's an attractive estuarial landscape with lots of wading birds and ducks and things of that sort."[833] Patience also had fond memories of Quietways: "At the weekend we all went down to Waldringfield and there he relaxed. We used to go on long walks and go bird-watching on the marshes. He bought a lovely little wooden sailing boat called 'The Tortoise' in which he 'taught us' to sail. He found a lot of time to teach me to drive a car. All the time he would try to concentrate but then he'd shout, 'By Jove, there's a redstart', or whatever it was and we'd both be distracted. Luckily we never had an accident."[834] Bragg got Patience to drive him to Diss, where he had been posted in the First World War.

Sailing, gardening, sketching, and bird-watching—weekends and holidays at Quietways proved the perfect antidote to weekday living in central London. On one occasion at Quietways, two American servicemen who had been billeted there during the Second World War came back to see the house again. Bragg, who was in the garden, wearing an old hat and a jacket with leather patches on the elbows, showed the visitors around the premises. Later, at the local pub, the Americans mentioned to the landlord what had transpired: "The owners of the house were out, but the gardener showed us around."[835]

There was a gardener, though, who looked after things when the Braggs were not in residence. As Bragg wrote to the previous owner of Quietways, "We have got the garden in good trim. Mr. Edgar, a retired income tax collector, gives me ten hours a week, he says he likes working in our garden because retired income tax collectors are not at all popular and he finds it a secluded place where no one sees him."[836]

Quietways housed a growing brood of grandchildren. Stephen and Maureen had three children: Nigel Lawrence, born 1952; Charles David, born 1956; and Andrew Christopher, born 1957. After graduating from Oxford, Margaret obtained a certificate in social science and administration from the London School of Economics. She married the diplomat Mark Heath in September 1954, and then joined him at his posting in Indonesia. The Heaths also had three children: John Nicholas, born 1956; Clare Penelope Margaret, born 1957; and William Mark, born 1959. When Patience graduated from Cambridge in 1957, she took a job in the library of the Foreign Office. She lived at the RI until 1959, when she married the banker David Thomson, son of Bragg's old friend George Thomson and grandson of J. J. The Thomsons had four children: Hugh David Bragg, born 1960; Benjamin John Paget, born 1963; Alice Mary Rose, born 1967; and Kathleen Anne Patience, born 1969. David Bragg married Elisabeth Bruno in April 1967; this union was without issue.[837]

In the autumn of 1955, Bragg returned to South Africa, this time with Alice. The itinerary—including a visit to a game reserve—was very similar to that of 1952. So were the lecture topics—Bragg spoke on "X-Rays and the Molecule," "The Discovery of Useful Electricity," "Proteins," and "The Work of the Royal Institution." Alice lectured on "Juvenile Courts in Britain" and "Women in Public Affairs."[838] Africa made a deep impression on Alice: She remembered "watching elephants, giraffes, and lions in a national park, the glories of the botanical gardens in Cape Town, a trip to watch a colony of night herons in the trees, while on the ground beneath were wild arum lilies each with a minute frog at its heart."[839]

It was a well-deserved break. After almost two years on the job, Bragg's rescue of the Royal Institution was almost complete. In December 1955, Bragg wrote to Joseph Gray, his former sound-ranging colleague and now Professor of Physics at Queen's University in Kingston, Ontario: "My main job for the last two years has been the reorganisation of the Royal Institution. The prospects are much brighter now. We are practically out of our financial difficulties and I have found a new function for the place in giving lectures for school children all the year round, illustrated by gorgeous experiments. This scheme is going very well."[840]

By the time Bragg took over as Director of the Davy–Faraday Laboratory, research at the Royal Institution had declined sadly from the glory days of WHB and his brilliant disciples: King had a small group on metal physics and Uli Arndt was studying proteins with X-rays. Bragg's first idea was to recruit Perutz or John Kendrew. Not surprisingly, neither wished to move to London, although they may have regretted that decision when the new Cavendish Professor, Nevill Mott, proved less broad-minded than his predecessor in his definition of "experimental physics."[841] Bragg then tried to entice Dorothy Hodgkin to the RI. Despite his entreaty that she would be "free of tiresome lecturing and tutoring," Hodgkin elected to stay in Oxford, where the analysis of vitamin B_{12} ($C_{63}H_{84}N_{14}O_{14}PCo$) was reaching a climax.[842]

There was nothing for it but to assemble a new group. Over his first couple of years at the RI, Bragg recruited Helen Scouloudi from John Desmond Bernal's department at Birkbeck; David Green, who had been a student of Perutz's; Tony North, who had worked on collagen with John Randall at King's; Jack Dunitz, a former student of Hodgkin's who had subsequently gone to the USA; and David Phillips, from the National Research Council laboratories in Ottawa.[843] All except Dunitz worked on proteins. Initially, the RI group worked as subcontractors on Perutz's analysis of hemoglobin and Kendrew's of myoglobin, Perutz and Kendrew both being appointed as Readers in the RI. Kendrew, in particular, made frequent trips to London to help Bragg get his new group going. As Bragg put it, "We got going by transferring a living part, as it were, of the Cambridge research to Albemarle Street."[844]

When Dunitz arrived in January 1956. "Uli Arndt and David Phillips were beginning to develop the linear diffractometer. There was an orthorhombic form of ox hemoglobin under investigation by David Green, and Helen Scouloudi was comparing diffraction patterns from various myoglobin crystals. Tony North was making X-ray photographs of lactoglobulin crystals."[845] The linear diffractometer made multiple simultaneous measurements of intensities, and was the first machine not to use X-ray film. It improved the "quality and speed of data acquisition by orders of magnitude."[846] Support for this research was obtained from the MRC and the Rockefeller Foundation.

One of the curious features of the RI was the juxtaposition of the Resident Professor's flat and the research laboratories in 20 Albemarle Street. Part of the flat was adjacent to the laboratories, connected by a passageway, and part was directly below them. On one occasion, a flood in one of the laboratories led to water dripping into Bragg and Alice's bedroom, which North remembers as "the only occasion that I remember him being angry."[847] Except for that incident, Bragg seems to have enjoyed being close to his researchers. As Phillips recalled: "This intimate arrangement enabled Bragg to visit the laboratories whenever he had a moment to spare or needed relief from the discussion of some tedious difficulty, and he would announce his imminent arrival by a characteristic stamp on the ancient and creaking floor boards."[848] According to Louise Johnson, then a graduate student, some of the students would use this warning as an opportunity to flee from the great man. In her experience, though, Bragg "was always most kind and interested in the work, even if it was quite a small problem."

Johnson's work was directed by Phillips but Bragg, as the only member of the Davy–Faraday Laboratory to have the appropriate status in the University of London, was her official supervisor. In that capacity, Bragg was one of the examiners of Johnson's Ph.D. thesis. He came to the oral examination dressed in the ornate robes and magnificent plumed hat he had worn to receive his honorary degree in Portugal in 1945: "The PhD exam never seemed to strike a serious note after that."[849]

Another research student who was impressed with the humility and enthusiasm of the Father of Crystallography was David Blow, who joined Perutz after Bragg left Cambridge. "When I had done enough to know what I was talking about, maybe summer 1956, Max once brought Bragg to my desk for me to show him my work. By this time I had a good knowledge of Bragg's incredible contributions to X-ray crystallography, and was overwhelmed that he would take the time to look at my insignificant little contribution. He sat beside me at my desk, exhibited tremendous interest and enthusiasm for what I showed him, and soon put me at my ease. I remember him making notes of what I was saying, which to my unsophisticated young man's mind seemed quite bizarre. I should be the one who made notes of everything he said!"[850]

The small size of the research laboratory (15–20 people), the proximity of the Director's flat, and above all Bragg's attitude meant that members of the RI crystallography group felt like family. Arndt recalls that they were invited to Bragg family skating trips on the pond at St James Park, to Patience's wedding and on an annual basis to Waldringfield.[851] The family atmosphere was strengthened in 1960 when Phillips married Bragg's secretary, Diana Hutchinson.

In April 1957, Perutz wrote to Bragg: "It now appears that the y parameter of our second mercury position is very close to that of the first, so that we may have difficulty in getting phase angles of reflexions with low orders of k. We still have no 3rd heavy atom position . . ."[852] After the initial breakthrough in 1953, when isomorphous replacement was first used to determine signs of reflections in hemoglobin, Perutz had quickly become bogged down again. As David Phillips told Max Blythe, "He tried hard to get more derivatives and a curious phenomenon got in the way. Every time he tried to produce derivative crystals, the cell dimensions of the space lattice of the crystal changed in some mysterious way so that the derivative wasn't compatible with the native structure."[853] Prevented from performing a three-dimensional Fourier synthesis by the paucity of heavy atom derivatives, he had to resort to a projection on the (010) plane at the higher resolution of 2.7 Å, but this provided little new structural information.

Myoglobin was proving far more amenable to X-ray analysis. Because it does not contain cysteine, the amino acid with which parachloromercuribenzoate reacts in hemoglobin, Kendrew asked two chemists working with him, Howard Dintzis and Gerhard Bodo, to come up with different methods of attaching heavy atoms. They succeeded in attaching mercury, gold, or silver at five different sites in the myoglobin molecule. The gods were not finished tormenting Perutz—it would be his former protégé who would achieve the honor of being the first to determine the structure of a protein molecule. As Perutz told Horace Judson, "There was Kendrew going ahead, things were going splendidly, he was probably solving his structure, and I was, having produced the method, getting absolutely nowhere!"[854]

The problem of locating the heavy atoms of protein crystals in three dimensions, critical to assigning structure factors to the heavy atoms and thereby determining phases, inspired Bragg to publish his last scientific paper. In this paper—published in *Acta Crystallographica*, a journal he had helped to found—Bragg developed a graphical method of determining the x, y, and z coordinates that circumvented the use of Patterson or Fourier methods. As Bragg realized, the problem was similar in principle to those faced in the X-ray analysis of simpler compounds: "In the early days of X-ray analysis . . . it was possible to determine coordinates to a high precision without using Fourier series. This was done by concentrating on spectra of high order, the calculated values of which were very sensitive to a small change in a coordinate, and it was a feasible method because the number of variables in the structure was so small. In the present case, we can use similar methods because generally we are only trying to find the coordinates of one atom, and they can be determined one at a time."[855]

This final paper of Bragg's was another demonstration that pen-and-paper methods, underpinned by a thorough understanding of optical principles, could still be used to solve problems at the frontiers of crystallographic research. Bragg accepted that the future belonged to the number-crunchers but felt no personal empathy for an approach that divorced crystal structure from X-ray optics. When computers replaced geometrical methods, Klug wrote, "Some of the romance was lost and with it, I believe, the need for the physical understanding so necessary in the early days of the subject."[856]

Bragg's graphical method of determining heavy atom coordinates was validated by comparison with values Kendrew had found using a Fourier method developed by Perutz. "Kendrew lent me a number of his measurements to mull over, and my great moment came when I convinced myself that the answer to both the above questions [could the positions of the heavy atoms be fixed in 3 dimensions and were the intensity changes large enough to determine the phases?] was 'yes'. My calculations had no influence on the steady march of Kendrew's analysis; they were like the 'colours' in a pan which show a miner that he has struck gold, but they proved that the answer to protein structure which we had sought for twenty-five years had at last been achieved. I could not resist testing one diffraction after another, seeing the answer come out in each case, till I worked myself to a complete mental standstill."[857]

Bragg's calculations confirmed Kendrew's values for the x, y, and z coordinates of the mercury atoms in sperm-whale myoglobin. This meant that structure factors could be calculated for these atoms and used to determine phase angles for all reflections. These phase values could in turn be used in Fourier syntheses that would determine the electron density throughout the three-dimensional space of the myoglobin unit cell. As Bragg later wrote of his last published research work, "it showed that the problem had been solved, and that final success was now certain."[858]

The phase problem had been solved for myoglobin, would be solved for hemoglobin once appropriate heavy-atom derivatives were found, and could in principle be solved for any protein. No wonder that Bragg was "moved to tears of emotion"[859]—the gigantic barrier that had, since the 1920s, prevented the full application of X-ray crystallography had finally been surmounted. No one could appreciate the significance of this more than Bragg. In his 1914 analyses of the alkaline halides, he had been the first to use the interference between waves diffracted by different atomic planes to gain insights into crystal structure. In the 1920s, he had pioneered the use of absolute intensity measurement and developed the concept of the structure factor. In 1929, he had performed the first two-dimensional Fourier analysis of a crystal structure. Ever since, the prospect of bringing together all these aspects of X-ray analysis to create an unambiguous way of determining crystal structure from the intensities of X-ray reflections had been severely limited by the phase problem. Now the full power of the X-ray method was finally unleashed. It was a fitting moment for Bragg to retire from research. As Perutz wrote: "If we think of Bragg as an artist and compare him to, say, Giotto, it is as though he had himself invented three dimensional representation, and then lived through all the styles of European painting from the Renaissance to the present day, to be finally confronted with computer art."[860]

It was one thing to have solved the phase problem, but quite another to apply that solution to a molecule more than 10 times larger than any that had been solved by X-ray methods. To construct a three-dimensional electron-density map of myoglobin at 6-Å resolution, Kendrew and his team measured the intensities of some 400 reflections from myoglobin and its five heavy-atom derivatives using a new microdensitometer rather than the previous method of visual inspection; corrections for the Lorentz and polarization effects were made with a desk calculator; to determine the phases, about 2000 circles had to be hand-drawn.

The calculation of the Fourier series, long the rate-determining step in construction of electron-density maps, was now the easy part. Kendrew had been foresighted in the application of computers to X-ray analysis. In 1949, Hugh Huxley had been told by Bragg that it would be "good for his soul" to calculate a two-dimensional Patterson projection of hemoglobin at moderate resolution using Beevers–Lipson strips. It took him about 2 weeks. His best friend was John Bennett, an engineering student at Christ's College. Bennett, who was working on the EDSAC (electronic delay storage automatic calculator) computer in the mathematics laboratory, told Huxley that his calculations were programmable, and wrote a program that did the hemoglobin analysis in half an hour (plus another half hour to print the results!).

When Huxley moved on to study muscle contraction, he turned the computer program over to his supervisor, Kendrew, who developed it further with Bennett.[861] Kendrew suggested to Bragg that a student be used to write programs for EDSAC, but Bragg was "not at first enthusiastic."[862] By a fortunate coincidence, one of the world's leading computer research groups was located about 100 yards from the Cavendish Laboratory. Using EDSAC, it took only 70 minutes to perform the Fourier synthesis of myoglobin—the first time a digital computer had been used in crystallographic analysis. As the results emerged from the machine, the electron-density contours were plotted on 16 sheets of transparent plastic and then stacked to create the model. As Maurice Wilkes, one of the members of the EDSAC team, observed, "No-one knew, until almost the very end, whether a clear structure for the molecule would emerge from an inspection of this stack or whether several years of sustained effort would have to be written off."[863]

The 6-Å electron-density map of myoglobin was published in March 1958, in a paper authored by five Cambridge researchers and Phillips from the RI; Bragg was thanked for "his interest and encouragement."[864] The map was in the form of 16 sections perpendicular to *y* and 2 Å apart. The myoglobin molecule appeared as a flat disk of ~43 Å × 35 Å × 23 Å. A smaller disk-shaped feature was obviously the heme group. As Perutz put it, "The first protein molecule beheld by man revealed itself as a long, winding visceral-looking object with a lump like a squashed orange attached to it."[865] At this low resolution, it was impossible to identify specific amino acids or even follow the course of the polypeptide chain. However, several sections of high-electron-density rods, 20–40 Å long, nearly circular in cross-section and 5 Å in diameter, could be distinguished—presumably the α-helices. Kendrew wrote: "Perhaps the most remarkable feature of the molecule are its complexity and its lack of symmetry. The arrangement seems to be almost totally lacking in the kind of regularities which one instinctively anticipates, and it is more complicated than has been predicated by any theory of protein structure."

The 6-Å map of myoglobin was only a staging post on the way to the real goal—an atomic-resolution structure. A resolution of 1.54 Å, the length of a C—C bond, would be ideal, but Kendrew decided that 2 Å was as much as the team could handle. Even then, they had to analyze 9600 reflections—in comparison to the 400 used for the 6-Å map—and perform intensity measurement on a quarter of a million spots because of the different derivatives and exposures required. Echoing Perutz's feelings about the three-dimensional Patterson synthesis of hemoglobin, Kendrew wrote in 1963, "I would not care to have to undertake such a task a second time."[866] A major contribution from Bragg's group was the automatic diffractometer designed by Arndt and Phillips, with which the intensities of successive reflections were recorded on punched tape that could be fed directly into a computer—EDSAC Mark II. Even with this faster machine, the calculation took 12 hours. Bragg estimated that the summation of the Fourier series for the 2-Å map of myoglobin, if done manually, would have taken "a skilled scientist" about two hundred years.[867]

The 12-hour run on EDSAC II was done one night in August 1959. David Phillips and Violet Shore from the RI, who had collected data from one of the heavy-metal derivatives, came for the occasion. According to Phillips, "the data was all fed in to the computer, and then the paper tape output began to chatter . . . some paper tape came out and we rushed over to the printer and printed out what it said, and it said 'Michael Rossman, Michael Rossman, Michael Rossman.' And this was just the sort of identification labels of the programme and nothing else came out. And we carried on there until one o'clock in the morning, two o'clock in the morning, three o'clock in the morning; Michael wrestling with the programme, people making sure the tape had gone in alright, everybody else rewinding paper tapes, taking care not to tear them and all the rest of it. And eventually, at some ridiculous hour in the middle of the night, the map started to come out . . . Now it was a 2-Å resolution map which isn't quite enough to show individual atoms, but it ought to show groups of atoms. So what we saw, to our delight, was something that looked a little like ridges of hills and these ridges of hills were the course of the polypeptide chain. In the middle of it one blob that was bigger than the rest, and that was the iron atom, and around that a rather plainer [planar?] group of structures and that was the haem group around the iron atom. So, that was the first thing that I concentrated on. And Vi and I produced a contour map showing the haem group and the iron in the middle of it. And we knew we were there at that point, though the map wasn't tremendously good."[868]

Most of the myoglobin team stayed up until they saw the α-helices, then went to bed.[869] The next day, according to Richard Dickerson, "We plotted the map sections on Plexiglass sheets, stacked the sheets over a light box, and threw a cocktail party at dusk on the Peterhouse [Kendrew's college] lawn to celebrate. I vividly remember Sir Lawrence Bragg, director of the Royal Institution and the man who had brought Perutz and Kendrew to Cambridge [sic], taking the elbow of guests at the party and propelling them to the light box, pointing at an α-helix that ran obliquely though the map sections, and saying excitedly: 'Look! See, it's hollow!' "[870]

It was not a true atomic-resolution map, but the internal structure of a protein had finally been cracked. The putative α-helices of the 6-Å analysis were now seen not only to be hollow, but also to be right-handed with an axial repeat of 5.4 Å. Sixty-five to 72% of the polypeptide chain was α-helical. Using information on the amino acid sequence of myoglobin provided by Allen Edmondson

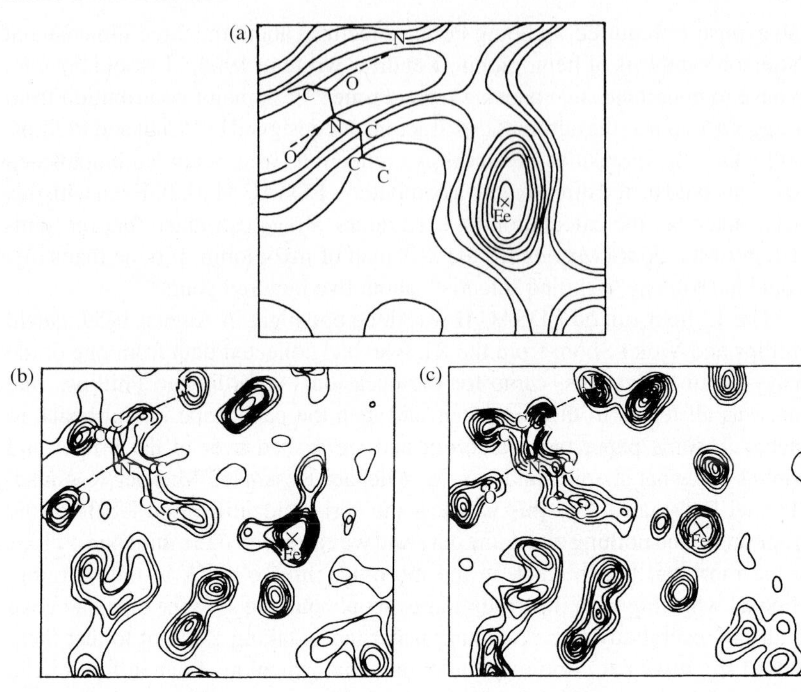

Fig. 9.1 Part of the electron-density map of myoglobin at different levels of resolution. (a) 6 Å; (b) 2 Å; (c) 1.5 Å. Reproduced, with permission, from Figure 8 of Kendrew, J. C. (1964). Myoglobin and the structure of proteins. In *Nobel Lectures in Chemistry, 1942–1962*, p. 690. Elsevier, Amsterdam. Copyright the Nobel Foundation, 1962

of the Rockefeller Institute for Medical Research, Kendrew was able to identify about two-thirds of the amino acids in the molecule (Figure 9.1).[871]

Perutz was on holiday and did not attend the party on the Peterhouse lawn. On August 5, he had written to Bragg: "There is no doubt now that we can safely go ahead with the calculation of a Fourier. I am going on holiday this evening (much to my regret). When I come back most of the measurements should have been turned into F's, and I hope that we can do the Fourier in September. Kendrew hopes to calculate his within the next two weeks. What a grand time we shall all have."[872] The technical problems that had plagued the hemoglobin project for several years had finally been overcome, and Perutz was able to calculate a three-dimensional Fourier—albeit at the relatively low resolution of 5.5 Å. In October 1959, Bragg wrote to Bernal, who had been Perutz's original supervisor, "I was in Cambridge last Thursday to see Perutz's haemoglobin. It really is thrilling. Haemoglobin is four myoglobins stuck together tetrahedrally, so like Kendrew's structure that the eye cannot detect any difference . . ."[873]

This analysis of horse oxyhemoglobin was published in February 1960, in a paper on which Tony North of the RI was one of the six coauthors.[874] The hemoglobin molecule was now seen to be a spheroid of 64 × 55 × 50 Å—very close to Bragg's 1952 prolate ellipsoid of 65 × 55 Å—consisting of four myoglobin-like subunits in a tetrahedral arrangement. The subunits were of two types that Perutz colored white or black in his plastic model (now termed α and β, respectively). Each of the white subunits was closely associated with one of the black ones, but the two half-molecules so formed did not make close contacts with one another. This explained the long-standing observation that the hemoglobin molecule could be dissociated into two identical halves. In confirmation of previous observations, a channel went through the center of the molecule and there were "dimples" where the two white and two black chains meet. The heme groups lay in pockets on the surface of the molecule.

In August 1960, the Fifth International Congress of the International Union of Crystallography was held in Cambridge. Such congresses were held every 3 years, the previous ones having been held in the United States, Sweden, France, and Canada. Bragg gave the Congress Discourse on "The Growth in the Power of X-Ray Analysis." Perutz presented one of five General Lectures, on the topic of "Structure of Crystalline Proteins." During the session on "Proteins and Related Compounds," Kendrew's 2-Å structure of myoglobin and Perutz's 5.5-Å structure of hemoglobin were presented. There were also papers on the X-ray analysis of insulin, ribonuclease, and viruses.[875] It was a triumphal coming-of-age for macromolecular crystallography. As Phillips later put it, "the scoffers were silenced."[876]

Looking back on the X-ray analysis of hemoglobin, Bragg wrote in 1965: "The progress of the work may be likened to the scaling of Mount Everest. A series of camps were established at ever-increasing heights, till finally a last camp was set up from which the successful assault on the summit was made." Mixing his metaphors somewhat, Bragg noted that "The 'nodes and loops' constituted a scaffolding which played a big part although it was knocked away when the building was completed."[877] It was certainly no exaggeration to say that hemoglobin was the Mount Everest of X-ray analysis. If Perutz was Edmund Hillary, Bragg was John Hunt, the expedition leader who did not make it to the summit.

Bragg could take great satisfaction from the fact that his new group at the RI had made solid contributions to the 2-Å structure of myoglobin and the 5.5-Å structure of hemoglobin. Led by David Phillips, the group was by 1962 strong enough to declare independence from Cambridge and study its "own" protein—lysozyme, from hen egg white. Crystals of this enzyme had been brought to the RI from MIT in 1960 by Roberto Poljak. The lysozyme project was led by David Phillips and included Poljak, Johnson, Tony North, and the chemist Colin Blake. Pauling and Corey, as noted above, had been working on lysozyme since the early 1950s. However, their efforts to solve the structure by comparing the amplitudes of reflections from niobium and tantalum salts of the protein were stymied when it turned out that the two forms were not isomorphous.[878]

In July 1960, the Royal Society celebrated its tercentenary. The celebrations included a exhibition "illuminating scientific research in progress in this country" which Bragg, as chair of the Royal Society's Soirée Committee, was asked to organize. Conveniently, this was to be held at the Royal Academy of Arts, just around the corner from the RI. During the planning for the tercentenary exhibition, Bragg had to referee a dispute which broke out between Phillips, who felt that the recently completed work on hemoglobin and myoglobin should have "pride of place," and Christopher Hinton, head of the Atomic Energy Authority, who wanted a model of a nuclear power station. In the end, Phillips had his way, and the centerpiece of the exhibition was a model of the myoglobin molecule.[879] Whether or not Bragg had any hand in this outcome, he must have been pleased to see protein crystallography triumph over nuclear physics!

The exhibit that featured the myoglobin model was entitled "Molecular Structure of Biological Systems"—the original name of Bragg and Perutz's group at Cambridge—and also contained displays on nucleic acid, virus, and muscle structure. Bragg must also have taken a paternal interest in two other exhibits featuring the work of former Cavendish colleagues—one on metal structure and the other on radioastronomy. The most distinguished visitor to the exhibition was Prince Philip, who was given a personal tour by Bragg. This was much more successful than Bragg's previous attempt to explain science to a royal visitor on the occasion of a tercentenary celebration—the Newton anniversary of 1946—Prince Philip took a keen interest and insisted on seeing every display.[880]

On September 21, 1960, Bragg gave the Rutherford Memorial Lecture at the University of Canterbury. Accompanied by Alice, he travelled to New Zealand via San Francisco, Hawaii, and Fiji. To illustrate his lecture, which was on "The Development of X-Ray Analysis," Bragg had brought Kendrew's model of myoglobin. Packed in a container the size of a large hatbox and carried as hand-luggage, this caused much bemusement to customs officers on the many stages of the journey to New Zealand. A pilot who wished to see the model for himself was treated to an in-flight presentation on protein structure![881]

Bragg also illustrated his Rutherford Lecture with a graph in which the logarithm of the number of parameters in a "typical" structure was plotted against year; the curve was hyperbolic until 1960, when it became an almost-vertical straight line. Bragg extrapolated that the million-parameter mark would be reached around 1965. "As I have worked for so long in this field, it will easily be understood how deep a gratification it is to me to witness the growing power of X-ray analysis and see how far it has progressed from the early days I remember so well."[882] However, conditions for the lecture were less than ideal. Bragg wrote to Phillips: "The proceedings were somewhat protracted with a honorary degree procedure first, an unveiling of a Rutherford portrait, a film of him lecturing, and by the time I came to speak everyone was frozen because the hall was not heated and the audience had been asked to come in evening dress!"[883] Rutherford was still a tough act for Bragg to follow.

The visit to New Zealand was unfortunately marred by a car accident in Wellington in which Alice suffered a bad concussion. On recovering from this, she and Bragg spent a month in Australia. For the Australian tour, which included stops in Brisbane, Canberra, Melbourne, Hobart, and Adelaide, Bragg had rounded up the usual suspects: He lectured on "Molecules of Living Matter," "The X-Ray Analysis of Biological Molecules," and "The Royal Institution." In Adelaide, Bragg addressed the sixth-form class at St Peter's College, his old school.[884] He and Alice visited the house his father had built and were pleased to find that it was now an old boys' club for the larger Adelaide schools. A grapevine that he had planted by the trellis at the back of the house was now "a noble vine, with a trunk as thick as a man's leg."[885] They also visited Aunt Lorna.

The only disappointment came when Bragg took Alice for a week's holiday to one of the coastal towns where he had stayed as a boy. He had looked forward to showing her the magnificent marine life of the St Vincent Gulf. Sadly, most of it had vanished, wiped out by the pollution of Adelaide's suburbs.[886]

The final stop on the Braggs' round-the-world trip was Hong Kong. On the flight back to England, they were surprised to receive VIP treatment. On arriving in London, they realized that they had been mistaken for the Governor of Hong Kong and his wife![887]

In November 1959, Bragg wrote to Rosalind Stubenberg of the Institute of Radio Engineers: "My main sciantific [sic] interests at the present time are research by X-ray analysis into the structure of biological molecules, and the development of the art of popular lecturing on scientific subjects."[888] By this time, however, Bragg's time as an active researcher was over. Indeed, even his time as leader of research groups had passed—he was now content to let Phillips and others run the research programs of the Davy–Faraday Laboratory. But Bragg still had an important role to play in the dispensation of scientific honors. His influence was not always exerted on behalf of distinguished scientists—Bragg, who had always acknowledged the important role of support personnel, persuaded Cambridge University to award George Crowe an M.A. degree when he retired from the Cavendish in 1959 after 52 years of service.[889] However, Bragg's main role in the dispensation of scientific honors was in two forums: The Royal Society and the Nobel Foundation.

As a Fellow of the Royal Society, Bragg was entitled to nominate new Fellows. Perutz had been elected in 1954, which left Kendrew and Crick as the main candidates from Bragg's former group in Cambridge (Watson, as an American, was only eligible to become a "foreign member," which he did in 1981). In the late 1950s, Bragg tried to get both men elected—over some opposition from those Fellows whom Crick had antagonized. Crick was elected in 1959 and Kendrew in 1960. Bragg appears to have taken a lesser role in the candidacies of other former Cavendish colleagues—in 1959, he ranked the loyal Will Taylor behind Kendrew and declined to support Bill Cochran.

Prestigious as an FRS may be, there could be no doubt about the ultimate scientific accolade—the Nobel Prize. As a Nobel Laureate himself, Bragg was each year asked to nominate candidates for the following year's awards. He could have nominated in any of the Nobel Prize categories, but for many years restricted himself to physics. In the period 1916–37, for which full records are available, Bragg exercised his nomination privilege on only five occasions: In 1922, he nominated Niels Bohr; in 1925, John Fleming; in 1926, Arthur Compton; in 1933, Erwin Schrödinger and Werner Heisenberg, with Paul Dirac as a second choice; in 1937, Ludwig Prandtl.[ll] Bohr, Compton, Schrödinger, Heisenberg, and Dirac won the Nobel Prize in physics the year of Bragg's nomination (Heisenberg was awarded the 1932 prize in 1933); Fleming and Prandtl never won a Nobel Prize.[890] In 1941, 1942, and 1943, Bragg nominated his former student Edward Appleton; no physics Nobels were awarded in the former two years, but Appleton won the Prize in 1947. Four years later, Bragg successfully nominated John Cockcroft and Ernest Walton. On at least one occasion, he apparently submitted a nomination for Cecil Powell, who won in 1950.

Therefore, of the 11 men Bragg is known to have nominated for the Nobel Prize in physics in the period 1916–51, nine won either that year or subsequently. Apart from Bragg's success in picking winners, the other notable feature of his list of nominations is its omissions. He did not nominate his old mentor C. T. R. Wilson, who shared the 1927 Prize with Compton, or his close friend George Thomson, who shared the 1937 Prize with Clinton Davisson.

The tremendous successes that had been achieved in the 1940s and 1950s by applying X-ray methods to the structures of organic molecules and macromolecules meant that Bragg now had to consider how—surely not whether—Nobel Prizes should be apportioned for work in the field of research he had founded. Linus Pauling had received the Nobel Prize in Chemistry in 1954, "for his research into the nature of the chemical bond and its application to the elucidation of the structure of complex substances."[mm] Perutz's work on hemoglobin was clearly an enormous achievement of vast significance; because Kendrew had actually been the first to solve a protein structure, however, a joint award would seem most appropriate. Among other X-ray crystallographers, Dorothy Hodgkin was probably the most deserving. Bragg was a great admirer of Hodgkin's work, in particular her 1957 structure of vitamin B_{12} revealing a hitherto-unknown chemical grouping.

The other achievement of X-ray analysis that was clearly of Nobel caliber was, of course, the structure of DNA. Here Bragg made the regrettable error of failing to recognize the importance of Rosalind Franklin's contribution. As early as January 1954, Bragg stated in an RI Discourse, "Very fine photographs of a form of nucleic acid (deoxyribonucleic acid), prepared by Dr M.F. Wilkins, have been obtained in Prof. J.T. Randall's laboratory at King's College, London."[891]

[ll] It is not clear why Bragg would nominate Prandtl, a mechanical engineer who worked on fluid dynamics.

[mm] On November 8, 1954, Bragg wrote to Pauling: "Alice and I were thrilled and delighted to see that you had got the Nobel Prize. I have long been a great admirer of your brilliant adventures into this field along quite new lines, and I am so glad your originality has received this well-deserved recognition. We send our warmest congratulations and all best wishes." (RI MS WLB24A/10). However, he apparently did not avail himself of the opportunity to nominate Pauling himself.

Likewise, Bragg referred in a 1957 article to "the structure of nucleic acids which we owe to Crick, Watson, and Wilkins."[892] On these occasions, and later, Bragg failed to acknowledge that it was Franklin's X-ray analysis—particularly her 1951 discovery of the "B" form of DNA—that had made the Watson–Crick model possible.

In view of Franklin's status as, in Brenda Maddox's words, "the Sylvia Plath of molecular biology,"[893] it is perhaps necessary to address the issue of whether Bragg's attitude to her was motivated by sexism or racism. For all his Edwardian sensibilities, Bragg's views on women in science—and the workplace in general—were highly progressive for the time. He antagonized his parents by insisting that Gwendy should go to university; he nominated Kathleen Lonsdale to be one of first female Fellows of the Royal Society; he appointed the first woman member of the Cavendish Laboratory, Helen Megaw; he had great admiration for Dorothy Hodgkin. Nor was Bragg, despite his occasional xenophobia, unwilling to advance the careers of Jews. As Aaron Klug, himself a Jew, put it: "If he had been anti-semitic he wouldn't have supported Perutz so strongly."[894] During the Nazi era, Bragg helped not only Perutz but also Paul Ewald and the Franco-Jewish physicist Adrienne Weill find refuge in Cambridge. In fact, Bragg helped Franklin herself get funding from the Medical Research Council in 1957.[895]

According to Klug, Franklin's scientific executor, Bragg's championing of Wilkins rather than the more deserving Franklin resulted from ignorance of the true facts of the case: "I don't think he knew about her at all, he didn't see her, he knew nothing about her, he didn't know the whole story."[896] Crick agreed: "He may not have been informed in detail about the position at King's, which, after all, was complex and I don't know that we would necessarily have gone into it at great lengths with him. So I don't think it was any malice on his part, I think it was just he thought Wilkins was the senior person or something like that."[897]

Certainly there was little or no communication between the heads of the two groups involved. Randall wrote to Bragg in 1968: "I don't think you and I were really closely in touch with each other about all this. If we had been—and as a former pupil I had some natural diffidence about raising such a potentially difficult matter—I think some of the repercussions that still reverberate might possibly have been avoided."[898] As noted above, Bragg had discouraged Randall from pursuing a Ph.D. at Manchester. On the other hand, Randall acknowledged that Bragg had helped him greatly in obtaining a Royal Society Warren Research Fellowship in 1937 and the chair of Natural Philosophy at St Andrews University in 1944. His ambivalent feelings towards Bragg played an important role in the discovery of the double helix.

Perhaps Bragg still felt guilty that he had allowed Watson and Crick to compete with the King's College group—after all, it had turned out that Pauling was a long way from being able to solve the DNA structure. Whatever the reason, Bragg "put every ounce of weight I could"[899] behind his effort to ensure that the appropriate King's College representative—in his view, Wilkins—shared in whatever glory Watson and Crick received. In December 1957, Bragg wrote to Gordon Sutherland, chairman of the Physics Selection Committee of the Royal Society and Director of the NPL: "I expect you know something about the nucleic acid story. Wilkins got good diffraction pictures of nucleic acid by developing new techniques in a very striking way. Crick and Watson, who knew of his results, had the inspiration which led to the solution of its structure. It was very hard luck for Wilkins that they put the key-stone in the arch and I always think that in fairness one ought to refer to the structure of nucleic acid as discovered by Crick, Watson, and Wilkins . . ."[900]

Watson was also keen that the King's group share any Nobel Prize awarded for the double helix; Kendrew recalled Watson telling him, around 1959, "If they ever ask your opinion about this, please make sure that Wilkins is put in too."[901]

Bragg's version of the DNA story reached Victor Rothschild, Chairman of the Agricultural Research Council, who tried unsuccessfully to set the record straight. Rothschild wrote to Bragg in January 1958: "As regards the Wilkins–Watson–Crick business, I have only heard about it from Perutz, whose account is somewhat different from yours. He described Wilkins as an 'amateur photographer' who was unable to interpret his pictures. He said that because of the work Cochran and Crick did on scattering from helical structures, Crick was able to interpret a photograph which Wilkins showed at some meeting (Wilkins was unable to interpret it). Perutz said that that woman crystallographer—I forget her name at the moment—whom Bernal is so keen on (crystallographically I mean) for years refused to accept Crick's interpretation. Perutz went on to say that Watson was responsible for doubling the helix."[902] Unfortunately, Rothschild's version of events appears to have done nothing to open Bragg's eyes to the contributions of "that woman crystallographer."[nn] In any event, Franklin's death in 1958 took her out of contention for the Nobel Prize, which cannot be awarded posthumously.

Bragg's view on the relative contributions of his own people to the double helix was also questionable. In March 1955, Bragg wrote to John Raper of Harvard University: "It was my impression that Watson was responsible for the brilliant and imaginative ideas in this work. He really has, I think, a touch of genius. Crick is a young man of great energy and wide reading who supplied knowledge about stereochemistry and symmetry and about X-ray diffraction which Watson lacked. But it seemed to me that Watson supplied the main girders of the structure, Crick knocked in the rivets which held it together. The trouble was that Crick, who is a very voluble young man, always did all the talking and Watson, being somewhat shy and sensitive, never had a chance to put his own case. I think he felt this very much and it led to certain difficulties between them . . . He is altogether a most interesting young man who much attracted me."[903] This letter was in response to Raper's request for a reference,

[nn] Rothschild's letter is probably a distortion of Perutz's opinion of Franklin. At least in later years, Perutz made it clear that he considered her work on DNA to be of central importance and highly deserving of Nobel honors.

so Bragg may have felt justified in playing up Watson's role with no prejudice to Crick. However, this idea of Watson doing the thinking while Crick did the talking was a fair reflection of Bragg's opinion on the subject. On a personal level, there is no doubt that Bragg much preferred Watson to Crick; this bias later influenced his attitude to the controversy over Watson's book *The Double Helix* (see below).

Bragg did not nominate anyone for the DNA work until 1960. In part, this may have been because he did not want Watson and Crick to receive a Nobel Prize before Perutz and Kendrew. In any case, it seems to have been the completion of the 2-Å structure of myoglobin in the summer of 1959 that decided Bragg that the time had come for Nobel nominations in macromolecular crystallography. In November of that year, he launched a lobbying campaign to enlist the support of fellow Nobel laureates. To Alan Moncrieff of the Institute for Child Health at the University of London, Bragg wrote: "I am invited to make a recommendation for the Physics [Nobel] Prize and I intend to suggest that it be awarded to the Cambridge group for its work on biological molecules . . . I propose to link together the names of Perutz, Kendrew, Crick and Watson."[904] Bragg appears not to have realized that Nobel Prizes could be shared only by a maximum of three people. A more surprising aspect of this proposal is that it omitted Wilkins, whose contribution to the double helix Bragg had earlier ranked with those of Watson and Crick. Around the same time, an alternative plan—again ignoring Wilkins—was floated to Cyril Hinshelwood, who had shared the 1956 Nobel Prize in chemistry: "The ideal would be a Physics prize for Perutz and Kendrew and and a Chemistry or Medical prize for Watson and Crick."[905]

Perhaps the most influential nomination—apart from Bragg's own—would be that of Linus Pauling. In response to Bragg's suggestion that Perutz, Kendrew, Watson, and Crick should share one or two Nobel Prizes, Pauling made the eminently reasonable point that an award for protein crystallography was premature, as "the most significant papers have not yet been published." He supported a nomination of Watson and Crick for the physiology or medicine Prize, and for chemistry suggested Dorothy Hodgkin together with Johannes Bijvoet of Utrecht University.[906] Bragg went along with most of these suggestions: He agreed to nominate Perutz, Kendrew, and Hodgkin for the physics Prize, but he felt that Bijvoet's work was not "sufficiently outstanding." He told Pauling he would write to Arne Westgren, chair of the Nobel Committee for Chemistry, suggesting consideration of Watson and Crick jointly by the Chemistry and Physiology or Medicine Committees, and asked Pauling to support this.[907]

In his nomination of Perutz, Kendrew, and Hodgkin for the 1960 physics Prize, the Father of Crystallography did not hesitate to pull rank: "may I remind your Committee that the Physics Prizes for 1914 and 1915 were bestowed for the work which started X-ray analysis. I have been associated with the science of X-ray analysis and seen it develop during the whole of my scientific career . . . "[908] By the time he submitted his nomination for the chemistry Prize, however,

Bragg had decided that Wilkins deserved to share the credit after all. He wrote to Westgren: "The solution [of the DNA structure] was their [Watson's and Crick's] own single brilliant piece of work, but it was both partly inspired by Wilkins' patient researches over many years in getting fine diffraction pictures of DNA, and has been confirmed since by Wilkins, who has analysed the structure in much greater detail. In my opinion, these three researchers ought to be grouped together."[909]

Under the unusual circumstances—in which all the research was done at King's but its successful interpretation was made at the Cavendish—it seemed only reasonable that someone from King's should share the rewards. As Crick put it, "The difficulty would have been if Rosalind had lived."[910] As Franklin was now dead, everyone involved—except the hard-nosed Pauling—seemed to think that Wilkins should be included in a DNA Prize, even if Franklin's work had been much more significant.

It is not clear who Pauling nominated for the 1960 Nobel Prizes. However, he obviously disapproved of Bragg's championing of Perutz and Kendrew, and belatedly decided to advance the case of his own collaborator, Corey. In March 1960, Pauling wrote to the Chemistry Committee giving advance notice of his intention to nominate Corey for the 1961 Prize: "I feel strongly, however [i.e., notwithstanding Bragg's nomination of Perutz and Kendrew], that no Nobel Prize should be awarded for the investigation of the detailed molecular structure of proteins without the inclusion of Robert Brainard Corey." Pauling suggested that the 1961 Chemistry Prize be 50% Corey, 25% Perutz, and 25% Kendrew.[911] The next day, Pauling wrote to the Physics Committee suggesting that Hodgkin, Perutz, and Kendrew share a Prize, but in chemistry rather than physics.[912] That same day, he wrote to the Chemistry Committee again, commenting on Bragg's nomination of Watson, Crick, and Wilkins, and claiming that the Watson–Crick structure "may to some extent have been stimulated" by his and Corey's triple-helical DNA structure! However, a Nobel Prize for DNA "might well be premature . . . because of existing uncertainty about the detailed structure of nucleic acid." "With respect to Wilkins, I may say that I recognize his virtuosity in having grown better fibers of DNA than any that had been grown before and in having obtained better x-ray photographs than were available before, but I doubt that this work represents a sufficient contribution to chemistry to permit him to be included among recipients of a Nobel Prize."[913]

It is likely that Westgren took a dim view of Pauling's attempts to nobble Bragg's nominees. In Cambridge, Pauling's manoeuvrings were perceived as a crude attempt to get a Nobel Prize for Corey as a quid pro quo for Pauling's support of the Cavendish workers. In June 1960, Perutz wrote to Bragg: "I have been thinking about what you told me about Pauling. If Pauling makes it a condition of his support that you support Corey, then we are not likely to receive any support from him. On the contrary."[914] The following month, Bragg wrote to Pauling pointing out the obvious flaw in his plan. Corey's greatest claim to fame was his role in the discovery of the α-helix—he had been a coauthor on all

Pauling's important papers on protein structure—but a Nobel Prize had already been awarded, in part for this work, to Pauling alone.[915]

In the event, the 1960 Nobel Prizes for physics and chemistry were awarded to the Americans Donald Glaser and Willard Libby, respectively. The campaign for 1961 reopened immediately thereafter. In November, 1960, Bragg wrote to Westgren, renewing his nominations of the previous year. Interestingly, he made it clear that he considered the work on proteins more deserving than that on DNA: "I give priority to a Nobel Prize shared between Perutz, Kendrew, and Mrs. Hodgkin." Because of the "borderline nature" of this work, it would be suitable for a Nobel Prize in chemistry or physics—or even physiology or medicine. Clearly, Bragg had been doing some discreet lobbying with Arne Tiselius, the President of the Nobel Foundation and a member of the Nobel Committee for Chemistry, as he told Westgren: "I had the opportunity of a very valuable talk with Tiselius . . ."[916] This apparently occurred when Tiselius visited the RI.

The campaign continued in July 1961, when Bragg wrote to Tiselius suggesting that a Nobel Prize to be awarded to Perutz, Kendrew, and Hodgkin.[917] He also sent Erik Hulthen, chair of the Physics Committee, new papers by Perutz and Kendrew—perhaps to address Pauling's criticism that "the most important papers have not yet been published."[918] However, the 1961 Nobel Prize in physics went to Robert Hofstadter of the United States and Rudolf Mössbauer of West Germany: The chemistry Prize was awarded to another American, Melvin Calvin.

It appears that Bragg did not submit nomination forms for the 1962 Nobel Prizes. His official position appears to have been that he adhered to his nominations for the 1960 Prizes—Perutz, Kendrew, and Hodgkin for physics, Watson, Crick, and Wilkins for chemistry—although this was qualified by Bragg's letters to Westgren and Hulthen and chats with Tiselius. Jacques Monod, a distinguished microbiologist at the Pasteur Institute in Paris, was invited by the Nobel Foundation to submit nominations for the 1962 Prizes. Monod asked Crick who should be credited with the DNA work. Crick felt strongly that Wilkins—who was, after all, an old friend—should be included. Like Bragg, Monod nominated Watson, Crick, and Wilkins for the Nobel Prize in chemistry.

However, the plot thickened in August 1962, when the organic chemist Sir Robert Robinson wrote to Bragg to tell him that he was re-thinking his nomination of Hodgkin for the chemistry Prize "years ago," particularly as he had learned that John Monteath Robertson used heavy metal derivatives to solve the structures of porphyrins and phthalocyanines before Hodgkin used them for penicillin. Robinson suggested a nomination of either Hodgkin and Robertson or Hodgkin, Robertson, and Perutz.[919] Apparently now worried that the inclusion of Hodgkin might spoil Perutz's and Kendrew's chances of a Nobel Prize, Bragg wrote to Westgren on October 1: "If the case is weakened by coupling the three of them, I would give first priority to awarding a prize to Perutz and Kendrew jointly."[920] Three days later, he nominated Perutz and Kendrew for the 1963 Chemistry Prize.[921] Perhaps the Chemistry Committee agreed with Robinson, perhaps they were influenced by Bragg, perhaps it was just a coincidence, but the Nobel Prize in chemistry for 1962 was awarded to John Kendrew and Max Perutz. The Prize for physiology or medicine was awarded to Francis Crick, James Watson, and Maurice Wilkins.

When the 1962 Nobel Prizes were announced, Bragg was in hospital, seriously ill after an operation for prostate cancer. According to Alice, he was so elated when he heard the news that he spent several hours trying to explain protein structure to the night-nurse. The following morning, Bragg's doctor told Alice: "Well, he's over the worst, but now I think he may die of excitement."[922]

In the end, Bragg's suggestions had, in general, prevailed. He wrote to thank Tiselius on December 13, 1962: "I think the news of the awards first of all for medicine and then for chemistry gave me the deepest pleasure of any event in my scientific life. I was delighted that Wilkins was included with Crick and Watson. He is of course overjoyed at the award, and it has removed all the bitterness he felt when he did so much work to get good figures of nucleic acid and then just missed their solution. I well remember our talk about my recommendations . . ."[923]

Wilkins knew that he owed his Nobel Prize to Bragg. He wrote to Bragg on November 30: "I am deeply appreciative of your understanding about the background to the original proposal of the DNA structure. It is good to have you refer openly to the bitterness of that time and to have your very sympathetic attitude to my position then. It was, of course, a trying time for me then but I tried not to show resentment. I think it was, in any case a very difficult situation for all of us, myself and Rosalind Franklin, Watson and Crick; I think all concerned had special difficulties. A scientific explosion was taking place and it created bigger problems of human behaviour than any of us knew how to cope with. Anyway that is now all finished for the award (as you imply) can be the last word said about it. I find it difficult to express my gratitude to you for what you have done for me in this. Clearly your determination in the matter must have been of the greatest importance, both because of your own position in science and because you were so closely in touch with what went on."[924]

With the macromolecular crystallographers taken care of, Bragg's influence on the Nobel process was now exerted on behalf of others. From 1963 until 1971, the year of his death, Bragg nominated Martin Ryle, the Cavendish Laboratory radioastronomer, for the physics Prize—for the first 4 years, jointly with Bernard Lovell, thereafter alone. Bragg undoubtedly felt a personal stake in Ryle's work, as he had persuaded the younger man to stick with radio waves rather than switch to nuclear physics. He also could relate to this kind of physics, as it, like X-ray crystallography, depended upon the interpretation of wave-interference phenomena. Ryle finally received the Nobel Prize in physics in 1974. Had Bragg lived to see this award, it would no doubt have given him great satisfaction. After protein crystallography, radioastronomy was the area of research he had most strongly supported—against the opposition of physicists in more traditional areas—during his tenure of the Cavendish Professorship. In the end, Bragg's post-war plan for Cambridge physics had been fully justified.

For the chemistry Prize, Bragg continued to nominate Hodgkin. There is little doubt that he felt a responsibility to ensure that the field of X-ray analysis, of which he was the grand old man, should receive the recognition that it deserved. Perhaps Bragg also felt some guilt that he had been willing to throw Hodgkin overboard in order to save Perutz and Kendrew. If so, he was no doubt relieved when Hodgkin was the sole recipient of the Nobel Prize in chemistry in 1964.

It would only be human if part of Bragg's satisfaction at the 1962 Nobel Prizes derived from the fact that his suggestions had prevailed over those of his old rival, Pauling. However, there was abundant compensation for Pauling the following year when the long-time campaigner against nuclear weapons was awarded the Nobel Peace Prize. In October 1963, Bragg offered his congratulations: "My wife and I were very thrilled to hear that you had been awarded the Nobel Peace Prize. We had always admired your courage in your attacks on nuclear war and testing and felt how well you deserved this tribute."[925] In the political arena, which he abhorred, Bragg felt no competition with his old rival and could offer sincere congratulations.

Bragg's recovery from his cancer surgery of late 1962 took many months, including a 3-week holiday in Teneriffe. In December 1963, Bragg and Alice embarked upon a two-month lecture tour of India, with stops in Madurai, Trivandrum, Mysore, Bangalore, Bombay, Delhi, Agra, Benares (Varanasi), and Calcutta. Bragg gave general lectures on a variety of subjects and also a course of lectures on X-ray analysis of minerals. The former were well-received but audiences dwindled during the latter—perhaps an inevitable consequence of Bragg's celebrity. Even the general lectures were somewhat more taxing than they would have been in a more-developed country: "The conditions are sometimes a little trying, a vast fan rotates over the lecture table and is apt to blow all one's notes away. A microphone and a loudspeaker is a status symbol even in the smallest classes and as one moves about pointing to slides etc., a student carefully moves along in a crouching position holding the microphone in front of one's mouth. People come in and out a great deal during the lecture . . ."[926]

The most important of the forty or so talks on Bragg's Indian tour was the Eighth Meghnad Saha Memorial Lecture, commemorating an astrophysicist who died in 1956, at the Saha Institute of Nuclear Physics in Calcutta. Bragg chose as his topic "The Difference Between Living and Non-Living Matter from a Physical Point of View." Many twentieth-century physicists, including Niels Bohr, Erwin Schrödinger, and Max Delbrück, had been deeply interested in the physical basis of life—Delbrück, Crick, and Wilkins had all given up physics for biology. But Bragg was only interested in biology as a source of interesting structures and was, in his own words, "not philosophically inclined."[927] Nonetheless, he now decided to tackle one of the most profound philosophical questions.

In his Saha Lecture, Bragg set himself the task of examining whether there is a fundamental difference between animal and vegetable on the one hand, and mineral on the other. In doing so, he argued, one must bear in mind that "the physical point of view" is limited to its space-time framework. The "detailed structure of living matter" is now much better understood through the solving of protein and nucleic acid structures. "Are all our human aspirations and fears, our art, our literature, our poetry, our religion, only dependent on the order in which two kinds of purine base, and two kinds of pyrimidine base, are arranged in a polymer chain in the nucleus." The scientist should confront this question, because it cuts to the core of scientific belief: "Science is allergic to mythology."

Bragg then described his view on the epistemological question of how the sciences were related to one another and to other forms of knowledge: "I think of the various forms of human interest as spread out in a series, mathematics, science, art, philosophy, religion. Mathematics is on the plain, where all is clear, one is as certain that two and two make four as is the Almighty. There is certainty, but it is completely devoid of any human interest. As one ascends the heights the mists close in, and on the summits of religion we cannot see and can only work with our intuition. Science is so fascinating because it is up in the foothills, above the plain of mathematics, and yet the mists have not closed in so as to obstruct all vision."

Bragg briefly described the cell, nucleus, and ribosome, and the relationship between nucleotide and amino acid sequences. "The haemoglobin molecule is four such [polypeptide] chains, it is a paragraph of four sentences . . . We write with letters of the alphabet which we place in a certain order. The Chinese write with thousands of special symbols each of which stands for a root idea such as 'house' or 'man'. Nature has chosen our way."

Minerals exist in the lowest-energy configuration so their structures are, in a way, obvious. A protein like myoglobin, however, could consist of amino acids in any order, but only one of these is compatible with life: "In the mineral race one horse is running and the chance of spotting the winner is very favourable. In the protein case there are 19 [types of amino acid] runners in 160 races [positions in the myoglobin polypeptide], and to win a prize one must guess them all correctly."[928]

This lecture put the question quite well but provided little in the way of original answers. For example, Bragg's analogy between the nucleotide sequence of a gene and a written sentence merely reiterates a point Schrödinger made in his 1944 book *What Is Life?*

Bragg had no deep insights into the meaning of life. However, he did have an instinctive understanding of the process of scientific discovery. In 1923, he wrote: "The history of science shows a series of alternating periods in which, on the one hand, discoveries that have been made are co-ordinated and brought into a scheme, and on the other hand, new discoveries are made that shatter this scheme and force us to consider it only a special case of a wider generalization."[929] This statement uncannily foreshadows Thomas Kuhn's 1962 book *The Structure of Scientific Revolutions*, in which Kuhn described periods of "science as usual" in which research is guided by a "paradigm," or set of mutually agreed-upon facts and approaches, until enough anomalies, or observations that do not fit within the paradigm, accumulate that a revolution occurs and a new paradigm adopted. In the same vein, Bragg wrote in 1961: "The situation [at the birth of a new science] may be compared to that of a supersaturated solution."[930]

Sivaramakrishna Chandrasekhar, who had left the RI in March 1961 to become Professor of Physics at Mysore, recorded that Bragg "kept the audience spellbound" during his lectures there, and afterwards was "practically mobbed by young undergraduates and schoolchildren for autographs."[931] Bragg "found Indian science a curious mixture of the inspired and the hopeless"[932] but was "intoxicated with the birds there."[933] India also made a strong impression on Alice: "There was Christmas Day in the South, in Trivandrum, where we went to service in a church packed with communicants, and dazzling with sunshine, great butterflies flying in and out, and the clergy bare-footed . . . Now I see myself in our High Commissioner's exquisitely tended garden in Delhi lecturing to a crowd of Indian women in beautiful saris, politely listening to my talk on Marriage Guidance, though their marriages were arranged, and they could not be divorced. Horrid memories also come of burning ghats along the Ganges, emaciated sacred cows wandering in the streets, the drive into Calcutta from the airport where shocking scenes of life and death were enacted along the packed road."[934]

Bragg's last teaching innovation at the RI was to offer courses on science to administrative civil servants. This arose from a suggestion made by one of the Managers, but was very much in line with Bragg's long-standing concern that political leaders were shockingly ignorant of scientific matters. The first course ran from October 1964 to June 1965 and involved 150 civil servants. There were six lectures, "The Properties of Matter," "Electricity and Magnetism," "Waves," "The Atomic Nucleus and Fundamental Particles," "Chemical Compounds and Chemical Reactions," and "Cells, Fundamental Units in Biology," each of which was followed by seminars given in groups of 50.

Reviewing the Civil Service Lectures at the end of the first year, Bragg realized that their content had been overly influenced by the Schools' Lectures. Intelligent adults, he concluded, could handle more advance concepts but had even less scientific background than school-children. Organizational problems in the timing of the lectures and size of classes were also identified. For the second year, enrolment was limited to 100 and nine topics were presented in two lectures each: "The Properties of Matter"; "Electricity and Magnetism"; "Waves and Vibrations" and "Waves, Conveyers of Energy and Information"; "The Interaction Between Radiation and Matter"; "The Atomic Nucleus" and "Atomic Energy"; "The Elements and Their Compounds" and "Molecules in Motion"; "Thermodynamics and Chemical Change" and "Biochemistry"; "Cells and Reproduction"; and "Communication Mechanisms in Biology." The philosophically minded immunologist Peter Medawar was also brought in to lecture on "Fact and Fiction About 'the Scientific Method.' " After each pair of lectures, the class split into "syndicates" of 12 to develop questions to ask the lecturer. "The methodological, historical and philosophic aspects were stressed much more, as against the attempt to 'teach' the basic facts of science."[935]

Even with this new orientation, the Civil Service Lectures were not a success. According to his son Stephen, Bragg "...said they were the most difficult audience to lecture to because they were taught to conceal their emotions and so don't appear to react to the lectures. Civil servants shouldn't have opinions, as it were, since they are trying to give factual statements to their ministers."[936]

March 31, 1965, was Bragg's 75th birthday. To honor the occasion, a party was held at the Davy–Faraday Laboratory. Bragg's research group gave him a very special present: A model of the 2-Å structure of lysozyme, which by good fortune had been completed just in time for the occasion. It was the first enzyme to have its structure determined, and contained the first evidence for the β-pleated sheet, a periodic arrangement of amino acids predicted by Pauling and Corey in 1951. It was also the sole work of the RI group. Phillips later wrote of Bragg's role in the lysozyme project in words that echoed the sentiments of Perutz and Kendrew: "The value of his constant advice, support and encouragement in this work can hardly be overestimated and his evident joy at the result gave the greatest possible pleasure to the people concerned."[937] Bragg was delighted for his team and made a drawing that accompanied the publication of the lysozyme structure in *Nature* (Figure 9.2).[938]

Bragg's joy at this, his last scientific project, is illustrated by a story told by Porter: "One night, in the time when he was still Director, he invited me after dinner, to accompany him to a room, called the Model Room, that was part of the Library store, in number 19 Albemarle Street. As we entered and closed the door behind us, it was clear that he had something very special to show me and was excited about it. On a table in the middle of the room, illuminated by one electric bulb, was a pile of perspex sheets, on each of which had been drawn the electron density cross sections of a very large molecule. It was lysozyme, the first enzyme structure to be worked out at high resolution, which had just been solved in the Davy–Faraday Laboratory by David Phillips and colleagues, under the encouraging and ever-helpful eye of Sir Lawrence. I listened to his happy story of how this had happened... how the molecule had been discovered in the first place by Alexander Fleming and many other wonderful things which infected the listener with the narrator's own excitement. When it became late and time to leave we found that we had locked ourselves into this uninhabited part of the building. We eventually escaped when Bragg remembered an internal telephone in some cupboard by which he was able to communicate with Jackson, the caretaker in his penthouse flat. Only those who have walked at night through the creaky corridors of the RI and imagined they saw some of its immortals, will fully appreciate the magic for me of that close encounter with one of them."[939]

In January 1966, Bragg took the lysozyme model to Berlin on a trip organized by the British Council. Patience, who accompanied her father, recalled: "I actually went to Berlin with my father with lysozyme on my lap on the plane and I was terribly excited because there was a pop group travelling with us. There

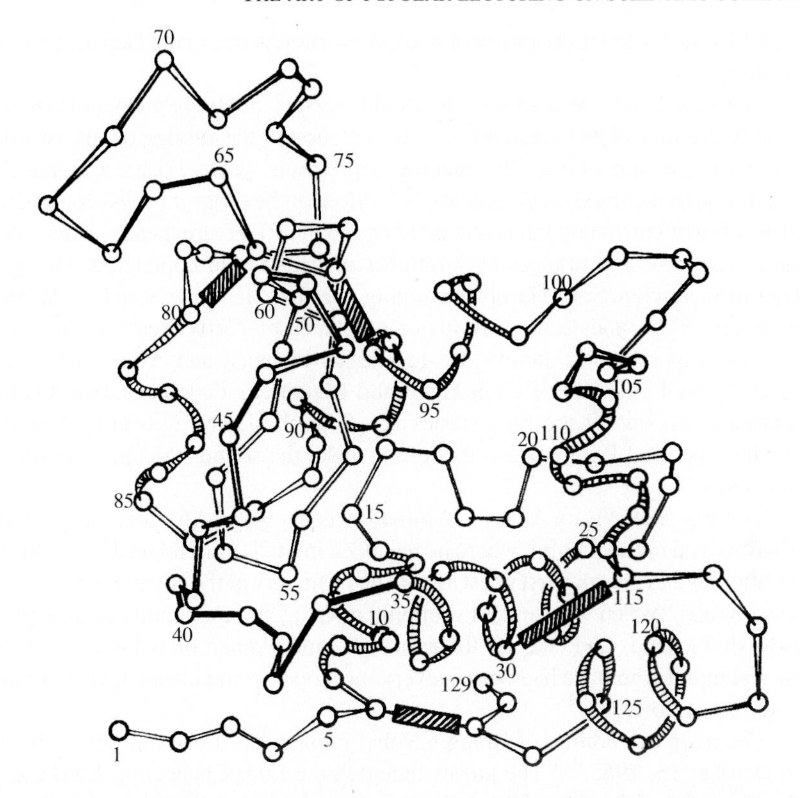

Fig. 9.2 Bragg's drawing of the polypeptide chain conformation in lysozyme. Reproduced, with permission, from Figure 5 of Blake, C. C. F., Koenig, D. F., Mair, G. A., North, A. C. T., Phillips, D. C., and Sarma, V. R. (1965). Structure of hen egg-white lysozyme. A three-dimensional Fourier synthesis at 2 Å resolution. *Nature* **206**, 757–61

were great crowds when we got off the aircraft, somebody with an enormous bunch of flowers and I was very impressed—because you know what it's like, you sort of take your father for granted—when I realized that the crowd and the flowers were for my father and not for the pop group."[940] Bragg used the model to illustrate his talk on "The X-Ray Analysis of Protein Structure" at the Fritz-Haber Institute.[941]

Bragg's described the lysozyme work as "the most recent experience which to me has been the greatest of all."[942] However, his delight was tinged with worry about the future of his protein crystallographers once he retired, as the research facilities of the RI went with the job. Oxford University was receptive to housing the group, but no college was interested in providing the fellowships normally used to supplement university salaries. Bragg enlisted the help of Sir Kenneth Lee, the industrialist who was godfather to his daughter Patience, and Harold Hemming, the former flash-spotter. With their help, an endowment fund was set up by three Oxford colleges. Phillips, North, Scouloudi, and others from the RI formed a new Laboratory of Molecular Biophysics in the Department of Zoology.[943]

The year 1965 was also the fiftieth anniversary of Bragg's Nobel Prize—he was the first Nobel Laureate to have celebrated the jubilee of his award. The BBC got into the act by making a program, "Fifty Years a Winner," which was transmitted on December 3.[944] Most of the program consisted of the BBC's Barry Westwood interviewing Bragg in the RI lecture theatre. However, there were also appearances by a number of friends and colleagues: George Thomson, Lucien Bull, Harold Hemming, Perutz, Kendrew, Crick, Watson, Hodgkin, Ryle, and Coates. A major emphasis of "Fifty Years a Winner" was the importance of family life to Bragg. Patience and Alice were featured, as well as Patience's son Hugh and Margaret's daughter Clare. Quite appropriately, one of the final scenes is a Schools' Lecture given by Bragg, in which he used Paget speech models to make the sounds "mama," "da-da," and "baby."

The day after "Fifty Years a Winner" was shown in Sweden, Bragg and Alice arrived in Stockholm, where he had been invited to attend the Nobel Prize ceremonies and give the first guest lecture in the history of the Nobel Foundation. Alice wrote: "When we went out shopping next day in Stockholm men stopped us to shake hands and take off their fur hats, and women blew kisses, and in the restaurant where we had coffee, everyone stood up and clapped. Our rooms were full of flowers."[945]

The main celebration of Bragg's Nobel jubilee was a party at the RI, held on October 15, 1965.[946] The guests included the Lord Chancellor, Lord Gardiner, and 20 of the other 28 British Nobelists, including Hodgkin, Kendrew, Perutz, Crick, Wilkins, Blackett, Cockcroft, and Howard Florey, who was also President of the Royal Society. The Swedish Ambassador represented the Nobel Foundation as Tiselius, recent ex-President of the Foundation and an Honorary Member of the RI, was unable to attend because of illness. Also present were Margrethe Bohr, wife of Niels Bohr, and Alexander Fleck, who had taken over from Brabazon as President of the RI two years earlier.

The evening began with a ceremony in the Lecture Theatre at which a congratulatory message from Queen Elizabeth was read out, and an illuminated address, signed by the Officers of the Royal Institution, the Lord Chancellor, and all the Nobel Prize winners, was presented to Bragg by Sir George Thomson. Fleck, Gardiner, Thomson, and Bragg then gave speeches. Fleck read congratulatory messages from several Nobelists unable to attend, including Sir Henry Dale and Alexander Todd (now Baron Todd of Trumpington). Gardiner paid tribute to Britain's scientific success. Instead of Bragg receiving the Nobel "as the coveted crown of a scientist's career," he had done so "at the absurdly early age of 25 . . . For 50 years he has had perhaps the rather difficult task of living up to it." One wonders whether Gardiner realized just how difficult living up to the Nobel Prize had been for Bragg.

In his speech, Thomson mentioned his qualifications for the job of making the presentation: Member of the RI, Nobel Laureate, and "close friend . . . of

Willie Bragg." He paid tribute to Bragg's unmatched lecturing skills and made indirect mention of his role in reforming the RI. Before Bragg, crystallography "was an affair of nice little labels attached to attractive-looking bits of stone all arranged neatly in glass-covered cabinets." In a more personal vein, Thomson referred to "the memories of long ago, memories of Cambridge, Trinity, the Cavendish Laboratory, memories even of days at sea and nights in quiet harbours."

The illuminated address with which Bragg was presented read: "On the occasion of the fiftieth anniversary of the award of the Nobel Prize for Physics jointly to yourself and to your father the Members of the Royal Institution of Great Britain wish to record their grateful recognition of your distinguished service in furthering the advance of science. By your zealous prosecution of original research you have added new lustre to the renown of the Royal Institution, and your enthusiasm and energy in expanding our educational activities have enhanced our long tradition for the diffusion and extension of useful knowledge. The British Nobel Laureates join in congratulating you on the jubilee of your award. They express their admiration of your pioneering work in X-ray crystallography and honour your inspiring leadership in the development of this science during the past fifty years."

Bragg then gave a gracious acceptance speech. In looking back to the events surrounding his Nobel award, however, he could not entirely conceal the traces of bitterness he still felt. Even on an occasion celebrating his own career, he felt it necessary to refute once again the old canard that the interpretation of Laue's X-ray diffraction pattern of zincblende was WHBs work: "It was, indeed, a joint effort between my father and myself, and I was not just following in my father's footsteps." Gwendy later wrote: "I do not think WL completely lost his reserve about his father until, in the years before his own death, his gathered wisdom and success and humour had dispelled the final shred of cloud that had hung between them . . ."[947] Only a few days before his death, Bragg wrote to Perutz: "I hope that there are many things your son is tremendously good at which you can't do at all, because that is the best foundation for a father–son relationship."[948]

Another ancient grudge unearthed was what Bragg considered his substandard reception by the Nobel Foundation in 1922: "We went over together and were fêted—but we did *not* have the grand party . . . The Nobel Foundation has been so kind as to invite my wife and myself to go to Stockholm on December 10th . . . this kind invitation which, though it has come late, has been none the less welcome." The wounds of those days were scarred over rather than completely healed.

A story from behind the scenes of the grand evening at the RI was recalled by Uli Arndt: "Bragg's gold medal was on display in the library; in the preparation of the exhibits the medal had been laid down in the preparation room on a drop of mercury and collected an unsightly stain. Messrs Johnson-Matthey, on consultation, prescribed the exact temperature of the heat-treatment needed to drive off the mercury. Bill Coates, the lecture assistant, claimed that he had lost several years of his life before the medal emerged in its pristine glory from the oven. I do not believe Sir Lawrence was ever told the story."[949]

Bragg's last full year at the RI had proven to be as personally satisfying as his last full year at the Cavendish Laboratory. Indeed, perhaps more so—he told Porter that "his time at the Royal Institution, in spite of all the difficulties, had been the happiest of his life."[950] Alice agreed that her husband had never been happier than in the period 1960–6.[951]

10
A very difficult affair indeed: Retirement, 1966–71

Following Bragg's retirement, he and Alice lived in London most of the year. Their London residence was the rented first floor of a house in The Boltons, South Kensington—a not entirely satisfactory arrangement, as the flat could be reached only through the owner's part of the house or by means of a fire escape.[952] Bragg continued to lecture at the RI, where he was an Emeritus Professor. He also did a lot of writing, including *The Development of X-ray Analysis*. Bragg appears to have viewed this book as some kind of counterpart to his autobiography, which contains very little about science. The majority, perhaps all, of the autobiography was written while Bragg was still at the RI. There are two references to the date of writing as being 1964, the latter occurring about three-quarters of the way through the text. However, that was probably not the date of completion. In February 1965, Bragg wrote to Philip Daly of B.B.C. Outside Broadcasts: "Here is the first part of my autobiography, up to the part just after World War I, about 1921. I have carried it on until 1949 but perhaps this beginning will show you whether it is any help to you [in preparing 'Fifty Year a Winner']."[953] As the final version ended with the events of 1951, Bragg must have still been working on it at the time this letter was written. The occurrence of two versions of the same incident in different places in the manuscript, as well as the different typefaces, suggest strongly that the autobiography was written in stages.

Bragg also told Daly: "Please regard it just for your eye only, it is not written with publication in mind, but as a family record." After his death, however, Alice tried to have the autobiography published.[954] Although it is not clear why this did not occur, it is safe to say that the manuscript would have required a great deal of editing. At least part of the text appears to have been dictated, as the names of many people Bragg knew well are mis-spelled (often phonetically), and only some of the many transcription errors have been corrected. Had he written his autobiography rather than dictating it, the result would doubtless have been much more satisfactory. Perutz wrote about his writing style: "He would illustrate his conclusions in a series of neatly drawn sketches, and then write the accompanying paper in a lucid and vivid prose. Some scientists produce such prose as a result of prolonged redrafting and polishing, but Bragg would do it in one evening, all ready to be typed the next day, rather like Mozart writing the overture to 'The Marriage of Figaro' in a single night."[955]

In November 1966, Bragg was awarded the highest honor of the Royal Society—the Copley Medal. Perhaps the timing was coincidental, but perhaps Bragg's retirement signalled an end to the feud between the RI and the Royal Society. In announcing the award of the Copley Medal, the President of the Royal Society, Bragg's old friend Patrick Blackett, hit the nail on the head: "The striking characteristic of Bragg as a scientist has been his direct and simple approach to complicated physical situations; his solutions of problems have a lucidity and simplicity which, in retrospect, make one forget how baffling they often seemed in advance." On November 23, Bernal wrote to Bragg: "Crystal structure may seem now an old story, and it is, but you, its only begetter, are still with us. Three new subjects, mineralogy, metallurgy, and now molecular biology, all first sprang from your head, firmly based on applied optics." In the New Year's list of 1967, Bragg was made a Companion of Honour. This order is restricted to 65 members who are appointed for services of national importance. Bragg told Bill Coates and Bruce Morris, who also worked in the RI workshop, that this award meant more to him than any other.[956]

The peace of Bragg's retirement was soon shattered when he became caught up in the violent controversy over the manuscript that became James Watson's book *The Double Helix*. However, Bragg could not complain about being sucked into a very acrimonious and difficult dispute—the book had been, in a sense, his idea. In May 1965, Bragg wrote to Watson: "Do you remember that I once told you how much I should like to see your version of the DNA discovery written up as part of an historical record?. . . Do write your version while your recollection is still fresh." He suggested that Watson's notes be filed together with his and Perutz's ("and if possible something from Crick, but whether I will get it is doubtful") in the archives of the Royal Society.[957] He also wrote to Warren Weaver: "We ought to have Watson's version of the story which is of such great importance in scientific history, and my recollections ought to be checked by his. Cannot you persuade him to write it? I have tried and he seems willing . . ."[958] It may seem curious that Bragg put so much more emphasis on Watson's version than Crick's—or, for that matter, Wilkins'—but this was the result of his continuing belief that Watson was the true brains behind the double helix. As Watson later wrote, "he told me that he wanted me to tell my side of the story since, given Francis Crick's brilliance, my contributions might well be thought those of a minor contributor."[959]

When he received Bragg's letter, Watson was already at work on his memoir. He had first conceived of it as a series of articles for *The New Yorker*, with the title "Annals of Crime." The first chapter was written in 1962.[960] The manuscript, now entitled "Honest Jim," was completed during a sabbatical spent in Cambridge in the autumn of 1965. Thomas Wilson, director of Harvard University Press, told Watson that the press would publish *Honest Jim* on condition that no-one was libeled and the major figures involved—including Crick, Wilkins, and Bragg—agreed to its publication.

At a lunch in London in January 1966 with Louise Johnson, Peter Pauling, and Watson, Tony North suggested that Bragg write the foreword to *Honest Jim*. Watson liked this idea, but feared Bragg's reaction to the manuscript, in which he initially portrayed Bragg as he had heard about him through Crick, rather than as he subsequently learned him to be. When Alice read *Honest Jim*, she was "appalled; so many of those people that he had met in his host country, and especially in Cambridge, were held up for criticism and ridicule, and none more than his Professor, W.L.B."[961] Bragg took a more liberal position on the rules of hospitality and professorial dignity; according to Patience, he merely chortled and told Alice: "But that's Watson, darling."[962]

Watson claimed that Bragg agreed to write the foreword to show that he was "acting in a magnanimous . . . way" and "above flattery."[963] This seems uncharacteristic. Certainly Bragg would not have stood in the way of a book that portrayed himself in an unflattering light, but his main motivation throughout the *Double Helix* controversy was to ensure that Watson's side of the DNA story was told rather than to flaunt his magnanimity.

Bragg's approval had been easily gained. Not so those of Crick and Wilkins. When they saw the manuscript of *Honest Jim*, which now had been given the third—equally tactless—title of *Base Pairs*, both Wilkins and Crick reacted with violent disapproval. Wilkins wrote to Watson in October 1966: "Most top scientists are fairly civilised, but your book, though you may not intent [sic] it, would give many people an impression of Francis as a feather-brained hyperthyroid, me an overgentlemanly mug and you an immature exhibitionist!" To his credit, Wilkins also disapproved of Watson's portrayal of Rosalind Franklin.[964] Crick objected to publication because the manuscript contained too much gossip and violated his privacy.[965] No doubt realizing that it would be very difficult to prevent Watson from publishing *Base Pairs* in the USA, Wilkins and Crick wrote to Bragg asking him to use his influence with Watson to ensure that the manuscript was withdrawn.[966]

It is important to realize that these early drafts of the book were much more inflammatory than the version that was eventually published. For example, in October 1966, Pauling wrote to Watson objecting to several phrases in *Honest Jim/Base Pairs*, including "his [Pauling's] infantile mistake," "the latest Linus' nonsense," and "Pauling's stooges."[967]

Bragg was in a difficult situation. He could hardly ask Watson to junk a manuscript he himself had solicited. He told Wilkins and Crick that he was inclined to approve the publication of Watson's book.[968] However, he took responsibility for having, however inadvertently, sown the dragon's teeth. Bragg solicited the opinions of Harold Himsworth and the eminent American protein chemist, John Edsall, and continued to agonize over the situation. He was put on the spot in December 1966, when Crick asked him to withdraw his foreword. On December 21, 1966, while he was in the USA, Bragg wrote a chronology of his involvement, concluding: "History must be published. Am worried harm due to scandal. Rewrite book strictly censoring all malice and anything unnecessary to story."[969]

Later in the trip, Bragg discussed the controversy during a lunch meeting with Edsall, Watson, and Thomas Wilson of Harvard University Press. Following that meeting, Wilson wrote to Nathan Pusey, President of Harvard University, stating that the book should be published but that it should be edited to remove "unintentional rudeness or unkindness." This decision enraged Crick, who was not mollified when he saw the revised manuscript. He wrote to Watson that *Base Pairs* "shows such a naive and egotistical view of the subject as to be scarcely credible."[970]

Bragg enlisted the help of the diplomatic John Kendrew. Together they went through the manuscript of what had now been re-renamed *The Double Helix* and sent Watson several revisions that Bragg insisted must be made as a condition of using his foreword. To Crick, Bragg wrote: "I hope I am taking the right line but it has been a very difficult affair indeed."[971]

As news of Watson's book leaked out, support for Crick and Wilkins increased. In January 1967, Pauling wrote to Watson: "I have decided that I should tell you my impression of your book. I think that it is a disgraceful exhibition of egocentricity and malevolence . . . I did not like some of the statements that you made about Sir Lawrence Bragg. I felt especially that your continued attack on Crick was abominable."[972] Four months later, Pauling wrote to Bragg: "I must say that I was shocked to read the book, perhaps one of the earlier drafts, after I had read your preface. I was indignant about the insinuations about my wife and the statements about other people, but also indignant about Watson's treatment of you. I do not think that you should give the book the support and validification [sic] that would be implied by your having written a preface, even despite your disclaimer."[973] Both Perutz and Edsall told Watson off for his grotesque portrayal of Franklin.[974]

Harvard University had had enough. In June, Watson wrote to Kendrew: "Pusey told Harvard University Press that the University should not become involved in a fight among scientists."[975] However, this decision was not to delay publication further. Thomas Wilson had retired from Havard University Press and he took Watson's manuscript with him. In February 1968, *The Double Helix* was published by Athenaeum Press. Watson later wrote: "I do not know whether I would have had the courage to see the publication of *The Double Helix* through to its end without Sir Lawrence's backing."[976]

To understand Bragg's bias towards Watson in the controversy over *The Double Helix* and his conviction that Watson rather than Crick had been the brains behind the 1953 DNA structure, it is illustrative to consider the parallel between Watson and the young Bragg. Both were natives of former British colonies who came to the mother country as young men; both participated in scientific breakthroughs of enormous significance in collaboration with an older, more-established man. Bragg's identification with Watson comes across clearly in the 1955 letter to John Raper quoted from above: "Watson, being somewhat shy and sensitive, never had a chance to put his own case. I think he felt this very much and it led to certain difficulties between them."[977] Substitute "Bragg" for "Watson," and this statement would apply exactly to the former's situation in the period 1912–14—and its lingering emotional aftermath. Considering the psychological baggage that Bragg carried as a result of his long struggle to gain scientific recognition independent of his father, he could scarcely help siding—perhaps subconsciously—with Watson.

In late 1966 and early 1967, Bragg, accompanied by Alice, made a 2-month lecture tour of the USA. He spoke at the American Association for the Advancement of Science meeting in Washington, DC, and gave about a dozen other lectures during stops at Buffalo, Boston, New Haven, Philadelphia, New Jersey, Atlanta, and Baltimore.[978] In Baltimore, Bragg delivered the Redding Lecture of the Franklin Institute on "Reminiscences of Fifty Years of Research." This talk illustrated how his lecture style had been changed by his time at the RI—the analogies were more vivid than ever. Bragg had often pointed out the similarities between the diffraction of visible light by a line grating and diffraction of X-rays by a crystal—this relationship had inspired his first and greatest discovery, and had influenced his thinking ever since. In the Redding Lecture, however, Bragg came up with a far more arresting image than he ever had before: "In the early days of cinema, when all these stunts were new, I remember one that we always used to laugh at very much: a man 'uneating' a chicken. He took pieces of chicken out of his mouth and put them on the plate and built up the chicken. Imagine taking the spectra [of light diffracted by a grating], turning them round, making them run backwards, then 'building up the chicken', with the light actually going through the grating. That's the way we now do all our x-ray work."

When Bragg reached the point in his lecture where he described the X-ray analysis of lysozyme, the metaphors and analogies were flying thick and fast: The enzyme's job is "to bite the skins of bacteria and so kill them"; the bacterial cell wall is made of "material rather like corduroy"; the oligosaccharide inhibitor of lysozyme is "rather like Hercules' sop to Cerberus on entering Hell, so that Cerberus shouldn't bite him!" He liked telling jokes in private discourse and used humor very effectively in public speaking. The Redding Lecture ended with a typically self-deprecating joke—showing an illustration of a tiny rock salt unit cell next to the giant myoglobin molecule, Bragg said: "This is interesting evidence of the way in which standards for Nobel Prizes have gone up in fifty years."[979] The Braggs concluded their American tour with a well-earned 2-week holiday in Arizona during which they identified 76 different species of bird.[980]

In 1967, Bragg was awarded the James Scott Prize Lectureship of the Royal Society of Edinburgh, which involved delivering a lecture in the area of "fundamental concepts of Natural Philosophy." Bragg chose as his subject "The Spirit of Science," based on a 1944 radio broadcast with the same title that examined "a scientist's view of the significance of the advances in science, during the last century or two." The central theme was the relationship between basic and applied research. Bragg began by noting that the Royal Society (dining)

Club has a toast to "the arts and sciences"; by "arts" is meant craftsmanship, or "project orientated" (practical) science, as opposed to "understanding orientated" (fundamental). His "main point" was that the rapid growth of science occurred because humans pursued knowledge for its own sake: "It is a strange paradox. As long as the knowledge was sought for the sake of technical advance, progress in technical advance was extremely slow. When technological advantage was ignored, a vastly greater body of knowledge about nature was quickly acquired which in turn made it possible for technology to race ahead." For example, Gilbert White became the father of ornithology because he was interested in birds for their own sake, not as food or an aid to hunting. An outstanding example was Galvani's finding that a frog's muscles twitched when a circuit was formed with two dissimilar metals, leading to Volta's invention of the pile, Faraday's discovery of induction, etc.: "If promise of usefulness had been the criterion of whether to continue with experiments, who would have pursued an observation on the kicking of frog's legs?"

The use of fundamental science for practical purposes is a slow process, Bragg argued, typically 40–50 years. This figure was derived from the time between Oersted's discovery of the magnetic effect of a current and the development of the telegraph, between Faraday's discovery of induction and the use of power stations, between Becquerel and Rutherford's discoveries of nuclear disintegration and the development of nuclear power. Because of this time lag, it is impossible to tell of what use a piece of fundamental research will be, or to ensure that it will only be used for "good" purposes. The "spirit of science" is international and ("this has interested me greatly") classless. What is it, Bragg asked, that gives the scientist "such profound aesthetic pleasure"? It is the realization that Nature, not man, "pronounces the verdict," and "there is the feeling that one is being given a glimpse of something far more enduring and fundamental than the ephemeral and local affairs of men."[981]

Unlike his previous foray into philosophy, the 1964 Saha Lecture on "The Difference Between Living and Non-Living Matter from a Physical Point of View," Bragg's Scott Lecture had something substantial to say. Bragg's "project orientated" and "understanding orientated" sciences are merely what Francis Bacon described over three hundred years earlier—and far more elegantly—as *experimenta fructifera* and *experimenta lucifera*. However, his view of their inter-dependence was forcefully argued. Bragg had long believed that fundamental research was a good investment for society even though the return on the investment could never be predicted.

Bragg returned to this theme in a December 1968 lecture to civil servants on "What Makes a Scientist?": "the scientific advances which have so increased technical powers have been almost entirely made without any thought of their possible practical use . . . When Charles II was moved to such mirth by the spectacle of the Fellows [of the Royal Society] studying the weighing of air, I do not think they were pursuing this research with an industrial end in view."[982]

Throughout the period of the Cold War, the social responsibility of scientists was a hot issue. As noted above, Bragg was not a political man; he usually declined to add his name to political causes, although he occasionally made exceptions in situations where he felt a sense of personal involvement, such as the treatment of non-whites in South Africa. Wisely, Bragg realized that the responsible use of scientific knowledge was not a matter of personal conscience on the part of scientists, but a matter for society as a whole. As he wrote in 1963, "since the scientist cannot envisage how his results will be used I do not think he can be told only to find out things which can be used for good purposes. It is rather like telling a boot maker that his boots must only be worn by parsons and not by criminals."[983] Rather, "the choice of what shall be done is surely a moral responsibility which we all share equally."[984] Unfortunately, the progress of science had outstripped the wisdom of our stewardship over nature: "Science has acquired so much knowledge about nature that it has enabled industry to do almost anything it likes. This has made us trustees of the world, which is in our power, and there can be no doubt this trusteeship has been widely abused."[985]

In his final years, Bragg lost none of his enthusiasm for science, although the scope of his interest had narrowed to the last area in which he had worked, the crystallography of proteins. Freed from the day-to-day worries of research, he was able to enjoy the spectacular flowering of the field that he had founded. But the rapid progress of X-ray analysis had to some extent alienated the Father of Crystallography from his offspring. Crowther wrote: "While he recognised the importance for large-scale, highly organised science, he had little taste for it."[986] For an artistically inspired classical physicist like Bragg, the use of computers and Fourier methods—ironically championed by Bragg himself—had robbed X-ray analysis of its romance.

In 1967, he attended a meeting in Oxford of the X-Ray Analysis Group at which no fewer than three new structures were presented—pancreatic ribonuclease, solved independently by groups in New York and London, and two much larger enzymes: Carboxypeptidase, solved by William Lipscomb's group at Harvard; and chymotrypsin, solved by David Blow in Cambridge. According to Blow, "Bragg was in fine form. 'Enough swallows', he said, 'to make a summer.' "[987]

Perhaps a greater pleasure for Bragg was a workshop on protein crystallography held at the Austrian village of Hirschegg/Kleinwalsertal in March 1968. Perutz presented his 2.8-Å Fourier analysis of horse methemoglobin.[988] It was, strictly speaking, not an atomic-resolution structure, but the amino acid sequence of hemoglobin was now known and with this the backbone and many of the amino acid side-chains could be located. After 30 years, Perutz had reached Ithaca at last.

For Bragg's 80th birthday in March 1970, Will Taylor organized a symposium at the RI on "X-Ray Analysis—Past, Present and Future." The first Bragg Lecture was given by Bert Warren—his diopside collaborator from the 1920s—on "The X-Ray Analysis of Glass Structures." The lecture was chaired by another old friend, Kathleen Lonsdale, now a Dame. The participants in the conference included crystallographers from many countries. According to Phillips, "Bragg was himself the liveliest participant."[989]

One of the last scientific conferences Bragg attended was a 1971 Royal Society Soirée. When Bragg came to his display on imperfections in solids, a post-doctoral fellow from Aberystwyth innocently asked him: "How much can I assume that you know about diffraction?" With a twinkle in his eye, Bragg replied "A little."[990]

Despite a tendency to succumb to respiratory infections, Bragg had enjoyed good health throughout his life. Now in his eighties, however, the signs of physical degeneration were unmistakeable. For example, back pain had incapacitated him for two days in 1968.[991] Nonetheless, the cancer surgery Bragg had undergone in 1962 was his only serious medical problem. That cancer recurred in the summer of 1971. On June 13, Bragg wrote to Phillips: "I saw my surgeon in Ipswich yesterday, and he says he must operate without delay on the old prostate wound which is giving trouble. This means going to hospital on Tuesday evening 15th and being there for an estimated month."[992] He survived the surgery, but suffered a relapse soon after and never left the hospital. Bragg died on July 1, 1971.

A service of thanksgiving was held at St James Church, Piccadilly, on September 23. On the back of the invitation to Blackett, Alice wrote: "Willie was very active and cheerful till the very end . . . We had a wonderful 50 years together."[993]

David Phillips took on the responsibility of completing Bragg's scientific testament, *The Development of X-Ray Analysis*, mainly an editorial task at this point. It was published in 1975. As early as March 1972, Phillips was asked by a representative of the publisher Macmillan to suggest a possible biographer. If this was an oblique suggestion that Phillips volunteer for the job, he did not take the hint.[994] He did, however, accept a commission from the Royal Society to write a biographical memoir. This extensive and generally very accurate account of Bragg's life was published in 1979.[995]

In 1974, the Cavendish Laboratory left Free School Lane after a century there. Part of the New Cavendish Laboratory, on Madingley Road on the outskirts of Cambridge, was the Bragg Building, which contains the administrative offices, lecture theaters, and library.[996] Bragg was also honored in the land of his birth. In 1984, both Adelaide houses he lived in as a child were added to the Register of South Australian State Heritage Items.[997]

Despite Bragg's lifelong wish to be thought of as a physicist, and despite his tenure of numerous senior positions in British and international physics, it must be said that his work had no great influence on the physics of his time—or after. Considered against the great achievements of the first half of the twentieth century—quantum theory, relativity, and the structure of the atom—Bragg's contributions to the physics of the solid state seem relatively minor. He was not a theoretical physicist—Cochran was merely stating the obvious when he wrote: "I had no great regard for him as a theorist."[998] Nor, despite his early training in mathematics, was Bragg a mathematical physicist. Crick said, "I don't think he was very powerful mathematically, I think some of the other physicists rather looked down on him."[999] Bragg always enlisted the help of mathematically inclined collaborators, such as Douglas Hartree, Evan Williams, and Charles Darwin, for his more abstract work. Of course, the same could be said of many great experimental physicists—Rutherford and J. J. Thomson, for example, were less mathematically adept than Bragg. But Rutherford and Thomson both made discoveries that fundamentally changed our view of the nature of matter. Bragg's great achievement was the less-fundamental one of inventing means by which the atomic structures of inorganic compounds and organic molecules could be determined. Well before Bragg's work on the alkaline halides, it was understood that rock salt was a cubic lattice of sodium and chlorine atoms; before his work on minerals, it was understood that the silicates were complex crystals of silicon, metals, and oxygen; before his work on proteins it was understood that these were folded polypeptides.

The term "classical physicist," often applied to Bragg by his contemporaries, was not, of course, meant as a compliment. As a classical physicist in the age of quantum theory, he was a scientific dinosaur in a world now dominated by mammals. Just as his Edwardian sensibilities would not allow him to address Perutz, a close friend, by his first name, Bragg's classical training prevented him from fully absorbing the anti-intuitive implications of quantum mechanics.

Having stated the obvious—that Bragg did not make any major contribution to twentieth-century physics—let us move on to the real significance of his work. Right from the beginning, it was scientists in other disciplines who were more affected by the X-ray analysis of crystals. Bragg's earliest work had profound implications for both inorganic and organic chemistry. His 1913 structure of rock salt showed that inorganic crystals were electrostatically bonded assemblages of anions and cations, rather than assemblages of electrostatically bonded salt molecules. His and his father's 1914 structure of diamond proved the tetravalency of the carbon atom and provided the first measurements of the length of the C—C bond that is the basis of organic chemistry. Generally, the techniques initially developed by the Braggs allowed chemists to determine the structures not only of inorganic crystals but of complex organic molecules, to the point where the X-ray diffractometer became one of the most powerful weapons in the armamentarium of the structural chemist. For example, Walter Hamilton wrote in 1970: "Recent issues of *Inorganic Chemistry* . . . have had crystal structure determination at the heart of 20 per cent of the papers."[1000]

Biology was, if anything, even more profoundly affected than chemistry by X-ray analysis. Generally speaking, chemists had other means of establishing the structures of their substances of interest. The structures of many of the small organic molecules studied by X-ray methods during Bragg's time, such as strychnine, phthalocyanine, and vitamin B_{12}, had been at least partially characterized by the chemical techniques of spectroscopy, retrosynthesis, etc. Biochemists, however, had no such options. Proteins were known to be polypeptides, but no chemical technique could distinguish between the almost-infinite number of possible conformations into which the flexible polypetide could be folded. Bragg's interpretation of Laue's diffraction pattern of zincblende laid the foundation of all X-ray analysis, but his later work played an important part in applying X-ray methods to far larger and more complex molecules than were dreamed of in 1912. Very early on, Bragg realized that the limitations of the trial-and-error method of crystal analysis could only be overcome by using the Fourier approach proposed by his father to mathematically reverse the physical process of diffraction and thereby "build up the chicken." His 1929 two-dimensional Fourier synthesis of diopside pointed the way for Perutz, and later others, to the X-ray analysis of proteins. Bragg not only supported and encouraged Perutz in what any reasonable person would have concluded was a fool's quest, but also contributed new approaches at key stages of the hemoglobin odyssey. Therefore, Bragg, the Father of Crystallography, also became one of the founders of molecular biology.

The irony is that Bragg was quite ignorant of chemistry and biology, and had little interest in either discipline. His lack of understanding of chemistry let him down most spectacularly in 1950, when he missed the α-helix, but throughout the time that he worked on proteins he never had more than a rudimentary grasp of their chemical structure. Nor did he have any interest in the nature of life, even during a period in which physicists were virtually stampeding into biology. Kendrew hit the nail on the head when he wrote in 1990: "I never found him very interested in biology—not even in the chemical structure of the compounds the structures of which he analysed . . . Basically he was a puzzle-solver; to him the great fascination was to interpret the complicated diffraction pattern, say of a protein crystal, in terms of its three-dimensional structure."[1001]

To Bragg, all crystalline substances were atomic crosswords that could be solved using X-ray methods, as long as sufficient ingenuity was put into the analysis. Whether the study material was common salt or the material of heredity, rock or precious stone made little difference to him. Teleology played little part in his thinking. Another quote from the article by Hamilton cited above would serve as a fitting epitaph for Bragg: "The crystallographer is a solid state scientist, and his interests are as broad as all of science."

Bragg had artistic and pragmatic sides to his personality. His artistry was inspired by the aesthetics of symmetrical atomic patterns. Very revealing is a comment he made in a 1967 lecture to the Royal Society of British Sculptors: "I do not believe anyone can be a scientist without being an artist at heart."[1002] His wife wrote: "In fact he told me, other things being equal, which of course they were not, he would have been as happy to be an artist as a scientist."[1003] Bragg's pragmatic side related to a form of research in which the correct solution was there, if only one could see it. A crystalline material has a lattice structure and its X-ray diffraction pattern is a direct and unique function of that structure—all the investigator has to do is solve the crossword puzzle. As Brian Pippard put it, "For him beauty and economy were the touchstones of a physical argument or an experiment, and unless one sympathized with his quest for these ideals one missed his intellectual power and subtlety."[1004]

Finally, one may ask why Bragg was so successful. Few would suggest that he had one of the great intellects of his time. Compared to Albert Einstein, Robert Oppenheimer, or Richard Feynman, Bragg's intellectual gifts seem quite humble. His diffidence towards Pauling seems to have been motivated, at least in part, by his realization that he could not match the brain-power of the Wizard of Pasadena.

Bragg had several weapons that compensated for his lack of transcendent intellectual brilliance. The first was his mastery of classical optics, which allowed him unmatched insights into the relationship between an object and its diffraction pattern, and inspired the "fly's eye" camera, the "X-ray microscope" and the minimum-wavelength principle. The second was his ability to visualize three-dimensional objects in space. James Crowther wrote: "During the war, when confronted with an aptitude test for fitting things together, he solved it in 33 seconds, when no one previously had done it in less than 4 minutes."[1005] The provenance of this story is not clear, but it is certainly true that Bragg solved the enigma of Laue's diffraction pattern of zincblende faster than vastly more experienced physicists. The visualization of the three-dimensional object that corresponds to a two-dimensional diffraction pattern is the obverse of the problem faced by the artist who tries to represent a three-dimensional world on a two-dimensional canvas. Bragg's ability to solve crystal structures may therefore owe more to the genetic (and epigenetic) influence of his mother than those of his father. A third factor is pointed out by two men who worked closely with Bragg. Henry Lipson wrote in 1970: "His outstanding quality was . . . his ability to see the essential point of a problem and to strip away the inessentials."[1006] This point was echoed by Crick: "I learned a lot from Bragg in the sense of grasping for the essence of a problem."[1007]

What of Bragg the man? Many great scientists are highly egotistical—indeed, it could be argued that a highly developed self-esteem is necessary in order to overturn the established scientific order. According to the Kuhnian view, new scientific theories do not add to the existing ones, but rather replace them. The proponent of a significant new theory must therefore believe that he is right and everyone else is wrong—not normally regarded as a healthy state of mind. Among his scientific peers, however, Bragg stands out for his humility—he was not too proud to talk to students as equals, or to lecture to children.

Phillips, who knew him well, summed up Bragg's personality in the following words: "Bragg had an artistic temperament with strong emotions normally kept in check by stern self-control . . . Bragg was a private family man . . . Forgetful of names, uneasy on committees, reluctant to face personal problems or angry scenes, he depended a great deal on his wife who sustained him through all the triumphs and difficulties of a long public life. There is no doubt that he found peace at the last."[1008]

References

"RI MS WLB," "RI MS WHB," and "RI MS RCB" refer to documents in the William Lawrence Bragg, William Henry Bragg, and Robert Charles Bragg archives at the Royal Institution of Great Britain, respectively. RI MS WLB 87 is W. L. Bragg's autobiography; RI MS 95 is Alice Bragg's memoir, "The Half Was Not Told."

1. RI MS WLB23B/25
2. Taylor, C. A. (2001). *A non-mathematical introduction to X-ray crystallography*, p. 5. International Union of Crystallography and University College Cardiff Press, Cardiff, Wales.
3. Taylor, C. A. (2001). *A non-mathematical introduction to X-ray crystallography*, p. 8. International Union of Crystallography and University College Cardiff Press, Cardiff, Wales.
4. Gibbs, R. M. (1984). *A History of South Australia: From Colonial Days to the Present*. Southern Heritage, Blackwood, South Australia.
5. Thomson, A. (1999). *The Singing Line: Tracking the Australian Adventures of my Intrepid Victorian Ancestors*. Doubleday, New York.
6. RI MS WHB14F/2
7. Andrade, E. N. da C. (1943). William Henry Bragg. *Obituary Notices of Fellows of the Royal Society*, 277–300; Caroe, G. M. (1978). *William Henry Bragg, 1862–1942: Man and Scientist*, p. 28. Cambridge University Press, Cambridge.
8. Tanner, J. R., ed. (1917). *The Historical Register of the University of Cambridge*, pp. 348–54; 542. Cambridge University Press, Cambridge.
9. Duncan, W. G. K. and Leonard, R. A. (1973). *The University of Adelaide, 1874–1974*, pp. 1–10; 15–16. Rigby, Adelaide.
10. Duncan, W. G. K. and Leonard, R. A. (1973). *The University of Adelaide, 1874–1974*, p. 187. Rigby, Adelaide.
11. Jenkin, J. G. (1985). The appointment of W. H. Bragg, F. R. S., to the University of Adelaide. *Notes and Records of the Royal Society of London* **40**, 75–99.
12. RI MS WHB37A/1/1
13. RI MS WHB37A/1/3; RI MS WHB37A/1/4; RI MS WHB37A/1/7
14. RI MS WLB87, pp. 1–4

15. Ewald, P. P. (1962). X-rays. In *Fifty Years of X-ray Diffraction* (P. P. Ewald, ed.), pp. 6–16. International Union of Crystallography, Utrecht.
16. RI MS WLB87, p. 7
17. RI MS WLB87, pp. 8–12
18. RI MS WLB87, p. 17
19. RI MS WLB87, pp. 12–15
20. Bragg, W. L. (1967). The start of X-ray analysis. *Chemistry* **40**, 8–13.
21. Jenkin, J. (1986). *The Bragg Family in Adelaide: a Pictorial Celebration*, p. 51. University of Adelaide Foundation, Adelaide.
22. RI MS WLB87, pp. 15–16
23. Jenkin, J. (1986). *The Bragg Family in Adelaide: a Pictorial Celebration*, p. 51. University of Adelaide Foundation, Adelaide.
24. RI MS WLB87, pp. 7–8
25. RI MS WLB54A/27
26. RI MS WLB87, p. 17
27. RI MS WLB87, p. 5
28. RI MS WHB30A/1
29. RI MS WLB87, p. 20
30. Jenkin, J. (1986). *The Bragg Family in Adelaide: a Pictorial Celebration*, p. 63. University of Adelaide Foundation, Adelaide.
31. Phillips, D. (1979). William Lawrence Bragg. *Biographical Memoirs of Fellows of the Royal Society of London* **25**, 75–143.
32. Quoted in Thomas, J. M. (1991). Bragg reflections. *Notes and Records of the Royal Society of London* **45**, 243–52.
33. Caroe, G. M. (1978). *William Henry Bragg, 1862–1942: Man and Scientist*, p. 78. Cambridge University Press, Cambridge.
34. Caroe, G. M. (1978). *William Henry Bragg, 1862–1942: Man and Scientist*, p. 39. Cambridge University Press, Cambridge.
35. Interview with Stephen Bragg, March 13, 2001.
36. RI MS WLB87, p. 18
37. RI MS WLB54A/27
38. Jenkin, J. (1986). *The Bragg Family in Adelaide: a Pictorial Celebration*, p. 47. University of Adelaide Foundation, Adelaide.
39. RI MS WLB54A/27
40. Jenkin, J. (1986). *The Bragg Family in Adelaide: a Pictorial Celebration*, p. 73. University of Adelaide Foundation, Adelaide.
41. Quoted in Caroe, G. M. (1978). *William Henry Bragg, 1862–1942: Man and Scientist*, p. 51. Cambridge University Press, Cambridge.
42. RI MS WLB87, p. 20
43. RI MS WLB87, p. 21
44. Bragg, W. L. and Caroe, G. M. (1962). Sir William Bragg, F. R. S. *Notes and Records of the Royal Society of London* **17**, 169–82.
45. Trevelyan, G. M. (1990). *Trinity College: a Historical Sketch.* Trinity College, Cambridge.

46. RI MS WLB87, p. 21
47. RI MS WLB37A/2/24
48. Caroe, G. M. (1978). *William Henry Bragg, 1862–1942: Man and Scientist*, p. 63. Cambridge University Press, Cambridge.
49. Interview with Patience Thomson, March 15, 2001.
50. Brooke, C. N. L. (1993). *A History of the University of Cambridge*, p. 600. Cambridge University Press, Cambridge.
51. RI MS WLB37A/2/24
52. RI MS WLB37A/2/25
53. RI MS WLB87, p. 21
54. Tanner, J. R., ed. (1917). *The Historical Register of the University of Cambridge*, pp. 598–600. Cambridge University Press, Cambridge.
55. RI MS WLB87, p. 22
56. RI MS WLB54A/198
57. RI MS WLB52A/85
58. Blackett, P. M. S. (1960). Charles Thomson Rees Wilson. *Biographical Memoirs of Fellows of the Royal Society of London* **6**, 269–95.
59. RI MS WLB87, p. 22–27
60. RI MS WLB10A/27
61. RI MS WLB87, p. 28
62. Crowther, J. G. (1974). *The Cavendish Laboratory, 1874–1974*. Science History Publications, New York.
63. Bragg, W. L. (1965). Reginald William James, 1891–1964. *Biographical Memoirs of Fellows of the Royal Society of London* **11**, 115–25.
64. RI MS WLB54A/345
65. Ewald, P. P. (1962). X-rays. In *Fifty Years of X-ray Diffraction* (P. P. Ewald, ed.), pp. 6–16. International Union of Crystallography, Utrecht.
66. Thomson, J. J. (1925). *The Structure of Light: the Fison Memorial Lecture, 1925*. Cambridge University Press, Cambridge.
67. Bragg, W. L. (1953). The discovery of X-ray diffraction by crystals. *Proceedings of the Royal Institution of Great Britain* **35**, 552–9.
68. Forman, P. (1969). The discovery of the diffraction of X-rays by crystals; a critique of the myths. *Archive for History of Exact Sciences* **6**, 38–71.
69. Burke, J. G. (1966). *Origins of the Science of Crystals*. University of California Press, Berkeley; Ewald, P. P. (1962). Crystallography. In *Fifty Years of X-ray Diffraction* (P. P. Ewald, ed.), pp. 17–30. International Union of Crystallography, Utrecht.
70. Forman, P. (1969). The discovery of the diffraction of X-rays by crystals; a critique of the myths. *Archive for History of Exact Sciences* **6**, 38–71.
71. Ewald, P. P. (1962). Laue's discovery of X-ray diffraction by crystals. In *Fifty Years of X-Ray Diffraction* (P. P. Ewald, ed.), pp. 31–56. International Union of Crystallography, Utrecht.

72. Friedrich, W., Knipping, P. and Laue, M. (1912). Interferenz-Erscheinungen bei Röntgenstrahlen. *Sitzungberichte der Bayerische Akademie der Wissenschaften* **42**, 303–22.
73. Laue, M. (1912). Eine quantitative Prüfung der Theorie für die Interferenzerscheinungen bei Röntgenstrahlen. *Sitzungberichte der Bayerische Akademie der Wissenschaften* **42**, 363–73.
74. Jenkin, J. (2001). A unique partnership: William and Lawrence Bragg and the 1915 Nobel Prize in physics. *Minerva* **39**, 373–92.
75. Bragg, W. L. (1945). X-ray analysis: past, present and future. *Proceedings of the Royal Institution of Great Britain* **33**, 393–400.
76. RI MS WLB49B/42
77. Bragg, W. L. (1967). The start of X-ray analysis. *Chemistry* **40**, 8–13.
78. RI MS WLB49B/42
79. RI MS WLB95, p. 131
80. Ewald, P. P. (1969). The myth of myths; comments on P. Forman's paper on "The discovery of the diffraction of X-rays in crystals". *Archive for History of Exact Sciences* **6**, 72–81.
81. von Laue, M. (1962). My development as a physicist: an autobiography. In *Fifty Years of X-Ray Diffraction* (P. P. Ewald, ed.), pp. 278–307. International Union of Crystallography, Utrecht.
82. Perutz, M. F. (1971). Sir Lawrence Bragg. *Nature* **233**, 74–6.
83. Bragg, W. L. (1949). Acceptance of the Roebling medal of the Mineralogical Society of America. *The American Mineralogist* **34**, 238–241.
84. Quoted in Heilbron, J. L. (1974). *H.G.J. Moseley: The Life and Letters of an English Physicist, 1887–915*, p. 195. University of California Press, Berkeley.
85. RI MS WLB94C/21
86. RI MS WLB49B/42
87. Bragg, W. L. (1914). The diffraction of short electromagnetic waves by a crystal. *Proceedings of the Cambridge Philosophical Society* **17**, 43–57.
88. Forman, P. (1969). The discovery of the diffraction of X-rays by crystals; a critique of the myths. *Archive for History of Exact Sciences* **6**, 38–71.
89. Ewald, P. P. (1962). The immediate sequels to Laue's discovery. In *Fifty Years of X-Ray Diffraction* (P. P. Ewald, ed.), pp. 57–80. International Union of Crystallography, Utrecht.
90. Bragg, W. L. (1966). Reminiscences of fifty years' research. *Proceedings of the Royal Institution of Great Britain* **41**, 92–100.
91. RI MS WLB94A/1
92. RI MS WHB26B/56
93. Bragg, W. L. (1912). The specular reflection of X-rays. *Nature* **90**, 410.
94. RI MS WLB49B/42

95. Bragg, W. L. (1962). Personal reminiscences. In *Fifty Years of X-ray Diffraction* (P. P. Ewald, ed.), pp. 531–9. International Union of Crystallography, Utrecht.
96. RI MS WLB87, p. 30
97. Bragg, W. L. (1954). Mr. Fred Lincoln. *Nature* **174**, 953.
98. Larson, E. (1962). *The Cavendish Laboratory: Nursery of Genius*, p. 42. Edmund Ward, London.
99. Bragg, W. L. (1966). Reminiscences of fifty years' research. *Proceedings of the Royal Institution of Great Britain* **41**, 92–100.
100. RI MS WLB49B/42
101. Barlow, W. and Pope, W. J. (1907). The relation between crystalline form and the chemical constitution of simple inorganic substances. *Journal of the Chemical Society* **91**, 1150–214.
102. RI MS WLB44C/41
103. Bragg, W. L. (1961). The development of X-ray analysis. *Proceedings of the Royal Society of London A* **262**, 145–158.
104. Ewald, P. P. (1962). Laue's discovery of X-ray diffraction by crystals. In *Fifty Years of X-Ray Diffraction* (P. P. Ewald, ed.), pp. 31–56. International Union of Crystallography, Utrecht.
105. Bragg, W. L. (1969). The early history of intensity measurements. *Acta Crystallographica* **25**, 1–3.
106. Bragg, W. H. and Bragg, W. L. (1913). The reflection of X-rays by crystals. *Proceedings of the Royal Society London Series A* **88**, 428–38.
107. Schuster, A. (1909). *An Introduction to the Theory of Optics*, p. 114. Edward Arnold, London.
108. Bragg, W. H. and Bragg, W. L. (1915). *X-Rays and Crystal Structure*, p. 81. G. Bell and Sons, Ltd., London.
109. Moseley, H. G. J. (1914). The high-frequency spectra of the elements. Part II. *Philosophical Magazine* **27**, 703–13.
110. RI MS WHB26B/60
111. Bragg, W. L. (1960). William Henry Bragg. *The New Scientist* **7**, 718–20.
112. RI MS WLB94A/7
113. Ewald, P. P. (1962). Laue's discovery of X-ray diffraction by crystals. In *Fifty Years of X-Ray Diffraction* (P. P. Ewald, ed.), pp. 31–56. International Union of Crystallography, Utrecht.
114. RI MS WLB44C/46
115. RI MS WLB94C/17
116. RI MS WLB94A/12
117. Bragg, W. L. (1914). The structure of some crystals as indicated by their diffraction of X-rays. *Proceedings of the Royal Society of London A* **89**, 248–77.
118. Bragg, W. H. and Bragg, W. L. (1914). The structure of the diamond. *Proceedings of the Royal Society of London A* **89**, 277–91.

119. von Laue, M. (1965). Historical introduction. In *International Tables for X-Ray Crystallography*, pp. 1–5. Kynoch Press, Birmingham.
120. Bragg, W. L. (1953). The discovery of X-ray diffraction by crystals. *Proceedings of the Royal Institution of Great Britain* **35**, 552–9.
121. Bragg, W. L. (1965). The history of X-ray analysis. *Contemporary Physics* **3**, 295–300.
122. Bragg, W. L. (1967). Reminiscences of fifty years of research. *Journal of The Franklin Institute* **284**, 211–28.
123. Groenewege, M. P. and Peerdeman, A. F. (1983). Johannes Martin Bijvoet. *Biographical Memoirs of Fellows of the Royal Society of London* **29**, 27–41.
124. Barlow, W. (1883). Probable nature of the internal symmetry of crystals. *Nature* **29**, 186–8; 205–7.
125. Sohncke, L. (1883). Probable nature of the internal symmetry of crystals. *Nature* **29**, 383–4.
126. Bragg, W. L. (1960). British achievements in X-ray crystallography. *Science* **131**, 1870–4.
127. Armstrong, H. E. (1927). Poor common salt! *Nature* **120**, 478.
128. Keeble, F. (1941). Henry Edward Armstrong. *Obituary Notices of Fellows of the Royal Society* **3**, 229–45.
129. Bragg, W. L. (1967). The start of X-ray analysis. *Chemistry* **40**, 8–13.
130. von Laue, M. (1965). Historical introduction. In *International Tables for X-Ray Crystallography*, pp. 1–5. Kynoch Press, Birmingham.
131. Bragg, W. L. (1961). The development of X-ray analysis. *Proceedings of the Royal Society of London A* **262**, 145–58.
132. Bragg, W. L. (1967). Reminiscences of fifty years of research. *Journal of The Franklin Institute* **284**, 211–228.
133. Bragg, W. L. (1914). The analysis of crystals by the X-ray spectrometer. *Proceedings of the Royal Society of London A* **89**, 468–89.
134. Bragg, W. L. (1949). Acceptance of the Roebling medal of the Mineralogical Society of America. *The American Mineralogist* **34**, 238–41.
135. RI MS WLB94A/13
136. Phillips, D. (1979). William Lawrence Bragg. *Biographical Memoirs of Fellows of the Royal Society of London* **25**, 75–143.
137. RI MS WLB94E/1
138. RI MS WLB87, p. 31
139. RI MS WLB95F/3
140. Interview with Stephen Bragg, March 13, 2001.
141. Caroe, G. M. (1978). *William Henry Bragg, 1862–1942: Man and Scientist*, p. 78. Cambridge University Press, Cambridge.
142. Jenkin, J. (1986). *The Bragg Family in Adelaide: a Pictorial Celebration*, p. 25. University of Adelaide Foundation, Adelaide.
143. Letter from Gwendolen Caroe to David Phillips, May 17, 1979. D. C. Phillips Collection, Bodleian Library, Oxford University.

144. Bragg, W. H. and Bragg, W. L. (1915). *X-Rays and Crystal Structure*, p. vii. G. Bell and Sons, Ltd., London.
145. Phillips, D. (1979). William Lawrence Bragg. *Biographical Memoirs of Fellows of the Royal Society of London* **25**, 75–143.
146. RI MS WLB94A/8
147. RI MS WLB37A/5/2
148. RI MS WHB28A/4
149. RI MS WLB37A/4/1
150. Phillips, D. (1979). William Lawrence Bragg. *Biographical Memoirs of Fellows of the Royal Society of London* **25**, 75–143.
151. RI MS WLB37A/6/3
152. RI MS WLB87, pp. 32–3.
153. Letter from Gwendolen Caroe to David Phillips, May 17, 1979. D. C. Phillips Collection, Bodleian Library, Oxford University.
154. RI MS WLB37A/6/2
155. RI MS WLB95E/1
156. RI MS WLB37A/6/39
157. Bragg, W. H. and Bragg, W. L. (1915). *X-Rays and Crystal Structure*. G. Bell and Sons, Ltd., London.
158. RI MS WLB94D/1
159. RI MS WLB37A/6/7
160. Andrade, E. N. da C. (1943). William Henry Bragg. *Obituary Notices of Fellows of the Royal Society* 12, 277–300.
161. RI MS WLB87, p. 34
162. RI MS RCB16
163. RI MS WLB37A/6/47
164. RI MS WLB87, p. 34
165. Jenkin, J. (1986). *The Bragg Family in Adelaide: a Pictorial Celebration*, p. 79. University of Adelaide Foundation, Adelaide.
166. RI MS WLB87, p. 28
167. RI MS WLB37B/3/29
168. Letter from Gwendolen Caroe to David Phillips, May 17, 1979. D. C. Phillips Collection, Bodleian Library, Oxford University.
169. RI MS WLB37B/4/4
170. RI MS WLB37A/6/22
171. RI MS WLB37A/6/30
172. RI MS WLB37A/6/31
173. Crawford, E. (1984). *The Beginnings of the Nobel Institution. The Science Prizes, 1901–1915*. Cambridge University Press, Cambridge.
174. RI MS WHB11A/8
175. RI MS WHB11A/9
176. Jenkin, J. (2001). A unique partnership: William and Lawrence Bragg and the 1915 Nobel Prize in physics. *Minerva* **39**, 373–92.

177. Crawford, E., Heilbron, J. L., and Ullrich, R. (1987). *The Nobel Population 1901–1937*, pp. 58–63. Office for History of Science and Technology, University of California, Berkeley.
178. RI MS WLB37A/6/25
179. RI MS WLB56B/20
180. RI MS WLB65A/104
181. RI MS WLB87, p. 35
182. RI MS WLB37A/6/34
183. RI MS WLB37A/6/49
184. RI MS WLB87, p. 40
185. RI MS WLB37B/1/64
186. RI MS WLB37B/2/26
187. Andrade, E. N. da C. (1943). William Henry Bragg. *Obituary Notices of Fellows of the Royal Society* **12**, 277–300.
188. RI MS WLB37B/1/29
189. RI MS WLB37B/1/31
190. RI MS WLB37B/1/36
191. RI MS WLB65A/104
192. RI MS WLB87, pp. 35–6
193. RI MS WLB37B/1/49
194. RI MS WLB32G/52
195. RI MS WLB32G/53
196. RI MS WLB37B/1/69
197. RI MS WLB87, p. 37
198. RI MS WLB87, p. 39
199. RI MS WLB37B/3/4
200. RI MS WLB37B/3/4
201. RI MS WLB87, p. 37
202. RI MS WLB37B/2/5
203. RI MS WLB65A/104
204. RI MS WLB37B/1/47
205. RI MS WLB87, p. 27
206. RI MS WLB37B/3/1
207. RI MS WLB37B/3/2
208. RI MS WLB37B/3/3
209. RI MS WLB37B/3/13
210. Interview with Stephen Bragg, March 13, 2001.
211. RI MS WLB23B/133
212. RI MS WLB37B/3/20
213. RI MS WLB32G/54
214. RI MS WLB37B/3/35
215. RI MS WLB37B/3/34
216. RI MS WLB37B/3/42
217. RI MS WLB87, p. 40
218. RI MS WLB37B/3/31

219. RI MS WLB94A/9
220. RI MS WLB37B/3/48
221. RI MS WLB37B/3/51
222. RI MS WLB37B/3/53
223. RI MS WLB37A/4/7
224. RI MS WLB87, pp. 41–2
225. RI MS WLB95, p. 112
226. Phillips, D. (1979). William Lawrence Bragg. *Biographical Memoirs of Fellows of the Royal Society of London* **25**, 75–143.
227. Minutes of the Senate of the University of Manchester, May 14, 1919.
228. RI MS WLB95, p. 134
229. RI MS WLB95, p. 139
230. RI MS WLB95, pp. 115–17
231. RI MS WLB95, p. 121
232. Crowther, J. G. (1974). *The Cavendish Laboratory, 1874–1974*, p. 169. Science History Publications, New York.
233. Peierls, R. (1985). *Bird of Passage: Recollections of a Physicist*, p. 101. Princeton University Press, Princeton.
234. Bragg, W. L. (1970). Manchester days. *Acta Crystallographica A* **26**, 173–7.
235. RI MS WLB87, p. 42
236. Phillips, D. (1979). William Lawrence Bragg. *Biographical Memoirs of Fellows of the Royal Society of London* **25**, 75–143.
237. Letter from Gwendolen Caroe to David Phillips, May 17, 1979. D. C. Phillips Collection, Bodleian Library, Oxford University.
238. Peierls, R. (1984). Reminiscences of Cambridge in the thirties. In *Cambridge Physics in the Thirties* (J. Hendry, ed.), pp. 195–200. Adam Hilger, Ltd., Bristol.
239. Wilson, D. (1983). *Rutherford: Simple Genius*, p. 228. MIT Press, Cambridge.
240. RI MS WLB95, p. 159
241. Bragg, W. L. (1970). Manchester days. *Acta Crystallographica A* **26**, 173–7.
242. Letter from Alice Bragg to David Phillips, June 8, 1979. D. C. Phillips Collection, Bodleian Library, Oxford University.
243. Letter from Gwendolen Caroe to David Phillips, May 17, 1979. D. C. Phillips Collection, Bodleian Library, Oxford University.
244. RI MS WLB87, p. 43
245. Letter from Alice Bragg to David Phillips, June 8, 1979. D. C. Phillips Collection, Bodleian Library, Oxford University.
246. Bragg, W. L. (1961). Mr W. A. Kay. *Nature* **189**, 621.
247. RI MS WLB95, p. 156
248. Thewlis, J. (1970). Personal reminiscences. *Acta Crystallographica A* **26**, 182–3.
249. RI MS WHB28A/5

250. Bragg, W. L. (1949). Acceptance of the Roebling medal of the Mineralogical Society of America. *The American Mineralogist* **34**, 238–41; Letter from Anthony North, March 14, 2002.
251. Bragg, W. L. (1920). The arrangement of atoms in crystals. *Philosophical Magazine* **40**, 169–89.
252. RI MS WHB28A/9
253. RI MS WHB28A/16
254. Darwin, C. G. (1914). The theory of X-ray reflexion. *Philosophical Magazine* **27**, 315–33; Darwin, C. G. (1914). The theory of X-ray reflexion. Part II. *Philosophical Magazine* **27**, 675–90.
255. Ewald, P. P. (1962). Laue's discovery of X-ray diffraction by crystals. In *Fifty Years of X-Ray Diffraction* (P. P. Ewald, ed.), pp. 31–56. International Union of Crystallography, Utrecht.
256. Bragg, W. L., James, R. W., and Bosanquet, C. H. (1921) The intensity of reflexion of X-rays by rock-salt. *Philosophical Magazine* **41**, 309–37.
257. Bragg, W. H. (1914). The intensity of reflexion of X rays by crystals. *Philosophical Magazine* **27**, 881–99.
258. Bragg, W. L., Darwin, C. G., and James, R. W. (1926). The intensity of reflexion of X-rays by crystals. *Philosophical Magazine* **1**, 897–922.
259. Bragg, W. L., James, R. W., and Bosanquet, C. H. (1921). The intensity of reflexion of X-rays by rock-salt. II. *Philosophical Magazine* **42**, 1–17.
260. Bragg, W. L., James, R. W., and Bosanquet, C. H. (1922). The distribution of electrons around the nucleus in the sodium and chlorine atoms. *Philosophical Magazine* **44**, 433–49.
261. RI MS WLB87, pp. 43–5
262. RI MS WHB28A/20
263. RI MS WLB37B/6/9
264. RI MS WLB37B/6/21
265. RI MS WLB95, pp. 141–2
266. Letter from Alice Bragg to David Phillips, June 8, 1979. D. C. Phillips Collection, Bodleian Library, Oxford University.
267. RI MS WLB95, p. 146
268. RI MS WLB87, pp. 45–6
269. Phillips, D. (1979). William Lawrence Bragg. *Biographical Memoirs of Fellows of the Royal Society of London* **25**, 75–143.
270. RI MS WLB10C/7
271. RI MS WLB95, p. 149
272. RI MS WLB95, p. 100
273. RI MS WLB95, p. 150
274. RI MS WLB87, p. 47
275. RI MS WLB95, pp. 163–5
276. RI MS WLB87, pp. 43–8
277. Interview with Patience Thomson, March 15, 2001.
278. RI MS WLB95, p. 143

279. Letter from Gwendolen Caroe to David Phillips, May 17, 1979. D. C. Phillips Collection, Bodleian Library, Oxford University.
280. RI MS WLB94B/2
281. RI MS WLB94B/9
282. RI MS WLB94B/9
283. Bragg, W. L. (1967). The diffraction of X-rays by crystals. In *Nobel Lectures in Physics, 1901–1921*, pp. 370–82. Elsevier, Amsterdam.
284. Jenkin, J. (2001). A unique partnership: William and Lawrence Bragg and the 1915 Nobel Prize in physics. *Minerva* **39**, 373–92.
285. Granqvist, G. (1967). Physics 1915. In *Nobel Lectures in Physics, 1901–1921*, pp. 363–6. Elsevier, Amsterdam.
286. Bragg, W. L. (1969). What makes a scientist? *Proceedings of the Royal Institution of Great Britain* **42**, 397–410.
287. RI MS WLB87, pp. 49–51
288. Letter from Gwendolen Caroe to David Phillips, May 17, 1979. D. C. Phillips Collection, Bodleian Library, Oxford University.
289. Interview with Patience Thomson, March 15, 2001.
290. RI MS WLB87, p. 51
291. RI MS WHB28A/4
292. Cleaveland, P. (1816). *An Elementary Treatise on Mineralogy and Geology*, p. 180. Cummings and Hilliard, Boston.
293. RI MS WLB94B/9
294. Bragg, W. L. (1923). The structure of aragonite. *Proceedings of the Royal Society of London A* **105**, 16–39.
295. Bragg, W. L. (1975). *The Development of X-Ray Analysis*, p. 73. G. Bell and Sons Ltd., London.
296. Wyckoff, R. W. G. (1922). *The Analytical Expression of the Results of the Theory of Space-Groups*. Carnegie Institution of Washington, Washington, D. C.
297. Bragg, W. L. (1924). The refractive indices of calcite and aragonite. *Proceedings of the Royal Society of London A* **105**, 370–86.
298. RI MS WLB87, p. 55
299. Bragg, W. L. (1938). Forty years of crystal physics. In *Background to Modern Science* (W. Pagel and J. Needham, eds.), pp. 77–92. Cambridge University Press, Cambridge.
300. Bragg, W. L. (1924). The influence of atomic arrangement on refractive index. *Proceedings of the Royal Society of London A* **106**, 346–68; Bragg, W. L. (1924). The refractive indices of calcite and aragonite. *Proceedings of the Royal Society of London A* **105**, 370–86.
301. Bragg, W. L. and Chapman, S. (1924). A theoretical calculation of the rhombohedral angle of crystals of the calcite type. *Proceedings of the Royal Society of London A* **106**, 369–77.
302. RI MS WLB87, pp. 52–4
303. RI MS WLB95, pp. 179–81
304. RI MS WLB87, pp. 55–6

305. Bragg, W. L. (1954). X-ray studies of biological molecules. *Nature* **174**, 55–9.
306. Wyart, J. (1962). Personal reminiscences. In *Fifty Years of X-ray Diffraction* (P. P. Ewald, ed.), pp. 685–90. International Union of Crystallography, Utrecht, Holland.
307. Klug, A. (1990). Reminiscences of Sir Lawrence Bragg. In *Selections and Reflections: the Legacy of Sir Lawrence Bragg* (J. M. Thomas and D. Phillips, eds.), pp. 129–33. Science Reviews Ltd, Northwood, U.K.
308. Bragg, W. L. (1967). The start of X-ray analysis. *Chemistry* **40**, 8–13.
309. Bethe, H. A. and Hildebrandt, G. (1988). Peter Paul Ewald. *Biographical Memoirs of Fellows of the Royal Society of London* **34**, 135–76.
310. Bragg, W. L., Darwin, C. G. and James, R. W. (1926). The intensity of reflexion of X-rays by crystals. *Philosophical Magazine* **1**, 897–922.
311. Bragg, W. L. and Brown, G. B. (1925). The crystalline structure of chrysoberyl. *Proceedings of the Royal Society of London A* **110**, 34–63.
312. Pauling, L. (1962). Early work on X-ray diffraction in the California Institute of Technology. In *Fifty Years of X-Ray Diffraction* (P. P. Ewald, ed.), pp. 623–8. International Union of Crystallography, Utrecht, Holland.
313. Horace Freeland Judson's interview with Linus Pauling, December 23, 1975. American Philosophical Society.
314. Pauling, L. (1990). My indebtedness to and my contacts with Lawrence Bragg. In *Selections and Reflections: the Legacy of Sir Lawrence Bragg* (J. M. Thomas and D. Phillips, eds.), pp. 86–8. Science Reviews Ltd, Northwood, U.K.
315. Bragg, W. L. (1961). The development of X-ray analysis. *Proceedings of the Royal Society of London A* **262**, 145–58.
316. Bragg, W. L. and West, J. (1926). The structure of beryl, $Be_3Al_2Si_6O_{18}$. *Proceedings of the Royal Society of London A* **111**, 691–714.
317. Bragg, W. L. (1929). Atomic arrangements in the silicates. *Transactions: Faraday Society* **25**, 291–314.
318. Bragg, W. H. (1915). X-rays and crystal structure. *Philosophical Transactions of the Royal Society A* **215**, 253–74.
319. Duane, W. (1925). The calculation of the X-ray diffracting power at points in a crystal. *Proceedings of the National Academy of Sciences USA* **11**, 489–93; Havighurst, R. J. (1925). The distribution of diffracting power in sodium chloride. *Proceedings of the National Academy of Sciences USA* **11**, 502–7.
320. Bragg, W. L. (1975). *The Development of X-Ray Analysis*, pp. 176–93. G. Bell and Sons Ltd., London.
321. RI MS WLB94B/22
322. RI MS WLB94B/23
323. RI MS WLB95, pp. 171–2

324. Bragg, W. L. (1926). The structure of phenacite, Be_2SiO_4. *Proceedings of the Royal Society of London A* **113**, 642–57.
325. Bragg, W. L. and West, J. (1927). The structure of certain silicates. *Proceedings of the Royal Society of London A* **114**, 450–73.
326. RI MS WLB87, p. 57
327. Interview with Stephen Bragg, March 13, 2001.
328. RI MS WLB87, p. 58
329. RI MS WLB95, p. 187
330. RI MS WLB87, pp. 57–9
331. Interview with Patience Thomson, March 15, 2001.
332. RI MS WLB13A/223
333. Bragg, W. L. (1927). Some views on the teaching of science. *Manchester Memoirs* **71**, 119–23.
334. RI MS WLB54A/475
335. RI MS WLB87, pp. 60–1
336. Warren, B. E. (1962). Personal reminiscences. In *Fifty Years of X-Ray Diffraction* (P. P. Ewald, ed.), pp. 667–71. International Union of Crystallography, Utrecht, Holland.
337. Letter from William Lawrence Bragg to Linus Pauling, March 1, 1928. Ava Helen and Linus Pauling Papers, Oregon State University Library.
338. RI MS WLB87, pp. 61–3; RI MS WLB95, p. 183
339. Warren, B. and Bragg, W. L. (1928). The structure of diopside, $CaMg(SiO_3)_2$. *Zeitschrift für Kristallographie und Kristallgeometrie* **69**, 168–93.
340. Bragg, W. L. (1967). Reminiscences of fifty years of research. *Journal of The Franklin Institute* **284**, 211–28.
341. Bragg, W. L. (1949). Acceptance of the Roebling medal of the Mineralogical Society of America. *The American Mineralogist* **34**, 238–41.
342. Anonymous (1946). Royal medals. *Nature* **158**, 842.
343. Bragg, W. L. and West, J. (1928). A technique for the X-ray examination of crystal structures with many parameters. *Zeitschrift fur Kristallographie Kristallgeometrie* **69**, 118–48.
344. RI MS WLB77C/52
345. Bragg, W. L. (1929). An optical method of representing the results of X-ray analysis. *Zeitschrift fur Kristallographie Kristallgeometrie* **70**, 475–92.
346. Bragg, W. L. (1929). The determination of parameters in crystal structures by means of Fourier series. *Proceedings of the Royal Society of London A* **123**, 537–59.
347. Bragg, W. L. (1961). The development of X-ray analysis. *Proceedings of the Royal Society of London A* **262**, 145–58.
348. Dodson, G. (2002). Dorothy Mary Crowfoot Hodgkin, O. M. *Biographical Memoirs of Fellows of the Royal Society of London* **48**, 3–41.

349. Lipson, H. (1990). The introduction of Fourier methods into crystal-structure determination. *Notes and Records of the Royal Society of London* **44**, 257–64.
350. Bragg, W. L. (1970). Manchester days. *Acta Crystallographica A* **26**, 173–7.
351. Bragg, W. L. (1929). Atomic arrangements in the silicates. *Transactions: Faraday Society* **25**, 291–314.
352. Pauling, L. (1929). The principles determining the structure of complex ionic crystals. *Journal of the American Chemical Society* **51**, 1010–26.
353. Phillips, D. (1979). William Lawrence Bragg. *Biographical Memoirs of Fellows of the Royal Society of London* **25**, 75–143.
354. Horace Freeland Judson's interview with Max Perutz, February 15, 1975. American Philosophical Society.
355. Interview with Aaron Klug, March 13, 2002.
356. Horace Freeland Judson's interview with Linus Pauling, December 23, 1975. American Philosophical Society.
357. Letter from Linus Pauling to William Lawrence Bragg, January 19, 1928. Ava Helen and Linus Pauling Papers, Oregon State University Library.
358. Letter from William Lawrence Bragg to Linus Pauling, August 24, 1928. Ava Helen and Linus Pauling Papers, Oregon State University Library.
359. Bragg, W. L. and Zachariasen, W. H. (1930). The crystalline structure of phenacite, Be2SiO4, and Willemite, Zn2SiO4. *Zeitschrift fur Kristallographie Kristallgeometrie* **72**, 518–28.
360. Bragg, W. L. (1929). The diffraction of short electromagnetic rays by a crystal. *Scientia* **45**, 153–62.
361. Burke, J. G. (1966). *Origins of the Science of Crystals*, p. 14. University of California Press, Berkeley.
362. Bragg, W. L. (1930). The structure of silicates. *Zeitschrift für Kristallographie und Kristallgeometrie* **74**, 237–305.
363. Bragg, W. L. (1949). Acceptance of the Roebling medal of the Mineralogical Society of America. *The American Mineralogist* **34**, 238–41.
364. Bragg, W. L. (1964). F. Machatschki. *Tschermaks mineralogische und petrographische Mitteilungen* **10**, 3.
365. RI MS WLB87, p. 63
366. RI MS WLB87, p. 66
367. Bragg, W. L. and Caroe, G. M. (1962). Sir William Bragg, F. R. S. *Notes and Records of the Royal Society of London* **17**, 169–82.
368. RI MS WLB87, p. 66
369. RI MS WLB87, p. 67
370. Letter from Gwendolen Caroe to David Phillips, May 17, 1979. D. C. Phillips Collection, Bodleian Library, Oxford University.
371. Caroe, G. M. (1978). *William Henry Bragg, 1862–1942: Man and Scientist*, p. 101. Cambridge University Press, Cambridge.
372. Peierls, R. (1985). *Bird of Passage: Recollections of a Physicist*, p. 99. Princeton University Press, Princeton.

373. Letter from Alexander Todd to Linus Pauling, May 30, 1938. Ava Helen and Linus Pauling Papers, Oregon State University Library.
374. Phillips, D. (1979). William Lawrence Bragg. *Biographical Memoirs of Fellows of the Royal Society of London* **25**, 75–143.
375. RI MS WLB87, pp. 67–8
376. Phillips, D. (1979). William Lawrence Bragg. *Biographical Memoirs of Fellows of the Royal Society of London* **25**, 75–143.
377. Pauling, L. (1990). My indebtedness to and my contacts with Lawrence Bragg. In *Selections and Reflections: the Legacy of Sir Lawrence Bragg* (J. M. Thomas and D. Phillips, eds.), pp. 86–8. Science Reviews Ltd, Northwood, U.K.
378. RI MS WLB77D/13
379. Phillips, D. (1979). William Lawrence Bragg. *Biographical Memoirs of Fellows of the Royal Society of London* **25**, 75–143.
380. RI MS WLB77C/128
381. RI MS WLB77C/136
382. RI MS WLB77D/69
383. Bragg, W. L. (1930). The structure of silicates. *Zeitschrift für Kristallographie und Kristallgeometrie* **74**, 237–305.
384. RI MS WLB77D/69
385. RI MS WLB77D/89
386. Warren, B. E. and Bragg, W. L. (1930). The structure of chrysotile $H_4Mg_3Si_2O_9$. *Zeitschrift für Kristallographie und Kristallgeometrie* **76**, 201–10.
387. RI MS WLB77D/156
388. RI MS WLB77D/141
389. RI MS WLB77D/143
390. Bragg, W. L. (1953). The discovery of X-ray diffraction by crystals. *Proceedings of the Royal Institution of Great Britain* **35**, 552–9.
391. RI MS WLB87, pp. 68–9
392. Interview with Patience Thomson, March 15, 2001.
393. Letter from Gwendolen Caroe to David Phillips, May 17, 1979. D. C. Phillips Collection, Bodleian Library, Oxford University.
394. Bragg, W. L. (1931). The architecture of the solid state. *Nature* **128**, 210–12; 248–50.
395. Phillips, D. (1979). William Lawrence Bragg. *Biographical Memoirs of Fellows of the Royal Society of London* **25**, 75–143.
396. Interview with Stephen Bragg, March 13, 2001.
397. RI MS WLB87, pp. 70–3; RI MS WLB95, pp. 190–4
398. RI MS WLB87, pp. 74–5
399. Brooke, C. N. L. (1993). *A History of the University of Cambridge*, p. 285. Cambridge University Press, Cambridge.
400. RI MS WLB95, pp. 196–202
401. Bragg, S. L. (1990). A personal view by his elder son. In *Selections and Reflections: the Legacy of Sir Lawrence Bragg* (J. M. Thomas

and D. Phillips, eds.), pp. 141–4. Science Reviews Ltd., Northwood, Middlesex.

402. RI MS WLB87, pp. 75–7
403. Bragg, W. L. (1933). Structure of alloys. *Nature* **131**, 749–53.
404. Bragg, W. L. (1937). Alloys. *Journal of the Royal Society of Arts* **85**, 431–47.
405. Lipson, H. (1990). The introduction of Fourier methods into crystal-structure determination. *Notes and Records of the Royal Society of London* **44**, 257–64.
406. Phillips, D. (1979). William Lawrence Bragg. *Biographical Memoirs of Fellows of the Royal Society of London* **25**, 75–143.
407. Letter from William Lawrence Bragg to Patrick Blackett, December 21, 1945. P. M. S. Blackett Collection, Royal Society Library (CSAC 63.1.79/H.21).
408. RI MS WLB87, p. 82
409. Bragg, W. L. (1962). Personal reminiscences. In *Fifty Years of X-ray Diffraction* (P. P. Ewald, ed.), pp. 531–9. International Union of Crystallography, Utrecht.
410. Bragg, W. L. (1935). Atomic arrangements in metals and alloys. *Journal of the Institute of Metals* **56**, 275–99.
411. Bragg, W. L. and Williams, E. J. (1935). The effect of thermal agitation on atomic arrangement in alloys—II. *Proceedings of the Royal Society of London A* **151**, 540–66.
412. Bragg, W. L. (1937). Alloys. *Journal of the Royal Society of Arts* **85**, 431–47.
413. Bragg, W. L. (1961). The development of X-ray analysis. *Proceedings of the Royal Society of London A* **262**, 145–58.
414. Bragg, W. L. and Williams, E. J. (1935). The effect of thermal agitation on atomic arrangement in alloys—II. *Proceedings of the Royal Society of London A* **151**, 540–66.
415. Bragg, W. L. (1934). The physical sciences. *Science* **79**, 237–40.
416. RI MS WLB87, pp. 77–81
417. RI MS WHB5A/14
418. Patterson, A. L. (1934). A Fourier series method for the determination of the components of interatomic distances in crystals. *Physical Review* **46**, 372–6.
419. Harker, D. (1936). The application of the three-dimensional Patterson method and the crystal structures of proustite Ag_3AsS_3, and pyrargyrite, Ag_3SbS_3. *Journal of Chemical Physics* **4**, 381–90.
420. Lipson, H. (1990). The introduction of Fourier methods into crystal-structure determination. *Notes and Records of the Royal Society of London* **44**, 257–64.
421. Phillips, D. (1979). William Lawrence Bragg. *Biographical Memoirs of Fellows of the Royal Society of London* **25**, 75–143.
422. Anonymous (1935). How dynamos work. *The Times*, January 2, p. 7.

423. RI MS WLB87, p. 84
424. Anonymous (1935). Electricity explained to children. *The Times*, January 4, p. 15.
425. Anonymous (1938). School speech days. *The Times*, July 2, p. 17.
426. Bragg, W. L. (1938). The examination evil. *The Times*, July 18, p. 8.
427. Bragg, W. L. (1938). The examination for honours in physics in the University of Manchester. *Yearbook of Education*, 351–5.
428. Bragg, W. L. (1935). The new crystallography. *Proceedings of the Royal Society of Edinburgh* **55**, 62–71.
429. Astbury, W. T. and Street, A. (1931). X-ray studies of the structure of hair, wool, and related fibres. *Philosophical Transactions of the Royal Society A* **230**, 75–101.
430. Bernal, J. D. and Crowfoot, D. (1934). X-ray photographs of crystalline pepsin. *Nature* **134**, 794–5.
431. Svedberg, T. and Fåhraeus, R. (1926). A new method for the determination of the molecular weight of the proteins. *Journal of the American Chemical Society* **48**, 430–8.
432. RI MS WLB77I/126–127
433. RI MS WLB77I/117
434. RI MS WLB77I/138
435. Andrade, E. N. da C. (1943). William Henry Bragg. *Obituary Notices of Fellows of the Royal Society*, 277–300.
436. RI MS WLB87, pp. 85–6.
437. Phillips, D. (1979). William Lawrence Bragg. *Biographical Memoirs of Fellows of the Royal Society of London* **25**, 75–143.
438. Bragg, W. L. and Lipson, H. (1936). The employment of contoured graphs of structure-factor in crystal analysis. *Zeitschrift für Kristallographie und Kristallgeometrie* **95**, 323–37.
439. RI MS WLB95, p. 203
440. RI MS WLB75A/30
441. RI MS WLB87, pp. 86–7; RI MS WLB95, pp. 204–6
442. Letter from Gwendolen Caroe to David Phillips, May 17, 1979. D. C. Phillips Collection, Bodleian Library, Oxford University.
443. Minutes of the Senate of the University of Manchester, November 4, 1937.
444. RI MS WLB95, p. 204
445. Phillips, D. (1979). William Lawrence Bragg. *Biographical Memoirs of Fellows of the Royal Society of London* **25**, 75–143.
446. RI MS WLB58A/153
447. RI MS WLB77K/97
448. Pyatt, E. (1983). *The National Physical Laboratory: a History*, p. 82. Adam Hilger Ltd., Bristol.
449. RI MS WLB87, pp. 87–8
450. RI MS WLB65A/111
451. RI MS WLB95, p. 211

452. RI MS WLB95, p. 159
453. Interview with Stephen Bragg, March 13, 2001.
454. RI MS WLB87, pp. 87–92
455. RI MS WLB95, pp. 211–12
456. RI MS WLB10A/11
457. RI MS WLB95, p. 210
458. RI MS WLB10A/8; interview with Stephen Bragg, March 13, 2001.
459. Pyatt, E. (1983). *The National Physical Laboratory: a History*, p. 70. Adam Hilger Ltd., Bristol.
460. RI MS WLB10A/27
461. RI MS WLB95, p. 210
462. RI MS WLB42B/154
463. Mott, N. (1990). Manchester and Cambridge. In *Selections and Reflections: the Legacy of Sir Lawrence Bragg* (J. M. Thomas and D. Phillips, eds.), pp. 96–7. Science Reviews Ltd, Northwood, U.K.
464. Pippard, B. (1990). Bragg—the Cavendish Professor. In *Selections and Reflections: the Legacy of Sir Lawrence Bragg* (J. M. Thomas and D. Phillips, eds.), pp. 97–100. Science Reviews Ltd, Northwood, U.K.
465. Cochran, W. and Devons, S. (1981). Norman Feather. *Biographical Memoirs of Fellows of the Royal Society of London* **27**, 255–82.
466. Crowther, J. G. (1974). *The Cavendish Laboratory, 1874–1974*, p. 250. Science History Publications, New York.
467. RI MS WLB42B/41
468. RI MS WLB10A/15
469. Anonymous (1938). Prof. W. L. Bragg, O.B.E., F.R.S. *Nature* **142**, 403.
470. RI MS WLB42B/198
471. RI MS WLB87, p. 90
472. Interview with Stephen Bragg, March 13, 2001.
473. RI MS WLB95, pp. 213–15
474. Larsen, E. (1962). *The Cavendish Laboratory: Nursery of Genius*, p. 55. Edmund Ward, London.
475. Arndt, U. (1990). The bird life of Albemarle Street. In *Selections and Reflections: the Legacy of Sir Lawrence Bragg* (J. M. Thomas and D. Phillips, eds.), pp. 105–8. Science Reviews Ltd, Northwood, U.K.
476. RI MS WLB87D/1
477. RI MS WLB77L/150
478. RI MS WLB87, p. 92
479. Perutz, M. F. (1970). Bragg, protein crystallography and the Cavendish Laboratory. *Acta Crystallographica A* **26**, 183–5.
480. Bernal, J. D. (1939). Structure of proteins. *Nature* **143**, 663–7.
481. RI MS WLB54A/282
482. Guinier, A. (1990). Golden rules. In *Selections and Reflections: the Legacy of Sir Lawrence Bragg* (J. M. Thomas and D. Phillips, eds.), pp. 108–9. Science Reviews Ltd, Northwood, U.K.
483. Interview with Aaron Klug, March 13, 2002.

484. Horace Freeland Judson's interview with Francis Crick, September 1, 1971. American Philosophical Society.
485. Interview with Max Perutz, March 12, 2001.
486. RI MS WLB13B/1
487. RI MS WLB13B/2
488. Crowfoot, D. (1938). The crystal structure of insulin. I. The investigation of air-dried insulin crystals. *Proceedings of the Royal Society of London A* **164**, 580–602.
489. Wrinch, D. M. (1936). The pattern of proteins. *Nature* **137**, 411–12.
490. RI MS WLB77L/6; RI MS WLB77M/18
491. RI MS WLB77M/60, RI MS WLB77M/71
492. Bragg, W. L. (1939). Patterson diagrams in crystal analysis. *Nature* **143**, 73–4.
493. Pauling, L. and Niemann, C. (1939). The structure of proteins. *Journal of the American Chemical Society* **61**, 1860–7.
494. Letter from Linus Pauling to William Lawrence Bragg, April 26, 1939. Ava Helen and Linus Pauling Papers, Oregon State University Library.
495. RI MS WLB78A/95
496. RI MS WLB78A/96
497. Bragg, W. L. (1939). Patterson diagrams in crystal analysis. *Nature* **143**, 73–4.
498. Robertson, J. M. (1939). Vector maps and heavy atoms in crystal analysis and the insulin structure. *Nature* **143**, 75–6.
499. Bernal, J. D. (1939). Structure of proteins. *Nature* **143**, 663–7.
500. RI MS WLB87, p. 91
501. RI MS WLB87, pp. 93–4
502. Bragg, W. L. (1946). The Austin wing of the Cavendish laboratory. *Nature* **158**, 326–7.
503. RI MS WLB95, p. 217
504. RI MS WLB87, pp. 137
505. RI MS WLB87, pp. 95–6
506. Letter from John Nye, December 20, 2002.
507. RI MS WLB87, p. 100
508. RI MS WLB14A/81
509. RI MS WLB39C/79
510. Phillips, D. (1979). William Lawrence Bragg. *Biographical Memoirs of Fellows of the Royal Society of London* **25**, 75–143.
511. RI MS WLB87, pp. 101–2
512. Phillips, D. (1979). William Lawrence Bragg. *Biographical Memoirs of Fellows of the Royal Society of London* **25**, 75–143.
513. RI MS WLB41A/17
514. RI MS WLB87, pp. 103–4
515. RI MS WLB41A/56
516. RI MS WLB37C/6/22
517. RI MS WLB37B/6/35

518. RI MS WLB37B/6/29
519. RI MS WLB37B/6/42
520. RI MS WLB37B/6/26
521. RI MS WLB37B/6/10
522. RI MS WLB37B/6/14
523. Interview with Patience Thomson, March 15, 2001.
524. RI MS WLB37B/6/43
525. RI MS WLB37B/6/44
526. RI MS WLB41B/126
527. RI MS WLB87, p. 104–5
528. RI MS WLB41B/5
529. RI MS WLB78A/224
530. RI MS WLB49B/3
531. Arndt, U. (1990). The bird life of Albemarle Street. In *Selections and Reflections: the Legacy of Sir Lawrence Bragg* (J. M. Thomas and D. Phillips, eds.), pp. 105–8. Science Reviews Ltd, Northwood, U.K.
532. Letter from Brian Pippard, January, 2003.
533. Pippard, B. (1990). Bragg—the Cavendish Professor. In *Selections and Reflections: the Legacy of Sir Lawrence Bragg* (J. M. Thomas and D. Phillips, eds.), pp. 97–100. Science Reviews Ltd, Northwood, U.K.
534. Wilson, D. (1983). *Rutherford: Simple Genius*, p. 609. MIT Press, Cambridge.
535. Letter from Antony Hewish, January 6, 2003.
536. Letter from John Nye, December 20, 2002.
537. Letter from Michael Whelan, January 28, 2002.
538. Letter from Peter Hirsch, November 20, 2002.
539. Letter from David Blow, January 16, 2002.
540. Phillips, D. (1979). William Lawrence Bragg. *Biographical Memoirs of Fellows of the Royal Society of London* **25**, 75–143.
541. RI MS WLB87, pp. 98–9
542. RI MS WLB87, p. 105
543. Interview with Patience Thomson, March 15, 2001.
544. RI MS WLB37B/6/46
545. RI MS WLB37B/6/47
546. RI MS WLB87, pp. 106–7
547. Bragg, W. L. and Caroe, G. M. (1962). Sir William Bragg, F. R. S. *Notes and Records of the Royal Society of London* **17**, 169–82.
548. Letter from Gwendolen Caroe to David Phillips, May 17, 1979. D. C. Phillips Collection, Bodleian Library, Oxford University.
549. Astbury, W. T. (1942). *Nature* **149**, 347–8.
550. RI MS WHB36/22
551. Quirke, V. (2002). "A big happy family": the Royal Institution under William and Lawrence Bragg, and the history of molecular biology. In *"The Common Purposes of Life": Science and Society at the Royal Institution of Great Britain* (F. A. J. L. James, ed.), p. 259. Ashgate Publishing Limited, Burlington, Vermont.

552. Bragg, W. L. (1939). A new type of X-ray 'microscope'. *Nature* **143**, 678.
553. Bragg, W. L. (1942). The X-ray microscope. *Nature* **149**, 470–1.
554. Bragg, W. L. (1944). Lightning calculations with light. *Nature* **154**, 69–72.
555. Bragg, W. L. and Stokes, A. R. (1945). X-ray analysis with the aid of the "fly's eye". *Nature* **156**, 332–3.
556. Perutz, M. F. (1942). X-ray analysis of haemoglobin. *Nature* **149**, 491–4.
557. Lomer, W. M. (1990). Blowing bubbles with Bragg. In *Selections and Reflections: the Legacy of Sir Lawrence Bragg* (J. M. Thomas and D. Phillips, eds.), pp. 115–18. Science Reviews Ltd, Northwood, U.K.
558. RI MS WLB33D/116
559. Letter from John Nye, December 20, 2002.
560. Bragg, W. L. (1942). A model illustrating intercrystalline boundaries and plastic flow in metals. *Journal of Scientific Instruments* **19**, 148–50.
561. Interview with Stephen Bragg, March 13, 2001.
562. Bragg, W. L. (1967). The spirit of science. *Proceedings of the Royal Society of Edinburgh, Section A (Mathematical & Physical Sciences)* **67**, 303–8.
563. Bragg, W. L. (1942). Physicists after the war. *Nature* **150**, 75–80 and 374.
564. RI MS WLB87, p. 110
565. Bragg, W. L. (1942). The balance of education. *The Times*, December 17, p. 5.
566. Phillips, D. (1979). William Lawrence Bragg. *Biographical Memoirs of Fellows of the Royal Society of London* **25**, 75–143.
567. Bragg, W. L. (1943). The exposition of science. *The Advancement of Science* **2:8**, 286–7.
568. Pickard, R. H., Findlay, A., and Bragg, W. L. (1943). Science and the state. *The Times*, June 19, p. 5.
569. Trevelyan, G. M. and Bragg, W. L. (1944). Specialization in schools. *The Times*, May 10, p. 5.
570. Bragg, W. L. (1944). Organization and finance of science in universities. *The Political Quarterly* **15**, 330–41.
571. RI MS WLB70B/1; RI MS WLB87, p. 107–10
572. RI MS WLB78B/182
573. Perutz, M. F. (1998). *I Wish I'd Made You Angry Earlier*, pp. 73–106. Cold Spring Harbor Laboratory Press, Plainsview, New York.
574. Letter from W. L. Bragg to Henry Dale, December 29, 1943 (93HD33.8.13); letter from F. M. R. Walshe to W. L. Bragg, December 4, 1943 (93HD33.8.14); letter from Bernard Hart to W. L. Bragg, December 21, 1943 (93HD33.8.15). H. H. Dale Collection, Royal Society Library.
575. RI MS WLB41B/230
576. Letter from W. L. Bragg to Henry Dale, November 18, 1943. H. H. Dale Collection, Royal Society Library (93HD33.8.6).
577. RI MS WLB62A/213

578. RI MS WLB95, p. 172
579. RI MS WLB87, p. 111
580. RI MS WLB95, p. 4
581. RI MS WLB58B/214
582. Interview with Stephen Bragg, March 13, 2001.
583. Horace Freeland Judson's interview with John Kendrew, January 18, 1971. American Philosophical Society.
584. "Fifty Years a Winner." BBC broadcast, December 3, 1965.
585. Mott, N. (1986). *A Life in Science*, pp. 71–2. Taylor and Francis, London.
586. Casimir, H. (1983). *Haphazard Reality: Half a Century of Science*, pp. 240–1. Harper & Row, New York.
587. Letter from Brian Pippard, January, 2003.
588. Horace Freeland Judson's interview with Max Perutz, December 6, 1970. American Philosophical Society.
589. Bragg, W. L. (1944). Organization and finance of science in universities. *The Political Quarterly* **15**, 330–41.
590. RI MS WLB87, p. 116
591. Bragg, W. L. (1944). Organization and finance of science in universities. *The Political Quarterly* **15**, 330–41.
592. Phillips, D. (1979). William Lawrence Bragg. *Biographical Memoirs of Fellows of the Royal Society of London* **25**, 75–143.
593. Perutz, M. F. (1970). Bragg, protein crystallography and the Cavendish Laboratory. *Acta Crystallographica A* **26**, 183–5.
594. Interview with Aaron Klug, March 13, 2002.
595. Pippard, B. (1990). Bragg—the Cavendish Professor. In *Selections and Reflections: the Legacy of Sir Lawrence Bragg* (J. M. Thomas and D. Phillips, eds.), pp. 97–100. Science Reviews Ltd, Northwood, U.K.
596. Crowther, J. G. (1974). *The Cavendish Laboratory, 1874–1974*, p. 391. Science History Publications, New York.
597. RI MS WLB87, p. 116
598. Bragg, W. L. (1943). Seeing ever-smaller worlds. *Nature* **151**, 545–7.
599. Arndt, U. (1990). The bird life of Albemarle Street. In *Selections and Reflections: the Legacy of Sir Lawrence Bragg* (J. M. Thomas and D. Phillips, eds.), pp. 105–8. Science Reviews Ltd, Northwood, U.K.
600. RI MS WLB49B/100
601. Bragg, W. L. (1945). Science and the nations. *The Times*, July 6, p. 5.
602. Bragg, W. L. (1949). Acceptance of the Roebling medal of the Mineralogical Society of America. *The American Mineralogist* **34**, 238–41.
603. RI MS WLB68A/76; RI MS WLB87, pp. 118–23
604. de Oliveira Marques, A. H. (1976). *History of Portugal.* Columbia University Press, New York; Anderson, J. M. (2000). *The History of Portugal.* Greenwood Press, Westport, Connecticut.

605. RI MS WLB87, pp. 118–19
606. RI MS WLB49C/148
607. Interview with Stephen Bragg, March 13, 2001.
608. RI MS WLB95, p. 237
609. Thomson, P. M. (1999). William Lawrence Bragg: the education of a scientist. *The Adelaidean*, May 31, p. 2.
610. RI MS WLB87, pp. 126–7
611. Letter from Antony Hewish, January 6, 2003.
612. Thomson, P. M. (1990). Thoughts about my father by his daughter. In *Selections and Reflections: the Legacy of Sir Lawrence Bragg* (J. M. Thomas and D. Phillips, eds.), pp. 144–6. Science Reviews Ltd, Northwood, U.K.
613. Interview with Patience Thomson, March 15, 2001.
614. Letter from Peter Hirsh, November 20, 2002.
615. "Fifty Years a Winner." BBC broadcast, December 3, 1965.
616. RI MS WLB95, pp. 167–8
617. Interview with Stephen Bragg, March 13, 2001.
618. Crowther, J. G. (1974). *The Cavendish Laboratory, 1874–1974*, p. 240. Science History Publications, New York.
619. Letter from Antony Hewish, January 6, 2003.
620. Arndt, U. (1990). The bird life of Albemarle Street. In *Selections and Reflections: the Legacy of Sir Lawrence Bragg* (J. M. Thomas and D. Phillips, eds.), pp. 105–8. Science Reviews Ltd, Northwood, U.K.
621. Interview with Patience Thomson, March 15, 2001.
622. Bragg, W. L. (1966). The art of talking about science. *Science* **154**, 1613–16.
623. Interview with Patience Thomson, March 15, 2001.
624. Interview with Patience Thomson, March 15, 2001.
625. Interview with Stephen Bragg, March 13, 2001.
626. Letter from John Nye, December 20, 2002.
627. RI MS WLB87, pp. 124–5
628. RI MS WLB31E/42
629. Anonymous (1946). Royal Medals. *Nature*, 842.
630. Phillips, D. (1979). William Lawrence Bragg. *Biographical Memoirs of Fellows of the Royal Society of London* **25**, 75–143.
631. Horace Freeland Judson's interview with John Kendrew, January 18, 1971. American Philosophical Society.
632. Crowther, J. G. (1974). *The Cavendish Laboratory, 1874–1974*, p. 306. Science History Publications, New York.
633. Kendrew, J. C. (1990). Bragg's broomstick and the structure of proteins. In *Selections and Reflections: the Legacy of Sir Lawrence Bragg* (J. M. Thomas and D. Phillips, eds.), pp. 88–91. Science Reviews Ltd, Northwood, U.K.
634. Kendrew, J. C. (1963). Myoglobin and the structure of proteins. *Science* **139**, 1259–66.

635. de Chadarevian, S. (2002). *Designs for Life: Molecular Biology After World War Two*, p. 126. Cambridge University Press, Cambridge.
636. Perutz, M. F. (1985). Early days of protein crystallography. *Methods in Enzymology* **114**, 3–18.
637. Boyes-Watson, J., Davidson, E., and Perutz, M. F. (1947). An X-ray study of horse methaemoglobin. I. *Proceedings of the Royal Society of London A* **191**, 83–132.
638. de Chadarevian, S. (2002). *Designs for Life: Molecular Biology After World War Two*, p. 61. Cambridge University Press, Cambridge.
639. Phillips, D. (1979). William Lawrence Bragg. *Biographical Memoirs of Fellows of the Royal Society of London* **25**, 75–143; Perutz, M. F. (1996). The Medical Research Council Laboratory of Molecular Biology. *Molecular Medicine* **2**, 659–62.
640. RI MS WLB87, p. 127
641. RI MS WLB56A/43
642. RI MS WLB87, pp. 128–30
643. Bragg, W. L. (1948). The Cavendish Laboratory. *Journal of the Institute of Metals.* **75**, 107–14; Bragg, W. L. (1948). Organisation and work of the Cavendish laboratory. *Nature* **161**, 627–8; Phillips, D. (1979). William Lawrence Bragg. *Biographical Memoirs of Fellows of the Royal Society of London* **25**, 75–143; Crowther, J. G. (1974). *The Cavendish Laboratory, 1874–1974*, pp. 269–90. Science History Publications, New York.
644. Cochran, W. (1990). Cavendish days. In *Selections and Reflections: the Legacy of Sir Lawrence Bragg* (J. M. Thomas and D. Phillips, eds.), pp. 103–5. Science Reviews Ltd, Northwood, U.K.
645. Phillips, D. (1979). William Lawrence Bragg. *Biographical Memoirs of Fellows of the Royal Society of London* **25**, 75–143.
646. Letter from John Nye, December 20, 2002.
647. Horace Freeland Judson's interview with Max Perutz, December 6, 1970. American Philosophical Society.
648. RI MS WLB95, p. 245
649. Phillips, D. (1979). William Lawrence Bragg. *Biographical Memoirs of Fellows of the Royal Society of London* **25**, 75–143.
650. Pippard, B. (1990). Bragg—the Cavendish Professor. In *Selections and Reflections: the Legacy of Sir Lawrence Bragg* (J. M. Thomas and D. Phillips, eds.), pp. 97–100. Science Reviews Ltd, Northwood, U.K.
651. Interview with Patience Thomson, March 15, 2001.
652. King, R. (1990). Of shoes—and ships—and sealing wax and string. In *Selections and Reflections: the Legacy of Sir Lawrence Bragg* (J. M. Thomas and D. Phillips, eds.), pp. 134–8. Science Reviews Ltd, Northwood, U.K.
653. Bragg, W. L. (1969). What makes a scientist? *Proceedings of the Royal Institution of Great Britain* **42**, 397–410.
654. Bragg, W. L. (1968). The white-coated worker. *Punch* **255**, 352–4.

655. Bragg, W. L. (1957). No time for science. *The Times*, January 3, p. 9.
656. Letter from Gwendolen Caroe to David Phillips, May 17, 1979. D. C. Phillips Collection, Bodleian Library, Oxford University.
657. RI MS WLB14E/1-2
658. Wilkins, M. H. F. (1987). John Turton Randall. *Biographical Memoirs of Fellows of the Royal Society of London* **33**, 493–535.
659. Huggins, M. L. (1943). The structure of fibrous proteins. *Chemical Reviews* **32**, 195–218.
660. Dunitz, J. D. (1990). Encounters with Bragg. In *Selections and Reflections: the Legacy of Sir Lawrence Bragg* (J. M. Thomas and D. Phillips, eds.), pp. 118–23. Science Reviews Ltd, Northwood, U.K.
661. Pauling, L. (1990). My indebtedness to and my contacts with Lawrence Bragg. In *Selections and Reflections: the Legacy of Sir Lawrence Bragg* (J. M. Thomas and D. Phillips, eds.), pp. 86–8. Science Reviews Ltd, Northwood, U.K.
662. Letter from Linus Pauling to Robert Corey, February 18, 1948. Ava Helen and Linus Pauling Papers, Oregon State University Library.
663. Quoted in Hager, T. (1995). *Force of Nature: the Life of Linus Pauling*, pp. 330–1. Simon & Shuster, New York.
664. Pauling, L. (1993). How my interest in proteins developed. *Protein Science* **2**, 1060–3.
665. RI MS WLB87, pp. 132–4
666. Bragg, W. L. (1948). Recent advances in the study of the crystalline state. *Science* **108**, 455–63.
667. RI MS WLB87, pp. 135–41; RI MS WLB56A/233
668. RI MS WLB69A/333
669. RI MS WLB79A/92
670. Perutz, M. F. (1949). An X-ray study of horse methaemoglobin. II. *Proceedings of the Royal Society of London A* **195**, 474–99.
671. Letter from Linus Pauling to Alexander Todd, July 19, 1939. Ava Helen and Linus Pauling Papers, Oregon State University Library.
672. Dodson, G. (2002). Dorothy Mary Crowfoot Hodgkin, O. M. *Biographical Memoirs of Fellows of the Royal Society of London* **48**, 3–41.
673. Olby, R. C. (1970). Francis Crick, DNA, and the central dogma. *Daedalus* **99**, 938–88.
674. Perutz, M. F. (1997). *Science is Not a Quiet Life: Unravelling the Atomic Mechanism of Haemoglobin*, p. 40. World Scientific Publishing Co., Singapore.
675. Olby, R. C. (1985). The "mad pursuit": X-ray crystallographers' search for the structure of haemoglobin. *History and Philosophy of Life Science* **7**.
676. Bragg, W. L. (1949). Giant molecules. *Nature* **164**, 7–10.
677. Thomson, P. M. (1990). Thoughts about my father by his daughter. In *Selections and Reflections: the Legacy of Sir Lawrence Bragg*

(J. M. Thomas and D. Phillips, eds.), pp. 144–6. Science Reviews Ltd, Northwood, U.K.

678. Interview with Patience Thomson, March 15, 2001.
679. RI MS WLB87, pp. 142–5
680. Bragg, W. L. (1948). Recent advances in the study of the crystalline state. *Science* **108**, 455–63.
681. Horace Freeland Judson's interview with Max Perutz, January 31, 1976. American Philosophical Society.
682. RI MS WLB13B/3
683. Bragg, W. L. (1965). First stages in the X-ray analysis of proteins. *Reports on Progress in Physics* **28**, 1–14.
684. Pauling, L. (1993). How my interest in proteins developed. *Protein Science* **2**, 1060–3.
685. Bragg, W. L., Kendrew, J. C., and Perutz, M. F. (1950). Polypeptide chain configurations in crystalline proteins. *Proceedings of the Royal Society of London A* **203**, 321–57.
686. Letter from Linus Pauling to Robert Olby, March 15, 1973. Ava Helen and Linus Pauling Papers, Oregon State University Library.
687. Horace Freeland Judson's interview with Francis Crick, September 10, 1975. American Philosophical Society.
688. Horace Freeland Judson's interview with John Kendrew, November 11, 1975. American Philosophical Society.
689. Interview with Max Perutz, March 12, 2001.
690. Dunitz, J. D. (1990). Encounters with Bragg. In *Selections and Reflections: the Legacy of Sir Lawrence Bragg* (J. M. Thomas and D. Phillips, eds.), pp. 118–23. Science Reviews Ltd, Northwood, U.K.
691. Kendrew, J. C. (1990). Bragg's broomstick and the structure of proteins. In *Selections and Reflections: the Legacy of Sir Lawrence Bragg* (J. M. Thomas and D. Phillips, eds.), pp. 88–91. Science Reviews Ltd, Northwood, U.K.
692. Bragg, W. L., Howells, E. R., and Perutz, M. F. (1952). Arrangement of polypeptide chains in horse methaemoglobin. *Acta Crystallographica* **5**, 136–41.
693. Bragg, W. L. (1967). Reminiscences of fifty years of research. *Journal of The Franklin Institute* **284**, 211–28.
694. Crowther, J. G. (1974). *The Cavendish Laboratory, 1874–1974*, p. 314. Science History Publications, New York.
695. Pauling, L. (1993). How my interest in proteins developed. *Protein Science* **2**, 1060–3.
696. Letter from Linus Pauling to David Harker, March 8, 1951. Ava Helen and Linus Pauling Papers, Oregon State University Library.
697. Pauling, L. and Corey, R. B. (1950). Two hydrogen-bonded spiral configurations of the polypeptide chain. *Journal of the American Chemical Society* **72**, 5349.
698. RI MS WLB87, p. 150

699. RI MS WLB70D/59
700. Interview with Patience Thomson, March 15, 2001.
701. RI MS WLB87, pp. 150–1
702. RI MS WLB69B/72
703. RI MS WLB95, p. 241
704. Crowther, J. G. (1974). *The Cavendish Laboratory, 1874–1974*, p. 285. Science History Publications, New York.
705. Pauling, L., Corey, R. B., and Branson, H. R. (1951). The structure of proteins: two hydrogen-bonded helical configurations of the polypeptide chain. *Proceedings of the National Academy of Sciences USA* **37**, 205–11.
706. Hager, T. (1995). *Force of Nature: the Life of Linus Pauling*, p. 378. Simon & Shuster, New York.
707. Todd, A. (1990). A recollection of Sir Lawrence Bragg. In *Selections and Reflections: the Legacy of Sir Lawrence Bragg* (J. M. Thomas and D. Phillips, eds.), pp. 95–6. Science Reviews Ltd, Northwood, U.K.
708. Horace Freeland Judson's interviews with Francis Crick, September 10, 1975, and John Kendrew, November 11, 1975. American Philosophical Society.
709. Letter from Max Perutz to Linus Pauling, August 17, 1951. Ava Helen and Linus Pauling Papers, Oregon State University Library.
710. Bragg, W. L. (1965). First stages in the X-ray analysis of proteins. *Reports on Progress in Physics* **28**, 1–14.
711. Interview with Max Perutz, March 12, 2001.
712. Letter from William Lawrence Bragg to Linus Pauling, June 13, 1951. Ava Helen and Linus Pauling Papers, Oregon State University Library.
713. Avery, O. T., MacLeod, C. M., and McCarty, M. (1944). Studies on the chemical nature of the substance inducing transformation of pneumococcal types. *Journal of Experimental Medicine* **89**, 137–59.
714. RI MS WLB54A/282
715. Horace Freeland Judson's interview with Max Perutz, December 4–5, 1970. American Philosophical Society.
716. RI MS WLB31E/52; Cochran, W. (1990). Cavendish days. In *Selections and Reflections: the Legacy of Sir Lawrence Bragg* (J. M. Thomas, and D. Phillips, eds.), pp. 103–5. Science Reviews Ltd, Northwood, U.K.; Interview with Francis Crick, January 11, 2002.
717. RI MS WLB53A/231
718. Memorandum from Linus Pauling to Robert Corey, July 24, 1951. Ava Helen and Linus Pauling Papers, Oregon State University Library.
719. Arndt, U. (1990). The bird life of Albemarle Street. In *Selections and Reflections: the Legacy of Sir Lawrence Bragg* (J. M. Thomas and D. Phillips, eds.), pp. 105–8. Science Reviews Ltd, Northwood, U.K.
720. RI MS WLB87, p. 154
721. Bragg, W. L. and Perutz, M. F. (1952). The external form of the haemoglobin molecule. I. *Acta Crystallographica* **5**, 277–83.

722. Bragg, W. L. and Perutz, M. F. (1952). The external form of the haemoglobin molecule. II. *Acta Crystallographica* **5**, 323–8.
723. Bragg, W. L., Howells, E. R., and Perutz, M. F. (1952). Arrangement of polypeptide chains in horse methaemoglobin. *Acta Crystallographica* **5**, 136–41.
724. Perutz, M. F. (1997). *Science is Not a Quiet Life: Unravelling the Atomic Mechanism of Haemoglobin*, pp. 44–5. World Scientific Publishing Co., Singapore.
725. Crick, F. H. C. (1988). *What Mad Pursuit: A Personal View of Scientific Discovery*, p. 47. Basic Books, New York.
726. Bragg, W. L. (1949). Giant molecules. *Nature* **164**, 7–10.
727. Bragg, W. L. and Perutz, M. F. (1952). The structure of haemoglobin. *Proceedings of the Royal Society of London A* **213**, 425–35.
728. Bragg, W. L. (1952). X-ray analysis of proteins. *The Proceedings of the Physical Society B* **65**, 833–46.
729. Letter from William Lawrence Bragg to Linus Pauling, May 5, 1952. Ava Helen and Linus Pauling Papers, Oregon State University Library.
730. Perutz, M. F. (1997). *Science is Not a Quiet Life: Unravelling the Atomic Mechanism of Haemoglobin*, p. 65. World Scientific Publishing Co., Singapore.
731. RI MS WLB68B/23, RI MS WLB68B/41
732. RI MS WLB95, pp. 241–6
733. Thewlis, J. (1953). Fortieth anniversary of the discovery of X-ray diffraction. *Nature* **171**, 106–7.
734. RI MS WLB12/3; Interview with Max Perutz, March 12, 2001; Interview with Francis Crick, January 11, 2002.
735. Interview with Max Perutz, March 12, 2001.
736. Watson, J. D. (1990). Bragg's Foreword to "The Double Helix". In *Selections and Reflections: the Legacy of Sir Lawrence Bragg* (J. M. Thomas and D. Phillips, eds.), pp. 111–13. Science Reviews Ltd, Northwood, U.K.
737. Horace Freeland Judson's interview with John Kendrew, November 11, 1975. American Philosophical Society.
738. RI MS WLB94A /9
739. RI MS WLB54A /282; RI MS WLB32E/7
740. Horace Freeland Judson's interview with John Kendrew, November 11, 1975. American Philosophical Society.
741. RI MS WLB12/90
742. Watson, J. D. and Crick, F. H. C. (1953). Molecular structure of nucleic acids. *Nature* **171**, 737–8; Wilkins, M. H. F., Stokes, A. R., and Wilson, H. R. (1953). Molecular structure of deoxypentose nucleic acids. *Nature* **171**, 738–40; Franklin, R. E. and Gosling, R. G. (1953). Molecular configuration in sodium thymonucleate. *Nature* **171**, 740–1.
743. Hager, T. (1995). *Force of Nature: the Life of Linus Pauling*, p. 428. Simon & Shuster, New York.

744. RI MS WLB87, p. 118
745. Bragg, W. L. (1961). The development of X-ray analysis. *Proceedings of the Royal Society of London A* **262**, 145–8.
746. Hodgkin, D. C. (1979). Crystallographic measurements and the structure of protein molecules as they are. *Annals of the New York Academy of Sciences* **325**, 121–48.
747. Perutz, M. F. (1997). *Science is Not a Quiet Life: Unravelling the Atomic Mechanism of Haemoglobin*, p. 67. World Scientific Publishing Co., Singapore.
748. Green, D. W., Ingram, V. M., and Perutz, M. F. (1954). The structure of haemoglobin. IV. Sign determination by the isomorphous replacement method. *Proceedings of the Royal Society of London A* **225**, 287–307.
749. Horace Freeland Judson's interview with William Lawrence Bragg, January 28, 1971. American Philosophical Society.
750. Bragg, W. L. and Perutz, M. F. (1954). The structure of haemoglobin. VI. Fourier projections on the 010 plane. *Proceedings of the Royal Society of London A* **225**, 315–29.
751. Kendrew, J. C., Bodo, G., Dintzis, H. M., Parrish, R. G., Wyckoff, H., and Phillips, D. C. (1958). A three-dimensional model of the myoglobin molecule obtained by X-ray analysis. *Nature* **181**, 662–6.
752. Kendrew, J. C. (1963). Myoglobin and the structure of proteins. *Science* **139**, 1259–66.
753. Horace Freeland Judson's interview with Max Perutz, February 7, 1976. American Philosophical Society.
754. Hodgkin, D. C. (1979). Crystallographic measurements and the structure of protein molecules as they are. *Annals of the New York Academy of Sciences* **325**, 121–48.
755. Perutz, M. F. (1985). Early days of protein crystallography. *Methods in Enzymology* **114**, 3–18.
756. Horace Freeland Judson's interview with Max Perutz, January 23, 1971. American Philosophical Society.
757. Perutz, M. F. (1997). *Science is Not a Quiet Life: Unravelling the Atomic Mechanism of Haemoglobin*, p. 67. World Scientific Publishing Co., Singapore.
758. Hager, T. (1995). *Force of Nature: the Life of Linus Pauling*, p. 436. Simon & Shuster, New York.
759. RI MS WLB88N/15
760. Crowther, J. G. (1974). *The Cavendish Laboratory, 1874–1974*, p. 328. Science History Publications, New York.
761. Perutz, M. F. (1970). Bragg, protein crystallography and the Cavendish Laboratory. *Acta Crystallographica A* **26**, 183–5.
762. Arndt, U. (1990). The bird life of Albemarle Street. In *Selections and Reflections: the Legacy of Sir Lawrence Bragg* (J. M. Thomas and D. Phillips, eds.), pp. 105–8. Science Reviews Ltd, Northwood, U.K.

763. Caroe, G. (1985). *The Royal Institution: an Informal History*. John Murray, London; James, F. A. J. L., ed. (2002). *"The Common Purposes of Life": Science and Society at the Royal Institution of Great Britain*. Ashgate Publishing Limited, Burlington, Vermont.
764. Eley, D. D. (1976). Eric Keightley Rideal. *Biographical Memoirs of Fellows of the Royal Society of London* **22**, 381–413.
765. Porter, G. (1990). W. L. B. at the R.I. In *Selections and Reflections: the Legacy of Sir Lawrence Bragg* (J. M. Thomas and D. Phillips, eds.), pp. 126–9. Science Reviews Ltd, Northwood, U.K.
766. Andrade, E. N. da C. (1943). William Henry Bragg. *Obituary Notices of Fellows of the Royal Society*, 277–300.
767. Cottrell, A. (1972). Edward Neville da Costa Andrade. *Biographical Memoirs of Fellows of the Royal Society of London* **18**, 1–20.
768. RI MS WLB37B/2/22
769. RI MS WLB58B/178
770. RI MS WLB56A/102
771. RI MS WLB87, pp. 152–3
772. RI MS WLB56A/478
773. RI MS WLB56A/479
774. Letter from Henry Dale to W. L. Bragg, January 19, 1952. H. H. Dale Collection, Royal Society Library (93HD33.8.52).
775. Letter from Edward Andrade to Linus Pauling, June 3, 1952. Ave Helen and Linus Pauling Papers, Oregon State University Library.
776. RI MS WLB11G/12
777. RI MS WLB11G/13
778. RI MS WLB15B/9
779. RI MS WLB95, p. 247
780. RI MS WLB95, p. 219
781. Letter from Alban Caroe to David Phillips, May 13, 1979. D. C. Phillips Collection, Bodleian Library, Oxford University.
782. Anonymous (1953). Royal Institution: Sir Lawrence Bragg, O. B. E., F. R. S. *Nature* **171**, 819.
783. Letter from Gwendolen Caroe to David Phillips, May 17, 1979. D. C. Phillips Collection, Bodleian Library, Oxford University.
784. RI MS WLB95, p. 255
785. Letter from William Lawrence Bragg to Patrick Blackett, undated (c. 1963?). P. M. S. Blackett Collection, Royal Society Library (CSAC 63.1.79/J.11).
786. RI MS WLB57C/243
787. RI MS WLB11F/5A
788. King, R. (1990). Of shoes—and ships—and sealing wax and string. In *Selections and Reflections: the Legacy of Sir Lawrence Bragg* (J. M. Thomas and D. Phillips, eds.), pp. 134–8. Science Reviews Ltd, Northwood, U.K.
789. RI MS WLB95, p. 248
790. RI MS WLB32G/16

791. King, R. (1990). Of shoes—and ships—and sealing wax and string. In *Selections and Reflections: the Legacy of Sir Lawrence Bragg* (J. M. Thomas and D. Phillips, eds.), pp. 134–8. Science Reviews Ltd, Northwood, U.K.
792. RI MS WLB95, p. 249
793. RI MS WLB11G/73
794. Letter from William Lawrence Bragg to Patrick Blackett, undated (c. 1963?). P. M. S. Blackett Collection, Royal Society Library (CSAC 63.1.79/J.11).
795. Letter from W. L. Bragg to Henry Dale, February 1, 1966. H. H. Dale Collection, Royal Society Library (93HD33.8.53).
796. King, R. (1990). Of shoes—and ships—and sealing wax and string. In *Selections and Reflections: the Legacy of Sir Lawrence Bragg* (J. M. Thomas and D. Phillips, eds.), pp. 134–8. Science Reviews Ltd, Northwood, U.K.
797. King, R. (1990). Of shoes—and ships—and sealing wax and string. In *Selections and Reflections: the Legacy of Sir Lawrence Bragg* (J. M. Thomas and D. Phillips, eds.), pp. 134–8. Science Reviews Ltd, Northwood, U.K.
798. Bragg, W. L. (1957). "Schools lectures" at the Royal Institution: a new venture. *Discovery* **18**, 66–7.
799. Coates, W. A. (1990). Sir William Bragg and his lecturer's assistant. In *Selections and Reflections: the Legacy of Sir Lawrence Bragg* (J. M. Thomas and D. Phillips, eds.), pp. 147–9. Science Reviews Ltd, Northwood, U.K.
800. Arndt, U. (1990). The bird life of Albemarle Street. In *Selections and Reflections: the Legacy of Sir Lawrence Bragg* (J. M. Thomas and D. Phillips, eds.), pp. 105–8. Science Reviews Ltd, Northwood, U.K.
801. Porter, G. (1990). W. L. B. at the R. I. In *Selections and Reflections: the Legacy of Sir Lawrence Bragg* (J. M. Thomas and D. Phillips, eds.), pp. 126–9. Science Reviews Ltd, Northwood, U.K.
802. Allibone, T. E. (1990). Christmas lectures to a juvenile auditory, 1959/60. In *Selections and Reflections: the Legacy of Sir Lawrence Bragg* (J. M. Thomas and D. Phillips, eds.), pp. 138–9. Science Reviews Ltd, Northwood, U.K.
803. Saunders, V. T. (1962). The Royal Institution Christmas Lectures, 1961. *Contemporary Physics* **3**, 469–73.
804. King, R. (1990). Of shoes—and ships—and sealing wax and string. In *Selections and Reflections: the Legacy of Sir Lawrence Bragg* (J. M. Thomas and D. Phillips, eds.), pp. 134–8. Science Reviews Ltd, Northwood, U.K.
805. Phillips, D. (1979). William Lawrence Bragg. *Biographical Memoirs of Fellows of the Royal Society of London* **25**, 75–143.
806. Phillips, D. C. (1970). W. L. Bragg at the Royal Institution, 1953–66. *Acta Crystallographica A* **26**, 186–8.

807. Porter, G. (1990). W. L. B. at the R. I. In *Selections and Reflections: the Legacy of Sir Lawrence Bragg* (J. M. Thomas and D. Phillips, eds.), pp. 126–9. Science Reviews Ltd, Northwood, U.K.
808. Caroe, G. (1985). *The Royal Institution: an Informal History*, p. 119. John Murray, London.
809. Bragg, W. L. (1965). The schools lectures at the Royal Institution. *Science* **150**, 1420–3.
810. Quoted in Caroe, G. (1985). *The Royal Institution: an Informal History*, p. 128. John Murray, London.
811. Bragg, W. L. (1958). Interpretation of science to the public. *Nature* **181**, 807–8.
812. Bragg, W. L. (1960). Atoms and molecules. *Contemporary Physics* **1**, 390–3.
813. Bragg, W. L. (1964). Minerals. *Proceedings of the Royal Institution of Great Britain* **40**, 64–81.
814. Interview with Stephen Bragg, March 13, 2001.
815. Bragg, W. L. (1957). The diffraction of short electromagnetic waves. *Journal Brit. I.R.E.*, 467–71.
816. Wain, R. L. (1990). As the barnacles stick to the rock. In *Selections and Reflections: the Legacy of Sir Lawrence Bragg* (J. M. Thomas and D. Phillips, eds.), p. 140. Science Reviews Ltd, Northwood, U.K.
817. RI MS WLB95, p. 250
818. Phillips, D. C. (1970). W. L. Bragg at the Royal Institution, 1953–66. *Acta Crystallographica A* **26**, 186–8.
819. RI MS WLB95, p. 253
820. RI MS WLB38C/76
821. Perutz, M. F. (1971). Sir Lawrence Bragg. *Nature* **233**, 74–6.
822. RI MS WLB25A/1; RI MS WLB25A/4; RI MS WLB25A/94; RI MS WLB25A/70; RI MS WLB25A/52.
823. Bragg, W. L. (1973). An awkward incident. In *A Random Walk In Science* (R. L. Weber and E. Mendoza, eds.), p. 192. The Institute of Physics, London.
824. Bragg, W. L. (1958). The contribution of the Royal Institution to the teaching of science. *The School Science Review* **40**, 240–5.
825. RI MS WLB33B/27
826. RI MS WLB33B/31
827. RI MS WLB2/31
828. Bragg, W. L. (1958). An international survey of recent scientific research. *New Scientist* **3**, 16–17; Max Blythe's interview with David Phillips, April 29, 1996. Oxford Brookes University Medical Sciences Video Archive MSVA 128; de Chadarevian, S. (2002). *Designs for Life: Molecular Biology After World War Two*, pp. 155–7. Cambridge University Press, Cambridge.
829. RI MS WLB56B/167; Interview with Patience Thomson, March 15, 2001.

830. Crick, F. H. C. (1988). *What Mad Pursuit: A Personal View of Scientific Discovery*, p. 53. Basic Books, New York.
831. Kendrew, J. C. (1990). Bragg's broomstick and the structure of proteins. In *Selections and Reflections: the Legacy of Sir Lawrence Bragg* (J. M. Thomas and D. Phillips, eds.), pp. 88–91. Science Reviews Ltd, Northwood, U.K.
832. Watson, J. D. (1990). Bragg's Foreword to "The Double Helix". In *Selections and Reflections: the Legacy of Sir Lawrence Bragg* (J. M. Thomas and D. Phillips, eds.), pp. 111–13. Science Reviews Ltd, Northwood, U.K.
833. Interview with Stephen Bragg, March 13, 2001.
834. Interview with Patience Thomson, March 15, 2001.
835. Interview with Stephen Bragg, March 13, 2001.
836. RI MS WLB57C/248
837. RI MS WLB95, p. 251
838. RI MS WLB68C/72; RI MS WLB68C/85
839. RI MS WLB95, p. 259
840. RI MS WLB56B/175
841. Mott, N. (1986). *A Life in Science*, p. 110. Taylor and Francis, London.
842. Letter from William Lawrence Bragg to Dorothy Hodgkin, May 6, 1955. D. M. C. Hodgkin Collection, Bodleian Library, Oxford University.
843. Phillips, D. (1979). William Lawrence Bragg. *Biographical Memoirs of Fellows of the Royal Society of London* **25**, 75–143.
844. Bragg, W. L. (1962). Personal reminiscences. In *Fifty Years of X-ray Diffraction* (P. P. Ewald, ed.), pp. 531–9. International Union of Crystallography, Utrecht.
845. Dunitz, J. D. (1990). Encounters with Bragg. In *Selections and Reflections: the Legacy of Sir Lawrence Bragg* (J. M. Thomas and D. Phillips, eds.), pp. 118–23. Science Reviews Ltd, Northwood, U.K.
846. Johnson, L. N. (1999). David Phillips and the origin of structural enzymology. *Trends in Biochemical Sciences* **24**, 287–9.
847. Letter from Anthony North, March 14, 2002.
848. Phillips, D. (1979). William Lawrence Bragg. *Biographical Memoirs of Fellows of the Royal Society of London* **25**, 75–143.
849. Letter from Louise Johnson, June 15, 2001.
850. Letter from David Blow, January 16, 2002.
851. Arndt, U. (1990). The bird life of Albemarle Street. In *Selections and Reflections: the Legacy of Sir Lawrence Bragg* (J. M. Thomas and D. Phillips, eds.), pp. 105–8. Science Reviews Ltd, Northwood, U.K.
852. RI MS WLB88M/21
853. Max Blythe's interview with David Phillips, April 29, 1966. Oxford Brookes University Medical Sciences Video Archive MSVA 128.
854. Horace Freeland Judson's interview with Max Perutz, December 4–5, 1970. American Philosophical Society.

855. Bragg, W. L. (1958). The determination of the coordinates of heavy atoms in protein crystals. *Acta Crystallographica* **11**, 70–5.
856. Klug, A. (1990). Reminiscences of Sir Lawrence Bragg. In *Selections and Reflections: the Legacy of Sir Lawrence Bragg* (J. M. Thomas and D. Phillips, eds.), pp. 129–33. Science Reviews Ltd, Northwood, U.K.
857. Bragg, W. L. (1966). Reminiscences of fifty years' research. *Proceedings of the Royal Institution of Great Britain* **41**, 92–100.
858. Bragg, W. L. (1965). First stages in the X-ray analysis of proteins. *Reports on Progress in Physics* **28**, 1–14.
859. Perutz, M. F. (1997). *Science is Not a Quiet Life: Unravelling the Atomic Mechanism of Haemoglobin*, p. 69. World Scientific Publishing Co., Singapore.
860. Perutz, M. F. (1971). Sir Lawrence Bragg. *Nature* **233**, 74–6.
861. Huxley, H. (1990). An early adventure in crystallographic computing. In *Selections and Reflections: the Legacy of Sir Lawrence Bragg* (J. M. Thomas and D. Phillips, eds.), pp. 133–4. Science Reviews Ltd, Northwood, U.K.
862. Crowther, J. G. (1974). *The Cavendish Laboratory, 1874–1974*, p. 306. Science History Publications, New York.
863. Wilkes, M. V. (1985). *Memoirs of a Computer Pioneer*, p. 192. MIT Press, Cambridge, Massachusetts.
864. Kendrew, J. C., Bodo, G., Dintzis, H. M., Parrish, R. G., Wyckoff, H., and Phillips, D. C. (1958). A three-dimensional model of the myoglobin molecule obtained by X-ray analysis. *Nature* **181**, 662–6.
865. Perutz, M. F. (1985). Early days of protein crystallography. *Methods in Enzymology* **114**, 3–18.
866. Kendrew, J. C. (1963). Myoglobin and the structure of proteins. *Science* **139**, 1259–66.
867. Bragg, W. L. (1961). What is life made of? *The Saturday Evening Post*, October 7, pp. 34–5, 54, 62, 64.
868. Max Blythe's interview with David Phillips, April 29, 1996. Oxford Brookes University Medical Sciences Video Archive MSVA 128.
869. Hodgkin, D. C. (1979). Crystallographic measurements and the structure of protein molecules as they are. *Annals of the New York Academy of Sciences* **325**, 121–48.
870. Dickerson, R. E. (1992). A little ancient history. *Protein Science* **1**, 182–6.
871. Kendrew, J. C., Dickerson, R. E., Strandberg, B. E., Hart, R. G., Davies, D. R., Phillips, D. C., and Shore, V. C. (1960). Structure of myoglobin: a three-dimensional Fourier synthesis at 2 Å resolution. *Nature* **185**, 422–7.
872. RI MS WLB88M/37
873. RI MS WLB88M/42
874. Perutz, M. F., Rossman, M. G., Cullis, A. F., Muirhead, H., Will, G., and North, A. C. T. (1960). Structure of haemoglobin: a three-dimensional

Fourier synthesis at 5.5-Å resolution, obtained by X-ray analysis. *Nature* **185**, 416–22.

875. Anonymous (1960). International Union of Crystallography Fifth General Assembly, International Congress and Symposia. *Acta Crystallographica* **13**, 965–71.
876. Law, J. (1973). The development of specialties in science: The case of X-ray protein crystallography. *Science Studies* **3**, 275–303.
877. Bragg, W. L. (1965). First stages in the X-ray analysis of proteins. *Reports on Progress in Physics* **28**, 1–14.
878. Horace Freeland Judson's interview with David Phillips, January 8, 1971. American Philosophical Society.
879. Max Blythe's interview with David Phillips, April 29, 1966. Oxford Brookes University Medical Sciences Video Archive MSVA 129.
880. Anonymous (1961). The Tercentenary Conversazione, 23 July 1960. *Notes and Records of the Royal Society of London* **16**, 25–30.
881. RI MS WLB95, p. 260
882. Bragg, W. L. (1961). The development of X-ray analysis. *Proceedings of the Royal Society of London A* **262**, 145–58.
883. Letter from William Lawrence Bragg to David Phillips, October 2, 1960. D. C. Phillips Collection, Bodleian Library, Oxford University.
884. RI MS WLB69E/1
885. RI MS WLB87, pp. 12–14
886. Phillips, D. (1979). William Lawrence Bragg. *Biographical Memoirs of Fellows of the Royal Society of London* **25**, 75–143.
887. RI MS WLB95, p. 261
888. RI MS WLB38C/32
889. Crowther, J. G. (1974). *The Cavendish Laboratory, 1874–1974*, p. 292. Science History Publications, New York.
890. Crawford, E., Heilbron, J. L., and Ullrich, R. (1987). *The Nobel Population 1901–1937*, pp. 89–155. Office for History of Science and Technology, University of California, Berkeley.
891. Bragg, W. L. (1954). X-ray studies of biological molecules. *Nature* **174**, 55–9.
892. Bragg, W. L. (1957). X-ray analysis. *The New Scientist* **3**, 19–21.
893. Maddox, B. (2002). *Rosalind Franklin: the Dark Lady of DNA*, p. xviii. HarperCollins, London.
894. Interview with Aaron Klug, March 13, 2002.
895. Maddox, B. (2002). *Rosalind Franklin: the Dark Lady of DNA*, p. 293. HarperCollins, London.
896. Interview with Aaron Klug, March 13, 2002.
897. Interview with Francis Crick, January 11, 2002.
898. RI MS WLB12/90
899. Horace Freeland Judson's interview with William Lawrence Bragg, January 28, 1971. American Philosophical Society.
900. RI MS WLB14D/38

901. Horace Freeland Judson's interview with John Kendrew, November 11, 1975. American Philosophical Society.
902. RI MS WLB14D/41
903. RI MS WLB59B/102
904. RI MS WLB13B/6
905. RI MS WLB13B/19
906. Letter from Linus Pauling to William Lawrence Bragg, December 15, 1959. Ava Helen and Linus Pauling Papers, Oregon State University Library.
907. RI MS WLB13C/3
908. RI MS WLB13B/28
909. RI MS WLB13E/1
910. Interview with Francis Crick, January 11, 2002.
911. RI MS WLB13C/6
912. RI MS WLB13C/7
913. RI MS WLB13C/5
914. RI MS WLB13C/11
915. RI MS WLB13C/12
916. RI MS WLB13B/17
917. RI MS WLB13B/48
918. RI MS WLB13B/47
919. RI MS WLB13B/55
920. RI MS WLB13B/57
921. RI MS WLB13B/58
922. "Fifty Years a Winner." BBC broadcast, December 3, 1965.
923. RI MS WLB32E/31
924. RI MS WLB32E/17
925. Letter from William Lawrence Bragg to Linus Pauling, October 11, 1963. Ava Helen and Linus Pauling Papers, Oregon State University Library.
926. RI MS WLB54A/486
927. RI MS WLB53A/147
928. Bragg, W. L. (1964). The difference between living and non-living matter from a physical point of view. *Science and Culture* **30**, 161–7.
929. Bragg, W. L. (1923). The new world of the atom. *The Yale Review* **12**, 755–72.
930. Bragg, W. L. (1961). The development of X-ray analysis. *Proceedings of the Royal Society of London A* **262**, 145–58.
931. Chandrasekhar, S. (1990). It's a black ibis. In *Selections and Reflections: the Legacy of Sir Lawrence Bragg* (J. M. Thomas and D. Phillips, eds.), pp. 123–5. Science Reviews Ltd, Northwood, U.K.
932. Letter from William Lawrence Bragg to Patrick Blackett, February 14, 1964, P. M. S. Blackett Collection, Royal Society Library (CSAC 63.1.79/J.11).
933. RI MS WLB33B/54

934. RI MS WLB95, p. 260
935. Bragg, W. L. (1966). The Royal Institution lectures in science for members of the administrative class of the civil service. *Contemporary Physics* **7**, 358–61.
936. Interview with Stephen Bragg, March 13, 2001.
937. Phillips, D. (1979). William Lawrence Bragg. *Biographical Memoirs of Fellows of the Royal Society of London* **25**, 75–143.
938. Blake, C. C. F., Koenig, D. F., Mair, G. A., North, A. C. T., Phillips, D. C., and Sarma, V. R. (1965). Structure of hen egg-white lysozyme: a three-dimensional Fourier synthesis at 2 Å resolution. *Nature* **206**, 757–61.
939. Porter, G. (1990). W. L. B. at the R. I. In *Selections and Reflections: the Legacy of Sir Lawrence Bragg* (J. M. Thomas and D. Phillips, eds.), pp. 126–9. Science Reviews Ltd, Northwood, U.K.
940. Interview with Patience Thomson, March 15, 2001.
941. RI MS WLB69L/17
942. Bragg, W. L. (1966). Reminiscences of fifty years' research. *Proceedings of the Royal Institution of Great Britain* **41**, 92–100.
943. Phillips, D. (1979). William Lawrence Bragg. *Biographical Memoirs of Fellows of the Royal Society of London* **25**, 75–143.
944. RI MS WLB23B/17
945. RI MS WLB95, p. 265
946. Anonymous (1965). Evening party at the Royal Institution. *Proceedings of the Royal Institution of Great Britain* **40**, 482–97.
947. Caroe, G. M. (1978). *William Henry Bragg, 1862–1942: Man and Scientist*, p. 177. Cambridge University Press, Cambridge.
948. Perutz, M. F. (1990). How Lawrence Bragg invented X-ray analysis. In *Selections and Reflections: the Legacy of Sir Lawrence Bragg* (J. M. Thomas and D. Phillips, eds.), pp. 71–85. Science Reviews Ltd., Northwood, U.K.
949. Arndt, U. (1990). The bird life of Albemarle Street. In *Selections and Reflections: the Legacy of Sir Lawrence Bragg* (J. M. Thomas and D. Phillips, eds.), pp. 105–8. Science Reviews Ltd, Northwood, U.K.
950. Porter, G. (1990). W. L. B. at the R. I. In *Selections and Reflections: the Legacy of Sir Lawrence Bragg* (J. M. Thomas and D. Phillips, eds.), pp. 126–9. Science Reviews Ltd, Northwood, U.K.
951. RI MS WLB95, p. 256
952. Letter from Stephen Bragg, December 15, 2002.
953. RI MS WLB23B/1
954. Letter from Alice Bragg to David Phillips, November 4, 1971. D. C. Phillips Collection, Bodleian Library, Oxford University.
955. Perutz, M. F. (1970). Bragg, protein crystallography and the Cavendish Laboratory. *Acta Crystallographica A* **26**, 183–5.
956. Morris, S. B. (1990). Memories of the workshop. In *Selections and Reflections: the Legacy of Sir Lawrence Bragg* (J. M. Thomas and

D. Phillips, eds.), pp. 149–50. Science Reviews Ltd., Northwood, Middlesex.
957. RI MS WLB53A/253
958. RI MS WLB53A/231
959. Watson, J. D. (1990). Bragg's Foreword to "The Double Helix". In *Selections and Reflections: the Legacy of Sir Lawrence Bragg* (J. M. Thomas and D. Phillips, eds.), pp. 111–13. Science Reviews Ltd, Northwood, U.K.
960. Horace Freeland Judson's interview with James Watson, October 5, 1973. American Philosophical Society.
961. RI MS WLB95, p. 263
962. Interview with Patience Thomson, March 15, 2001.
963. Watson, J. D. (1990). Bragg's Foreword to 'The Double Helix'. In *Selections and Reflections: the Legacy of Sir Lawrence Bragg* (J. M. Thomas and D. Phillips, eds.), pp. 111–13. Science Reviews Ltd, Northwood, U.K.
964. RI MS WLB12/13a
965. RI MS WLB12/12
966. RI MS WLB12/11; RI MS WLB12/13
967. Letter from Linus Pauling to James Watson, October 20, 1966. Ava Helen and Linus Pauling Papers, Oregon State University Library.
968. RI MS WLB12/14; RI MS WLB12/15
969. RI MS WLB/26
970. RI MS WLB12/42
971. RI MS WLB12/44
972. Letter from Linus Pauling to James Watson, January 4, 1967. Ava Helen and Linus Pauling Papers, Oregon State University Library.
973. Letter from Linus Pauling to William Lawrence Bragg, May 17, 1967. Ava Helen and Linus Pauling Papers, Oregon State University Library.
974. RI MS WLB12/35; Interview with Max Perutz, March 12, 2001.
975. RI MS WLB12/64
976. Watson, J. D. (1990). Bragg's Foreword to "The Double Helix". In *Selections and Reflections: the Legacy of Sir Lawrence Bragg* (J. M. Thomas and D. Phillips, eds.), pp. 111–13. Science Reviews Ltd, Northwood, U.K.
977. RI MS WLB59B/102
978. RI MS WLB69D/98, RI MS WLB69D/99
979. Bragg, W. L. (1967). Reminiscences of fifty years of research. *Journal of The Franklin Institute* **284**, 211–28.
980. RI MS WLB14A/7
981. Bragg, W. L. (1967). The spirit of science. *Proceedings of the Royal Society of Edinburgh, Section A (Mathematical & Physical Sciences)* **67**, 303–8.

982. Bragg, W. L. (1969). What makes a scientist? *Proceedings of the Royal Institution of Great Britain* **42**, 397–410.
983. RI MS WLB52B/241
984. Bragg, W. L. (1969). What makes a scientist? *Proceedings of the Royal Institution of Great Britain* **42**, 397–410.
985. RI MS WLB33D/146
986. Crowther, J. G. (1974). *The Cavendish Laboratory, 1874–1974*, p. 306. Science History Publications, New York.
987. Letter from David Blow, January 16, 2002.
988. RI MS WLB54A/324–325; RI MS WLB54A/326
989. Phillips, D. (1979). William Lawrence Bragg. *Biographical Memoirs of Fellows of the Royal Society of London* **25**, 75–143.
990. Thomas, J. M. (1991). Bragg reflections. *Notes and Records of the Royal Society of London* **45**, 243–52.
991. Letter from W. L. Bragg to Patrick Blackett, February 8, 1968. P. M. S. Blackett Collection, Royal Society Library (CSAC 63.1.79/F.42).
992. Letter from William Lawrence Bragg to David Phillips, June 13, 1971. D. C. Phillips Collection, Bodleian Library, Oxford University.
993. Card from Alice Bragg to Patrick Blackett (undated). P. M. S. Blackett Collection, Royal Society Library (CSAC 63.1.79/J.11).
994. Letter from Robert Maycock to David Phillips, March 27, 1972; letter from David Phillips to Robert Maycock, April 18, 1972. D. C. Phillips Collection, Bodleian Library, Oxford University.
995. Phillips, D. (1979). William Lawrence Bragg. *Biographical Memoirs of Fellows of the Royal Society of London* **25**, 75–143.
996. Crowther, J. G. (1974). *The Cavendish Laboratory, 1874–1974*, p. 422. Science History Publications, New York.
997. Jenkin, J. (1986). *The Bragg Family in Adelaide: a Pictorial Celebration*, p. 35. University of Adelaide Foundation, Adelaide.
998. Cochran, W. (1990). Cavendish days. In *Selections and Reflections: the Legacy of Sir Lawrence Bragg* (J. M. Thomas and D. Phillips, eds.), pp. 103–5. Science Reviews Ltd, Northwood, U.K.
999. Interview with Francis Crick, January 11, 2002.
1000. Hamilton, W. C. (1970). The revolution in crystallography. *Science* **169**, 133–41.
1001. Kendrew, J. C. (1990). Bragg's broomstick and the structure of proteins. In *Selections and Reflections: the Legacy of Sir Lawrence Bragg* (J. M. Thomas and D. Phillips, eds.), pp. 88–91. Science Reviews Ltd, Northwood, U.K.
1002. RI MS WLB4/11
1003. RI MS WLB95, p. 121
1004. Pippard, B. (1990). Bragg—the Cavendish Professor. In *Selections and Reflections: the Legacy of Sir Lawrence Bragg* (J. M. Thomas and D. Phillips, eds.), pp. 97–100. Science Reviews Ltd, Northwood, U.K.

1005. Crowther, J. G. (1974). *The Cavendish Laboratory, 1874–1974*, p. 313. Science History Publications, New York.

1006. Lipson, H. (1970). W. L. Bragg—an appreciation. *Acta Crystallographica A* **26**, 180–2.

1007. Horace Freeland Judson's interview with Francis Crick, September 1, 1971. American Philosophical Society.

1008. Phillips, D. (1979). William Lawrence Bragg. *Biographical Memoirs of Fellows of the Royal Society of London* **25**, 75–143.

Index